AF528552

AQUAPEDIA

A COMPREHENSIVE AQUACULTURE GUIDE

Dr. E. Prabu is currently working as Assistant Professor at Directorate of Incubation and Vocational training in Aquaculture, Tamil Nadu Dr.J.Jayalalithaa Fisheries University, Muttukadu, Chennai. He has good knowledge on aquaculture, aquaculture nutrition, especially in fish nutrigenomics and also, he has experienced in species specific feed formulation using feed formulation softwares. He has authored nearly 60 publications including research, review and popular articles. He has operated different research and establishment projects as principal and co-principal investigator with the total budget of Rs. 9.66 crores through the funding support from government and aqua feed industries.

Dr. N. Felix, the Vice Chancellor of Tamil Nadu Dr. J. Jayalalithaa Fisheries University, boasts an illustrious career spanning more than three decades in the realms of teaching, research and extension. His administrative service has been demonstrated through his capacities as Acting Vice-Chancellor, Registrar, Director and Dean over an impressive eight-year tenure. He has contributed significantly to academic works, having authored over 150 scientific and semi-scientific papers, 70 conference abstracts, 5 textbooks and an additional 80 publications. His leadership has steered the successful execution of 38 R&D, institutional entrepreneurship and industry linkage projects, collectively valued at Rs. 35 crores from various funding agencies.

Dr. K. Ravaneswaran is a distinguished academic and an expert in Fisheries Extension, currently serving as the Controller of Examinations i/c., at Tamil Nadu Dr. J. Jayalalithaa Fisheries University, Muttukadu, Chennai. With a robust academic foundation that includes a B.F.Sc., M.Sc. in Agricultural Extension, and a Ph.D. in Animal Husbandry Extension, Dr. Ravaneswaran has dedicated 34 years to the field of aquaculture extension, where he has made significant contributions to both research and practical training.

Dr. A. Uma, Director of Directorate of Incubation and Vocational training in Aquaculture at Tamil Nadu Dr. J Jayalalithaa Fisheries University, is an expert in Aquaculture and Aquatic Animal Health with 25 years of experience. She holds a Master's and PhD in Aquaculture from the Fisheries College and Research Institute, Thoothukudi. Dr. Uma began her career at CP Aquaculture India Pvt Limited and later contributed significantly to establishing a State-level Referral Laboratory for aquatic health. Her research includes fish health management and molecular diagnostics, with six technologies developed and two patents. She has published 80 research papers, authored three books, supervised 50 students, and led 15 major projects. Dr. Uma has received several awards, including the Tamil Nadu Young Woman Scientist Award.

Mr. A. Monishkumar is currently working as an Assistant professor who has completed post-graduation in the Department of Fish Nutrition and Feed Technology at Institute of Fisheries Post Graduate Studies of Tamil Nadu Dr. J. Jayalalithaa Fisheries University (TNJFU), India. His current researches focus on fish and shrimp nutrition, nutritional programming, and nutrigenomics. Also he is having well expertise in species specific feed formulation using feed formulation softwares.

AQUAPEDIA

A COMPREHENSIVE AQUACULTURE GUIDE

– Authors –

Dr. E. Prabu

Dr. N. Felix

Dr. K. Ravaneswaran

Dr. A. Uma

Mr. A. Monishkumar

2025

Daya Publishing House®

A Division of

Astral International Pvt. Ltd.

New Delhi – 110 002

ISBN: 978-93-5919-229-1

Published by : **Daya Publishing House®**
A Division of
Astral International Pvt. Ltd.
– ISO 9001:2015 Certified Company –
4736/23, Ansari Road, Ground Floor,
Darya Ganj, New Delhi-110 002
Ph. 011-43549197, 23278134
e-mail: info@astralint.com
Website: www.astralint.com

Printed at : **Replika Press Pvt Ltd**

PREFACE

Aquaculture has rapidly emerged as one of the most important sectors in global food production, contributing significantly to food security, employment, and economic growth. As the industry continues to evolve, the need for a comprehensive resource that captures both foundational knowledge and the latest innovations has never been greater. AQUAPEDIA: A Comprehensive Aquaculture Guide is designed to meet this need, providing an in-depth exploration of the diverse elements that define modern aquaculture practices.

This book is structured to offer a holistic understanding of aquaculture, starting with the fundamental Principles of Aquaculture, which lay the groundwork for students, researchers, and practitioners. The chapters on Water and Soil Quality Management and Farm Engineering emphasize the critical role of environmental stewardship and the infrastructure needed for sustainable operations.

With the growing interest in ornamental species, the book includes a dedicated section on Ornamental Fish Culture, while Finfish Nutrition and Feeding Management and Finfish Diseases and Health Management provide essential insights into the dietary and health aspects of aquatic species. The burgeoning shrimp industry is well-covered, with chapters on Shrimp Farming and Shrimp Diseases and Health Management.

As aquaculture expands into diverse environments, chapters on Freshwater Aquaculture, Coastal Aquaculture and Mariculture, and the Culture of Fish Food Organisms offer guidance on species management in different ecological contexts. The inclusion of Advances in Aquacultural Practices and Integrated Aquaculture reflects the latest technological developments and sustainable approaches.

We hope AQUAPEDIA serves as a valuable resource for anyone involved in the ever-evolving field of aquaculture.

Dr. E. Prabu | Dr. N. Felix
Dr. K. Ravaneswaran | Dr. A. Uma
Mr. A. Monishkumar

CONTENTS

CONTENTS

CHAPTER 1

PRINCIPLES OF AQUACULTURE

Aeration: Aeration is the process of increasing the oxygen content in aquaculture water, which is essential for maintaining high levels of dissolved oxygen (5-8 mg/L) necessary for the respiration of aquatic organisms and the breakdown of organic matter. Aeration techniques include mechanical aerators, such as paddle wheels and diffusers, which enhance oxygen transfer and circulation in the water column. Effective aeration supports high stocking densities and intensive production by preventing oxygen depletion, reducing stress, and minimizing the risk of fish kills.

Aquaculture Economics: Aquaculture economics focuses on the financial aspects of aquaculture production, including cost structures, profitability, market dynamics, and economic sustainability. It involves analyzing production costs (e.g., feed, labor, energy), pricing strategies, and the economic impacts of factors such as disease outbreaks or environmental regulations. Economic assessments help aquaculture businesses plan their operations, optimize resource use, and make investment decisions. Global trade and market access also play a significant role in the economic viability of aquaculture.

Aquaculture Equipment: Aquaculture equipment includes the various tools and devices used in the farming of aquatic species. This equipment ranges from simple nets and tanks to advanced aeration systems, water quality sensors, automatic feeders, and filtration systems. Equipment selection depends on the type of aquaculture system (e.g., pond, RAS, or cage culture) and the species being farmed. Proper maintenance and regular upgrading of equipment are essential for efficient farm operation, ensuring animal welfare, and optimizing production.

Aquaculture Policies: Aquaculture policies are the strategic guidelines and governmental regulations that shape the development and management of the aquaculture sector. These policies address issues such as licensing, environmental protection, resource allocation, and market access. National and international policies aim to balance the need for aquaculture expansion with sustainability and conservation goals. Policy frameworks also promote research, innovation, and training to improve industry practices, while supporting economic growth and food security.

Aquaculture Practices: Aquaculture practices encompass the methods and techniques used in the farming of aquatic organisms, including site selection, species selection, system design (e.g., ponds, cages, recirculating systems), stocking densities, feeding regimes, water quality management, and disease prevention. These practices vary depending on the species being cultured, the environmental conditions, and the production goals. Sustainable aquaculture practices aim to optimize production efficiency while minimizing environmental impacts, such as pollution, resource depletion, and biodiversity loss. Best management practices (BMPs) are often developed to guide farmers in implementing effective and environmentally responsible aquaculture operations.

Aquaculture Systems: Aquaculture systems refer to the controlled environments designed for the cultivation of aquatic organisms, including fish, shellfish, and aquatic plants. These systems can be broadly categorized into extensive, semi-intensive, and intensive systems based on the level of management and inputs such as feed, water exchange, and aeration. Optimal water quality parameters, including dissolved oxygen (5-8 mg/L), temperature (dependent on species, e.g., 24°-28°C for tilapia), and pH (6.5-8.5), are crucial for maintaining healthy production. Recirculating aquaculture systems (RAS), pond systems, and raceway systems are common examples, each tailored to specific species and production goals.

Aquaculture vs Agriculture: Aquaculture and agriculture are both forms of farming but differ significantly in their environments and practices. While agriculture involves cultivating crops and raising livestock on land, aquaculture focuses on the cultivation of aquatic organisms in water. Aquaculture can be more efficient in terms of feed conversion, as fish generally have higher feed conversion ratios (FCR) than land animals, meaning they require less feed to produce the same amount of protein. For example, the FCR for farmed fish like tilapia can be as low as 1.2:1, compared to 7:1 for beef cattle. Both systems require careful management of resources, but aquaculture faces unique challenges such as maintaining water quality, controlling diseases, and managing environmental impacts.

Aquaculture: Aquaculture is the controlled process of cultivating aquatic organisms, including fish, mollusks, crustaceans, and aquatic plants, for human consumption, recreation, or conservation. This practice is carried out in a range of water environments, including freshwater, brackishwater, and marine systems. The goal of aquaculture is to maximize the growth and yield of these species through optimized feeding, breeding, water quality management, and disease control. Typical water parameters for successful aquaculture include a temperature range of 22°C to 30°C, dissolved oxygen levels above 5 mg/L, and pH levels between 6.5 and 8.5, depending on the species and system used.

Aquafeeds: Aquafeeds are specially formulated feeds designed to meet the nutritional requirements of cultured aquatic species. These feeds typically contain a balanced mixture of protein, lipids, carbohydrates, vitamins, and minerals, with the specific composition varying depending on the species and life stage. For example, carnivorous fish such as salmon require high-protein diets (40-50% protein), while omnivorous species like tilapia require lower protein levels (25-35%). The quality of aquafeeds directly influences growth rates, feed conversion efficiency, and overall health of the cultured species.

Aquaponics: Aquaponics is a sustainable food production system that combines aquaculture (raising aquatic animals) with hydroponics (cultivating plants in water). In an aquaponic system, the waste produced by the aquatic animals provides nutrients for the plants, which in turn help filter and clean the water. This symbiotic relationship reduces the need for chemical fertilizers and minimizes water consumption. Aquaponics systems can be operated at various scales, from small backyard setups to commercial operations, and are often seen as a sustainable method for producing both fish and vegetables.

Aquatic Animal Health: Aquatic animal health in aquaculture focuses on the prevention, diagnosis, and treatment of diseases affecting farmed species. Maintaining aquatic animal health involves regular monitoring, proper water quality management, biosecurity measures, and the use of vaccines, probiotics, or other treatments when necessary. Disease outbreaks can have severe economic and ecological impacts, making the management of aquatic animal health critical for both farm productivity and sustainability. Collaborating with veterinarians and health experts is essential for maintaining healthy stocks.

Aquatic Ecosystems: Aquatic ecosystems encompass the complex interactions between living organisms, such as fish, plants, and microorganisms, and their non-living environment, including water chemistry, substrate, and temperature. These ecosystems can be freshwater, brackish, or marine, each characterized by distinct physical, chemical, and biological conditions. Key parameters influencing these ecosystems include salinity, pH, dissolved oxygen, and nutrient levels. The health and stability of aquatic ecosystems are vital for the sustainability of aquaculture practices, as they directly affect the productivity and welfare of cultured species.

Aquatic Plant Management: Aquatic plant management in aquaculture involves controlling the growth of submerged, emergent, and floating plants in production systems. While some aquatic plants, like algae, can provide natural food for species like carp, excessive plant growth can interfere with water circulation, reduce dissolved oxygen levels, and create habitat for disease vectors. Management techniques include mechanical removal, herbicide application (with caution), and biological control methods, such as introducing plant-eating fish. Proper management of aquatic plants helps maintain balanced ecosystems and supports optimal production conditions.

Artificial Propagation: Artificial propagation in aquaculture refers to the controlled breeding and reproduction of aquatic species through techniques such as hormone-induced spawning, in vitro fertilization, and larval rearing. These methods allow for the consistent production of large numbers of offspring, ensuring a reliable supply of seedstock for grow-out operations. Artificial propagation is essential for species that do not readily reproduce in captivity or require specific environmental triggers for spawning. This practice supports the expansion of aquaculture by reducing reliance on wild-caught seedstock and enabling selective breeding programs.

Biological Factors: Biological factors in aquaculture include the living components of the system that interact with the cultured species, such as plankton, aquatic plants, pathogens, predators, and the cultured species themselves. These factors play a crucial role in determining the productivity and health of the aquaculture environment. For instance, plankton serves as the primary food source in extensive and semi-intensive systems, while pathogens and predators represent potential threats that need to be managed. Understanding and managing biological interactions, such as the balance between beneficial and harmful organisms, is key to maintaining a healthy and productive aquaculture system.

Bioremediation: Bioremediation in aquaculture is the use of natural or engineered organisms, such as bacteria, fungi, or plants, to degrade, detoxify, or remove pollutants from water and sediments. This process helps manage waste, reduce the environmental impact of aquaculture, and improve water quality. Common bioremediation methods include the use of biofilters, constructed wetlands, and microbial inoculants that break down organic matter, nitrogen compounds, and other contaminants. Bioremediation supports sustainable aquaculture by minimizing the accumulation of harmful substances in the environment.

Biosecurity: Biosecurity in aquaculture encompasses the measures and protocols implemented to prevent the introduction and spread of infectious diseases and harmful pathogens within and between aquaculture facilities. These measures include the use of disease-free broodstock, disinfection of equipment, quarantine procedures for new stock, and strict water quality management. Effective biosecurity practices are essential to protect the health of cultured species, minimize economic losses, and ensure sustainable production. Regular monitoring and rapid response to potential disease threats are critical components of a robust biosecurity plan.

Brackishwater Aquaculture: Brackishwater aquaculture refers to the farming of aquatic organisms in water that has a salinity level between that of freshwater and seawater, typically ranging from 5 to 30 ppt. Species like shrimp (*Penaeus monodon*), milkfish (*Chanos chanos*), and mud crabs are commonly cultured in these environments. Brackishwater systems are often found in estuaries, coastal lagoons, and mangrove areas. Key management practices include monitoring and adjusting salinity, temperature, and pH, with optimal temperature ranges generally between 25°C to 30°C. Brackishwater aquaculture is essential for producing high-value species in regions where both freshwater and marine systems are accessible.

Brackishwater Species: Brackishwater species in aquaculture are aquatic organisms that can tolerate a range of salinities, typically from 0.5 to 30 ppt. These species are commonly cultured in estuaries, coastal lagoons, and mangrove areas, where salinity levels fluctuate due to tidal influences. Examples of brackishwater species include shrimp (e.g., *Penaeus monodon*, *Penaeus vannamei*), milkfish, and mud crabs. These species are often chosen for their economic value and adaptability to variable salinity conditions. Effective management of brackishwater systems, including salinity control, water exchange, and feeding practices, is essential for optimizing the growth and survival of these species.

Brackishwater Systems: Brackishwater systems in aquaculture are those where the salinity is intermediate between freshwater and seawater, typically ranging from 5 to 30 ppt. These systems are often found in estuaries, coastal lagoons, and mangrove areas, and are used to culture species adapted to such conditions, such as shrimp, milkfish, and mud crabs. Managing a brackishwater system requires careful monitoring of salinity levels, which can fluctuate due to tidal movements and freshwater inflow. The temperature in these systems generally ranges from 25°C to 30°C, with a pH between 7.5 and 8.5. Brackishwater aquaculture is significant for producing high-value species in coastal regions, contributing to both local economies and export markets.

Breeding Techniques: Breeding techniques in aquaculture refer to the methods used to propagate and improve aquatic species through controlled reproduction. Techniques include selective breeding, hybridization, and genetic engineering. Selective breeding involves choosing individuals with desirable traits, such as fast growth or disease resistance, to produce offspring with enhanced characteristics. Optimal breeding conditions, such as temperature, photoperiod, and water quality, are species-specific and must be carefully managed to ensure high reproductive success and healthy progeny.

Broodstock Management: Broodstock management in aquaculture involves the selection, care, and breeding of mature individuals used for reproduction. Proper broodstock management ensures the production of high-quality eggs and larvae with desirable traits such as fast growth, disease resistance, and high survival rates. Key practices include maintaining optimal water quality, providing balanced nutrition, and preventing stress. The genetic diversity of broodstock is also important to avoid inbreeding and maintain healthy populations. Regular monitoring of reproductive performance helps optimize breeding cycles and improve hatchery outputs.

Cage Culture: Cage culture is a form of aquaculture where fish or other aquatic species are raised in cages or net enclosures submerged in natural water bodies such as lakes, rivers, or the ocean. These cages can vary in size and are typically constructed from materials that allow free water exchange, providing a controlled environment that protects the cultured species from predators while optimizing feeding and growth conditions. Commonly cultured species include tilapia, salmon, and sea bass. Cage culture allows for high stocking densities and efficient space utilization, but it requires careful management of water quality, disease prevention, and feeding practices to minimize environmental impacts and ensure high productivity.

Candidate Species: Candidate species in aquaculture refer to aquatic organisms that are selected for cultivation based on specific criteria such as market demand, growth rate, adaptability to culture conditions, feed conversion efficiency, and resistance to diseases. These species are chosen for their potential to provide economic benefits to the farmer while being suited to the environmental conditions of the culture system. Common candidate species include tilapia, carp, shrimp, and salmon. The selection of candidate species is critical to the success of aquaculture operations, as it influences the production methods, management practices, and overall sustainability of the enterprise.

Certification and Standards: Certification and standards in aquaculture refer to voluntary or mandatory programs that ensure aquaculture products meet certain quality, environmental, and social criteria. Certification schemes, such as the Aquaculture Stewardship Council (ASC) or Global GAP, assess practices related to sustainability, animal welfare, feed sourcing, and worker rights. Certified products often carry a label that informs consumers about the farm's adherence to these standards, providing market advantages and promoting responsible aquaculture practices. Meeting certification requirements can open access to premium markets and enhance a farm's reputation.

Chemical Factors: Chemical factors in aquaculture refer to the water's chemical composition, which includes parameters such as pH, dissolved oxygen, salinity, hardness, and concentrations of nutrients like ammonia, nitrite, and nitrate. These factors directly influence the health, growth, and survival of aquatic organisms. For example, maintaining a pH range of 6.5 to 8.5 is crucial for most species, while dissolved oxygen levels should typically be kept above 5 mg/L. Ammonia and nitrite, which are toxic at high levels, must be carefully monitored and managed through biofiltration, water exchange, or chemical treatments. Maintaining balanced chemical conditions is vital for preventing stress and disease in cultured species.

Climate Change Adaptation: Climate change adaptation in aquaculture involves adjusting farming practices and systems to cope with the impacts of climate change, such as rising temperatures, sea level rise, and changes in precipitation patterns. Adaptation strategies include selecting climate-resilient species, modifying production systems (e.g., using recirculating aquaculture systems), and improving water and energy efficiency. Aquaculture operations may also need to relocate to areas less vulnerable to climate impacts or adopt new technologies to mitigate risks. Effective adaptation is crucial for maintaining productivity and resilience in the face of changing environmental conditions.

Cost-Benefit Analysis: Cost-benefit analysis in aquaculture involves comparing the costs of production (e.g., feed, equipment, labor, energy) with the expected benefits, such as revenue from sales. This analysis helps aquaculture businesses assess the financial feasibility of different production strategies and investment options. It includes evaluating short-term and long-term profitability, considering risks such as disease outbreaks, environmental changes, and market fluctuations. Effective cost-benefit analysis is essential for making informed decisions, optimizing resource allocation, and ensuring the sustainability of aquaculture operations.

Criteria for Species Selection: The criteria for species selection in aquaculture involve a comprehensive evaluation of various factors to determine the suitability of a species for cultivation. Key criteria include the species' growth rate, feed conversion ratio (FCR), tolerance to environmental conditions (e.g., temperature, salinity, pH), resistance to diseases, reproductive efficiency, market demand, and economic profitability. Additionally, the availability of seed or juveniles, the species' adaptability to culture systems (e.g., ponds, cages, tanks), and regulatory considerations are also critical. Selecting the right species based on these criteria is essential for optimizing production efficiency and ensuring the sustainability of aquaculture operations.

Disease Management: Disease management in aquaculture involves the prevention, diagnosis, and treatment of diseases that affect cultured aquatic species. It includes the implementation of biosecurity measures, vaccination, use of probiotics, and the application of therapeutic agents such as antibiotics and antiparasitics. Maintaining optimal water quality, reducing stress through proper stocking densities, and providing balanced nutrition are crucial for disease prevention. Early detection and appropriate treatment of diseases are vital to minimize mortality rates and economic losses in aquaculture operations.

Disease Prevention: Disease prevention in aquaculture involves implementing strategies and practices to reduce the risk of disease outbreaks in cultured species. These strategies include maintaining optimal water quality, controlling stocking densities, implementing biosecurity measures (e.g., quarantining new stock), and ensuring proper nutrition. Vaccination programs, the use of probiotics, and regular health monitoring are also part of disease prevention. Preventing disease is critical for minimizing economic losses, improving animal welfare, and ensuring the sustainability of aquaculture operations.

Domestication: Domestication in aquaculture is the process of adapting wild aquatic species to farming conditions through selective breeding and controlled reproduction. Over time, domesticated species become more suited to aquaculture environments, exhibiting traits such as faster growth, higher reproductive success, and increased resistance to disease. Examples of domesticated aquaculture species include Atlantic salmon, tilapia, and Pacific white shrimp. Domestication reduces reliance on wild-caught broodstock, improving the sustainability and efficiency of aquaculture operations.

Effluent Treatment: Effluent treatment in aquaculture refers to the processes used to treat wastewater discharged from aquaculture systems before it is released into the environment. Effluents typically contain organic matter, nutrients (nitrogen and phosphorus), and other pollutants that can lead to water quality degradation if not properly managed. Treatment methods include sedimentation, biological filtration, constructed wetlands, and biofloc technology. Effective effluent treatment systems help prevent eutrophication of surrounding water bodies, protect biodiversity, and comply with environmental regulations.

Environmental Impact: The environmental impact of aquaculture refers to the effects that aquaculture practices can have on surrounding ecosystems. Key concerns include water pollution from nutrient-rich effluents, habitat destruction (e.g., mangrove clearing for shrimp farms), and the potential for escaped farmed species to disrupt local ecosystems. Effective environmental management practices include water treatment systems, responsible stocking densities, and the use of sustainable feeds. Regulations and certifications, such as the Aquaculture Stewardship Council (ASC), aim to mitigate negative impacts and promote sustainable practices in the industry.

Environmental Monitoring: Environmental monitoring in aquaculture involves the continuous assessment of key environmental parameters, such as water quality (e.g., temperature, dissolved oxygen, pH, ammonia, and nitrates) and soil conditions, to ensure optimal growing conditions for cultured species. Monitoring also includes assessing the impacts of aquaculture activities on the surrounding ecosystem, such as nutrient runoff or habitat alteration. Real-time monitoring systems and data analysis help aquaculturists make informed decisions to maintain productivity, prevent environmental degradation, and comply with regulatory standards.

Extensive Aquaculture: Extensive aquaculture is characterized by low input and low-density farming, where aquatic organisms are reared in natural or semi-natural water bodies such as ponds, lakes, or coastal lagoons. This system relies primarily on the natural productivity of the water body, with minimal supplemental feeding or aeration. Stocking densities are low, typically less than

1,000 kg/ha, which allows the natural ecosystem to support the growth of the organisms. Extensive aquaculture is often practiced in rural areas with limited resources and infrastructure. While it is environmentally sustainable and has low operational costs, the production yields are generally lower compared to more intensive systems.

Feed Conversion Ratio (FCR): The feed conversion ratio (FCR) is a measure of the efficiency with which aquatic organisms convert feed into body mass. It is calculated as the amount of feed consumed divided by the weight gained. An FCR of 1.0 means that 1 kg of feed produces 1 kg of biomass, while a higher FCR indicates less efficient feed conversion. FCR values vary by species, feed quality, and management practices. Lower FCR values are desirable as they indicate better feed efficiency, reduced waste, and higher profitability. In well-managed systems, FCRs typically range from 1.0 to 1.5.

Feed Formulation: Feed formulation in aquaculture involves the scientific design of diets that meet the nutritional needs of specific aquatic species while optimizing growth, health, and feed conversion efficiency. The formulation process balances protein, lipids, carbohydrates, vitamins, and minerals, considering factors like species, life stage, and production goals. Ingredients often include fishmeal, plant proteins, oils, and binders. Advances in feed formulation focus on reducing reliance on fishmeal and incorporating sustainable alternatives like insect protein or algae, while maintaining optimal feed conversion ratios (FCR) of 1.0-1.5 in intensive systems.

Feeding Practices: Feeding practices in aquaculture involve the management of feed type, quantity, and frequency to ensure the optimal growth and health of cultured species. Feed formulations are species-specific, balancing protein, lipid, carbohydrate, vitamin, and mineral content to meet nutritional requirements. For instance, carnivorous fish like salmon require high-protein diets (35-50% protein), while herbivorous species like tilapia can thrive on lower-protein feeds (25-35% protein). Proper feeding practices aim to maximize feed conversion efficiency while minimizing waste and environmental impact.

Fisheries Management: Fisheries management in aquaculture involves the regulation and oversight of fishery resources to ensure their sustainable use. This includes managing wild fish stocks that are either directly harvested or used to supply aquaculture operations. Fisheries management practices include setting catch limits, protecting critical habitats, preventing overfishing, and ensuring that wild populations are not negatively impacted by aquaculture activities. The goal is to promote a balance between economic profitability and the conservation of aquatic ecosystems.

Freshwater Aquaculture: Freshwater aquaculture involves the farming of aquatic species in environments where the salinity is less than 0.5 ppt, such as rivers, lakes, ponds, and reservoirs. Key species include carps, tilapia, catfish, and freshwater prawns. This form of aquaculture requires careful management of water quality parameters, such as maintaining a temperature range of 25°C to 30°C, pH between 6.5 and 8.5, and dissolved oxygen above 4 mg/L. Freshwater aquaculture is highly prevalent, especially in regions with abundant freshwater resources, and it contributes significantly to global fish production.

Freshwater Species: Freshwater species in aquaculture refer to aquatic organisms that thrive in low-salinity environments, typically less than 0.5 ppt. These species are cultured in ponds, lakes, rivers, and reservoirs, and include popular fish such as carp (e.g., common carp, grass carp), tilapia, catfish, and freshwater prawns (e.g., *Macrobrachium rosenbergii*). Freshwater species are often chosen for their adaptability to varying water conditions, fast growth rates, and high market demand. The successful cultivation of freshwater species requires careful management of water quality parameters such as temperature (optimally between 25°C and 30°C), pH (6.5 to 8.5), and dissolved oxygen levels (above 4 mg/L).

Freshwater Systems: Freshwater systems in aquaculture refer to the cultivation of aquatic species in water bodies with low salinity levels, typically less than 0.5 ppt. These systems include ponds, lakes, rivers, and reservoirs where species such as carp, tilapia, catfish, and freshwater prawns are commonly cultured. The management of freshwater systems involves maintaining optimal water quality parameters, including a temperature range of 25°C to 30°C, pH levels between 6.5 and 8.5, and dissolved oxygen concentrations above 4 mg/L. Freshwater aquaculture systems are the most widespread globally and are essential for producing affordable and accessible protein sources, particularly in developing countries.

Genetic Improvement: Genetic improvement in aquaculture involves the selective breeding and genetic modification of aquatic species to enhance desirable traits such as growth rate, disease resistance, and feed efficiency. Genetic improvement programs use techniques like selective breeding, hybridization, and marker-assisted selection to accelerate the development of superior strains. Advances in genomics and biotechnology have enabled more precise and efficient genetic improvement, leading to more productive and resilient aquaculture species.

Global Aquaculture Scenario: Globally, aquaculture has become the fastest-growing food production sector, accounting for more than 50% of the fish consumed worldwide. Asia dominates the industry, with China being the largest producer, contributing approximately 60% of global aquaculture output. The industry encompasses a wide range of species and systems, from freshwater pond culture to marine cage farming. Key challenges include managing environmental impacts, disease control, and the sustainability of feed resources. Despite these challenges, global aquaculture production has surpassed 120 million metric tons annually, providing a vital source of protein and supporting millions of livelihoods.

Grow-out Systems: Grow-out systems in aquaculture refer to the phase where juvenile aquatic organisms are reared to market size. These systems can be extensive (e.g., ponds, cages) or intensive (e.g., recirculating systems, raceways) depending on the level of management and resource input. Key factors in grow-out systems include stocking density, water quality management, feeding strategies, and disease control. The success of a grow-out system is measured by parameters such as growth rate, feed conversion ratio (FCR), and survival rate. Optimal conditions vary by species but typically involve maintaining temperature, dissolved oxygen, and pH within species-specific ranges.

Harvesting Techniques: Harvesting techniques in aquaculture involve the methods used to capture and collect cultured aquatic organisms when they reach market size. Techniques vary depending on the species and culture system but commonly include netting, draining ponds, or using specialized equipment like seines and traps. The timing of harvest is crucial to ensure that the organisms are in prime condition for market, with considerations for stress minimization, water quality, and product quality. Proper handling during harvest is essential to prevent physical damage and reduce post-harvest mortality.

Hatchery Design: Hatchery design in aquaculture refers to the layout and construction of facilities for breeding and rearing aquatic species during their early life stages. Effective hatchery design includes considerations for water quality management, biosecurity, and efficient workflow. Key components include broodstock tanks, spawning areas, larval rearing tanks, and water filtration systems. The design must ensure optimal environmental conditions (e.g., temperature, dissolved oxygen) and facilitate easy handling of broodstock and larvae. A well-designed hatchery enhances productivity, reduces mortality, and supports the production of high-quality seedstock for grow-out operations.

Hatchery Management: Hatchery management involves the operations and practices required to successfully breed, hatch, and rear early life stages of aquatic organisms, such as fish and shellfish, in a controlled environment. This includes broodstock management, water quality control, feeding of larvae, and disease prevention. Optimal water temperature, dissolved oxygen levels, and appropriate live feed (e.g., Rotifers and Artemia) are critical factors in ensuring high survival rates and healthy larvae production. Effective hatchery management is essential for supplying high-quality seedstock for grow-out operations.

History of Aquaculture: The history of aquaculture dates back over 4,000 years, with origins in ancient China, where early forms of fish farming involved the cultivation of carp in ponds. The practice spread to other parts of Asia and later to Europe, with the Romans developing oyster farming techniques. The modern era of aquaculture began in the mid-20th century with advancements in hatchery technology, feed production, and water quality management, leading to the rapid expansion of the industry. Today, aquaculture is a global industry, producing nearly half of the world's fish supply for human consumption, and continues to evolve with technological innovations.

Hybridization: Hybridization in aquaculture refers to the crossing of different species or strains to produce hybrids with desirable traits. Hybridization can lead to improved growth rates, disease resistance, or tolerance to environmental conditions compared to either parent species. For example, hybrid catfish (crosses between channel catfish and blue catfish) are widely farmed due to their fast growth and resistance to certain diseases. However, hybridization must be carefully managed to avoid issues such as reduced fertility or negative impacts on wild populations.

Integrated Culture: Integrated culture, also known as Integrated Multi-Trophic Aquaculture (IMTA), is an advanced aquaculture system where multiple species from different trophic levels are cultivated together in a synergistic manner. This system mimics natural ecosystems, where the waste products of one species become the nutrients for another. For example, in a marine IMTA system, finfish are cultured alongside filter-feeding mollusks and seaweeds, with the latter absorbing excess nutrients from the fish. Integrated culture improves resource efficiency, enhances environmental sustainability, and reduces the environmental footprint of aquaculture operations by recycling nutrients and minimizing waste discharge.

Integrated Multi-Trophic Aquaculture (IMTA): Integrated Multi-Trophic Aquaculture (IMTA) is a system that involves the cultivation of multiple species from different trophic levels in the same environment. For example, fish, shellfish, and seaweeds can be farmed together, with each species playing a role in maintaining a balanced ecosystem. The fish produce organic waste that serves as nutrients for seaweeds, while shellfish filter the water. IMTA enhances sustainability by reducing waste and improving resource efficiency, promoting a circular economy in aquaculture.

Integrated Pest Management: Integrated Pest Management (IPM) in aquaculture is a holistic approach to controlling pests and diseases that minimizes the use of chemical treatments. IPM combines biological control (e.g., introducing predators or beneficial microorganisms), mechanical control (e.g., traps or barriers), and cultural practices (e.g., managing water quality) to prevent and manage pests. The goal is to reduce the reliance on pesticides, minimize environmental impacts, and promote the health of cultured species. IPM practices are tailored to specific pests and environmental conditions, ensuring effective and sustainable pest management.

Intensive Aquaculture: Intensive aquaculture is a high-input system where aquatic organisms are raised in a controlled environment with a focus on maximizing production. This system involves the use of artificial feeds, aeration, and frequent water exchange to support high stocking densities, often exceeding 10,000 kg/ha. Intensive aquaculture systems are typically used for high-value species such as shrimp, salmon, and tilapia. These systems require significant investment in infrastructure and technology, and they demand meticulous management to prevent issues such as oxygen depletion, disease outbreaks, and water quality deterioration. Despite these challenges, intensive aquaculture can yield substantial outputs, making it a key component of commercial aquaculture operations.

Larval Nutrition: Larval nutrition in aquaculture focuses on providing the necessary nutrients to support the rapid growth and development of larval stages. Early larval stages often rely on live feeds such as microalgae, rotifers, and Artemia, which provide essential proteins, lipids, and fatty acids. As larvae grow, they may transition to formulated feeds that are tailored to their nutritional needs. Ensuring proper larval nutrition is crucial for high survival rates and strong growth, as deficiencies can lead to developmental issues and increased susceptibility to disease.

Larviculture: Larviculture is the process of rearing the larval stages of aquatic organisms, such as fish, shrimp, or mollusks, from hatching until they reach the juvenile stage. This period is critical, as larvae are highly sensitive to environmental conditions and require specific water quality parameters, live feed (e.g., microalgae, rotifers, Artemia), and temperature controls. Successful larviculture requires meticulous management of water quality (e.g., ammonia levels <0.01 mg/L, dissolved oxygen >5 mg/L) and a constant supply of nutritious, appropriately-sized feed to support rapid growth and high survival rates.

Liming: Liming involves the application of lime (calcium carbonate, calcium hydroxide, or calcium oxide) to aquaculture ponds to raise soil and water pH, thereby enhancing the availability of nutrients and improving overall pond productivity. Optimal pH levels in aquaculture ponds range between 6.5 and 8.5. Lime application rates typically range from 1,000 to 2,000 kg/ha depending on soil acidity and buffering capacity. Regular liming ensures a stable pH environment, reduces the solubility of toxic metals, and promotes the growth of beneficial microorganisms.

Live Feed Culture: Live feed culture in aquaculture refers to the production of small, live organisms such as microalgae, rotifers, and Artemia (brine shrimp) that are used to feed the larvae of many aquatic species during their early life stages. These live feeds provide essential nutrients, including proteins, lipids, and fatty acids, that are crucial for larval growth and development. Culturing live feed requires maintaining specific environmental conditions, such as light intensity, temperature, and nutrient levels. Proper live feed management is essential for high larval survival rates and the successful rearing of many finfish and shellfish species.

Mariculture: Mariculture is a branch of aquaculture that involves the cultivation of marine organisms in ocean waters, coastal environments, or man-made systems such as tanks or ponds filled with seawater. Common species include finfish (e.g., salmon, sea bass), mollusks (e.g., oysters, mussels), and seaweeds. Mariculture is particularly significant for species that require high salinity, typically around 30-35 ppt. Water temperature, oxygen levels, and nutrient availability are carefully monitored to ensure the optimal growth of the species. Mariculture supports both food production and the restoration of endangered marine species.

Marine Aquaculture: Marine aquaculture, or mariculture, focuses on the cultivation of marine species in their natural or controlled oceanic environments. This includes species such as sea bass, mussels, oysters, and various seaweeds. Marine aquaculture systems maintain seawater salinity levels at approximately 35 ppt and temperatures suited to the specific species, often ranging from 10°C to 28°C. These systems are designed to mimic natural oceanic conditions as closely as possible, ensuring that the cultured species can thrive. Marine aquaculture contributes significantly to the seafood supply and helps alleviate pressure on wild fish stocks.

Marine Species: Marine species in aquaculture are aquatic organisms that thrive in high-salinity environments, typically around 35 ppt. These species are cultured in marine environments such as open seas, coastal bays, and man-made marine systems, including cages and tanks. Common marine species include sea bass, sea bream, grouper, salmon, and various shellfish like oysters and mussels. Marine species are often selected for their high market value and demand, particularly in export markets. Cultivating marine species requires careful management of water quality, including temperature (species-specific, often between 10°C and 28°C), oxygen levels, and waste management, to ensure optimal growth and productivity.

Marine Systems: Marine systems in aquaculture involve the farming of species in saltwater environments, including open oceans, coastal bays, and man-made structures like marine cages and tanks. The salinity in these systems is typically around 35 ppt, and species such as sea bass, mussels, oysters, and seaweeds are commonly cultured. Marine aquaculture requires the maintenance of stable water quality, with temperature ranges specific to each species, often between 10°C to 28°C, and careful management of oxygen levels, nutrient input, and waste removal. Marine systems are crucial for supplying seafood and have the potential for expansion through technologies like offshore farming and integrated multi-trophic aquaculture (IMTA).

Market Trends: Market trends in aquaculture refer to the patterns and dynamics in the global demand and supply of aquaculture products. Trends are influenced by factors such as consumer preferences, price fluctuations, sustainability concerns, technological advancements, and regulatory changes. For instance, there is a growing demand for sustainably farmed seafood, traceability, and aquafeeds that reduce reliance on fishmeal. Understanding market trends is critical for aquaculture producers to make informed decisions regarding species selection, production scale, and investment strategies.

Monoculture: Monoculture in aquaculture refers to the practice of cultivating a single species of aquatic organism in a specific system or water body. This approach allows for specialized management practices, including targeted feeding, disease control, and environmental management, optimized for the particular species. Common examples of monoculture include the farming of species like tilapia, shrimp, or salmon. While monoculture can lead to high yields and efficient production, it can also increase the risk of disease outbreaks and environmental impacts due to the lack of biodiversity and potential accumulation of waste products. Monoculture systems require careful monitoring and management to sustain production and minimize risks.

National Aquaculture Scenario: India ranks second in global aquaculture production, driven by the farming of freshwater species such as carp, catfish, and prawns, as well as brackishwater shrimp. The country's diverse climate and vast water resources provide favorable conditions for aquaculture, with major production centers in states like Andhra Pradesh, West Bengal, and Tamil Nadu. Government initiatives, such as the Pradhan Mantri Matsya Sampada Yojana (PMMSY), aim to further boost the sector's growth by improving infrastructure, promoting sustainable practices, and enhancing market access. India's aquaculture industry is a critical component of its economy, contributing significantly to food security and export earnings.

Nursery Rearing: Nursery rearing refers to the stage in aquaculture where juvenile aquatic organisms, such as fish or shellfish, are raised from the post-larval stage until they are large enough to be transferred to grow-out systems. During this stage, organisms are often fed formulated feeds that match their growing nutritional needs and are kept in environments with optimal water quality parameters to promote healthy development. The nursery phase typically lasts a few weeks to months and is critical for ensuring the survival and robust growth of the stock before it enters the final grow-out phase.

Nutrient Requirements: Nutrient requirements in aquaculture refer to the essential nutrients that aquatic species need for growth, reproduction, and overall health. These nutrients include macronutrients like proteins, lipids, and carbohydrates, as well as micronutrients such as vitamins and minerals. Specific nutrient requirements vary by species and life stage. For example, juvenile fish typically require diets containing 35-50% protein, while adult fish may need less. Nutrient deficiencies or imbalances can lead to poor growth, increased susceptibility to disease, and reduced reproductive success.

Pen Culture: Pen culture is a type of aquaculture where aquatic organisms are raised in pens or enclosures made of netting or other materials that allow water exchange with the surrounding environment. These pens are typically placed in natural water bodies such as lakes, rivers, or coastal areas, and can range in size from a few square meters to several hectares. Pen culture provides a semi-controlled environment, protecting the cultured species from predators while allowing them access to natural food resources. It is commonly used for species such as tilapia, carp, and certain marine fish. The success of pen culture depends on careful site selection, pen construction, and regular monitoring of water quality.

Physical Factors: Physical factors in aquaculture encompass the environmental conditions that influence the growth and health of aquatic organisms. These include temperature, light, water flow, turbidity, and substrate composition. Temperature, for example, affects metabolic rates, feeding behavior, and reproductive cycles, with most species thriving within a specific range, often between 20°C to 30°C. Light affects primary productivity and can influence the behavior of both cultured species and their prey. Water flow and turbidity impact oxygen levels, waste removal, and the distribution of feed. Proper management of physical factors is essential to create optimal conditions for aquaculture production.

Polyculture: Polyculture is an aquaculture practice where two or more compatible species are cultivated together in the same system or water body. The species selected for polyculture typically have different ecological niches or feeding habits, which allows for more efficient utilization of available resources, such as space, feed, and water quality. For example, in a carp polyculture system, species like silver carp, grass carp, and common carp are raised together, each utilizing different trophic levels-plankton, aquatic plants, and detritus, respectively. Polyculture can enhance overall productivity, improve water quality through species interactions, and reduce the risk of disease outbreaks by diversifying the cultured species.

Pond Culture: Pond culture is a common method in aquaculture where fish or other aquatic organisms are raised in controlled pond environments. These ponds are typically shallow, with depths ranging from 1 to 2 meters, and can vary in size from small backyard setups to large commercial ponds covering several hectares. Pond culture allows for easy monitoring and management of water quality parameters, such as temperature, dissolved oxygen, and pH. Fertilization, feeding, and aeration are often used to enhance productivity. Species such as carp, tilapia, and catfish are commonly farmed in ponds, which can be either freshwater or brackishwater systems, depending on the species and location.

Pond Fertilization: Pond fertilization is the practice of adding nutrients to aquaculture ponds to enhance the growth of natural food organisms like phytoplankton and zooplankton, which serve as a primary food source for fish and shrimp. Fertilizers can be organic (e.g., animal manure) or inorganic (e.g., urea and superphosphate). The application rates and types of fertilizers depend on the existing nutrient levels in the pond and the species being cultured. For instance, nitrogen and phosphorus are often applied at a ratio of 4:1, with typical application rates of 5-10 kg/ha of nitrogen weekly during the growing season.

Pond Preparation: Pond preparation in aquaculture involves a series of steps to create an optimal environment for the culture of aquatic species. This includes clearing debris, repairing pond structures, drying and liming the pond bottom to adjust pH, and applying fertilizers to stimulate the growth of natural food organisms. Proper pond preparation ensures good water quality, promotes the growth of beneficial microorganisms, and reduces the risk of disease outbreaks. The timing and intensity of these activities depend on factors like pond size, species being cultured, and local environmental conditions.

Pond Productivity: Pond productivity in aquaculture refers to the ability of a pond to support the growth and yield of cultured species, which is influenced by the pond's natural and managed inputs, including nutrient levels, primary and secondary productivity, and water and soil quality. High pond productivity is often achieved through proper fertilization, water management, and stocking practices that enhance the growth of both primary producers (e.g., phytoplankton) and secondary consumers (e.g., fish). Measuring and optimizing pond productivity is critical for maximizing yields and ensuring the sustainability of the aquaculture operation.

Post-Harvest Management: Post-harvest management in aquaculture refers to the handling, processing, and preservation of aquatic products after harvest to ensure quality and safety for consumers. This includes practices such as proper cleaning, chilling, and packaging of fish, shellfish, or other aquaculture products. The goal is to minimize spoilage, maintain freshness, and prevent contamination. Techniques such as cold storage (0-4°C) and rapid freezing are commonly used to extend shelf life. Effective post-harvest management is vital for maintaining the value and marketability of aquaculture products.

Post-stocking Management: Post-stocking management in aquaculture encompasses all activities performed after the aquatic organisms have been introduced into the culture system. This includes regular monitoring and maintenance of water quality parameters, such as temperature, dissolved oxygen, pH, and nutrient levels, to ensure a conducive environment for growth. Post-stocking management also involves feeding strategies, health management through disease prevention and treatment, and periodic sampling to assess growth rates and biomass. Effective post-stocking management ensures optimal growth, minimizes mortality, and maximizes production efficiency throughout the culture cycle.

Pre-stocking Management: Pre-stocking management in aquaculture involves a series of preparatory activities conducted before the introduction of aquatic organisms into the culture system. These activities include pond or tank preparation, such as drying, liming, and fertilization to enhance natural productivity and ensure optimal water quality. Water quality parameters like pH, alkalinity, and dissolved oxygen are adjusted to suitable levels for the species to be stocked. Pre-stocking management also involves the selection and acclimatization of healthy juveniles or fingerlings, ensuring they are disease-free and well-suited to the culture environment. Proper pre-stocking management is critical for the successful establishment and growth of the cultured species.

Primary Productivity: Primary productivity in aquaculture refers to the rate at which energy is converted by photosynthetic organisms, such as phytoplankton, into organic substances that form the base of the food web. This process is critical for supporting the growth of herbivorous and omnivorous species in the culture system. Primary productivity is influenced by factors such as light availability, nutrient concentrations (e.g., nitrogen and phosphorus), and water temperature. High primary productivity is often desirable in extensive and semi-intensive systems, as it can reduce the need for external feed inputs by naturally sustaining the food chain within the system.

Productivity Estimation: Productivity estimation in aquaculture involves assessing the potential yield of a culture system based on various factors such as species selection, stocking density, water quality, feeding practices, and management strategies. Estimation methods may include biomass sampling, growth rate analysis, and modeling of production parameters. Accurate productivity estimation is essential for planning and optimizing production cycles, managing inputs, and forecasting economic returns. It also helps in identifying potential bottlenecks or limiting factors in the culture system, allowing for adjustments to improve efficiency and sustainability.

Regulatory Framework: The regulatory framework for aquaculture encompasses the laws, regulations, and guidelines that govern the farming of aquatic species. This framework includes rules related to environmental protection, animal welfare, food safety, and the use of chemicals and antibiotics. Regulations vary by country and region but generally aim to promote sustainable practices, protect natural ecosystems, and ensure the safety of aquaculture products. Compliance with the regulatory framework is essential for maintaining a license to operate, accessing markets, and avoiding legal penalties.

Risk Management: Risk management in aquaculture involves identifying, assessing, and mitigating risks that could impact production, profitability, or environmental sustainability. These risks include disease outbreaks, environmental changes (e.g., temperature fluctuations, water quality deterioration), market volatility, and regulatory changes. Strategies for managing risk include diversifying species and production methods, implementing biosecurity measures, securing insurance coverage, and staying informed about emerging trends and technologies. Effective risk management is crucial for ensuring the long-term success and resilience of aquaculture businesses.

Running Water Culture: Running water culture, also known as flow-through systems, involves the continuous flow of water through tanks or raceways where aquatic species are cultured. This system ensures a constant supply of fresh water, which helps to maintain optimal water quality, particularly dissolved oxygen levels, and to remove waste products. Running water culture is commonly used for species such as trout and salmon, which thrive in well-oxygenated, cool water environments. The system's efficiency depends on the availability of a reliable water source and the ability to manage the flow rate to meet the specific needs of the cultured species, typically maintaining flow rates of 5-10 L/min per kg of fish.

Secondary Productivity: Secondary productivity in aquaculture is the generation of biomass by heterotrophic organisms, such as zooplankton, benthos, and fish, which consume primary producers (e.g., phytoplankton) or other secondary producers. This level of productivity is crucial for the growth of higher trophic level species in the aquaculture system. Secondary productivity depends on the availability of primary producers, water quality, and the overall health of the aquatic ecosystem. In aquaculture, managing secondary productivity is important for optimizing the growth of cultured species and ensuring the efficient transfer of energy through the food web.

Selective Breeding: Selective breeding in aquaculture is the process of choosing broodstock with desirable traits, such as fast growth, high survival rates, or disease resistance, to produce offspring with enhanced characteristics. This practice, similar to traditional livestock breeding, has led to significant improvements in aquaculture productivity. Selective breeding programs rely on careful genetic selection, record-keeping, and controlled breeding environments. Over multiple generations, selective breeding can result in strains that are better adapted to farming conditions and environmental challenges.

Semi-intensive Aquaculture: Semi-intensive aquaculture represents a middle ground between extensive and intensive systems, involving moderate input levels to enhance the natural productivity of the water body. Farmers may provide supplemental feeding, fertilization, and basic water management to support higher stocking densities, typically between 1,000 to 5,000 kg/ha. This system is commonly used in pond culture and can achieve higher yields than extensive aquaculture while maintaining relatively low environmental impacts. The success of semi-intensive aquaculture depends on careful management practices, including regular monitoring of water quality parameters such as dissolved oxygen, which should be kept above 4 mg/L, and pH levels maintained between 6.5 and 8.5.

Soil Analysis: Soil analysis in aquaculture is the process of evaluating the physical and chemical properties of pond soil to determine its suitability for supporting aquatic life. Key parameters include pH (optimal range 6.5 to 7.5), organic matter content, and nutrient levels, such as nitrogen and phosphorus. Soil analysis helps in identifying issues like acidification, nutrient deficiencies, or toxic substance accumulation, which can affect pond productivity and water quality. Based on the results, management practices such as liming, fertilization, or soil amendments can be implemented to improve the pond's suitability for aquaculture, enhancing the growth and health of the cultured species.

Soil Chemistry: Soil chemistry in aquaculture pertains to the study of soil properties, including pH, organic matter content, and nutrient availability, which influence the productivity of pond-based systems. The ideal soil pH for aquaculture ponds ranges between 6.5 and 7.5, promoting optimal nutrient availability and minimizing toxic metal solubility. High organic matter content supports the growth of beneficial microorganisms but requires careful management to prevent excessive oxygen consumption and the buildup of harmful substances such as hydrogen sulfide.

Soil Quality Management: Soil quality management in aquaculture is essential for maintaining the productivity and sustainability of pond-based culture systems. It involves the assessment and adjustment of soil parameters such as pH (optimal range 6.5 to 7.5), organic matter content, and nutrient levels (e.g., nitrogen, phosphorus). Proper soil management practices, including liming, fertilization, and regular monitoring, help to create a conducive environment for the growth of benthic organisms and the cultured species. Good soil quality supports the natural productivity of the pond, improves water retention, and reduces the risk of harmful conditions such as acidification or excessive nutrient buildup.

Soil Quality Parameters: Soil quality parameters in aquaculture ponds include factors such as pH (optimal range 6.5-7.5), organic matter content, and nutrient availability (e.g., nitrogen, phosphorus). Soil texture and permeability also influence water retention and aeration. Proper soil quality management involves liming to adjust pH, fertilization to enhance nutrient levels, and aeration to prevent anaerobic conditions that can lead to the accumulation of toxic compounds like hydrogen sulfide. Soil quality directly impacts water quality and the productivity of pond-based aquaculture systems.

Species Identification: Species identification in aquaculture is the process of accurately recognizing and classifying the species being cultured, which is critical for effective management, breeding programs, and disease control. Identification can be based on morphological characteristics, such as body shape, fin structure, and coloration, or through genetic methods, such as DNA barcoding or molecular markers. Correct species identification ensures that the right species are selected for specific culture environments and that hybrids or non-native species are avoided. It also plays a role in maintaining biodiversity and preventing the introduction of invasive species into natural ecosystems.

Stocking Density: Stocking density refers to the number of aquatic organisms, such as fish or shrimp, introduced into a given volume or area of water. It is a critical factor in aquaculture that influences growth rates, health, and overall production efficiency. Optimal stocking densities vary by species, production system, and management practices. For example, tilapia in intensive pond systems may be stocked at 2-3 fish/m^2, while shrimp in semi-intensive systems may be stocked at 10-20 individuals/m^2. Maintaining appropriate stocking densities ensures adequate space, oxygen availability, and waste management, reducing stress and disease outbreaks.

Super Intensive Aquaculture: Super intensive aquaculture is the most advanced and high-density form of aquaculture, often utilizing cutting-edge technologies such as recirculating aquaculture systems (RAS) and biofloc technology. These systems can support extremely high stocking densities, sometimes exceeding 100,000 kg/ha, with water quality maintained through continuous filtration, aeration, and biosecurity measures. Super intensive aquaculture is typically used for species with high market demand, such as shrimp and certain fish species. These systems are designed for maximum efficiency, with minimal water usage and waste production, making them suitable for regions with limited water resources. However, they require substantial investment, technical expertise, and energy inputs.

Sustainable Aquaculture: Sustainable aquaculture refers to farming practices that meet current production needs without compromising the ability of future generations to meet their needs. This includes minimizing environmental impact, conserving resources, promoting animal welfare, and ensuring social and economic viability. Key practices in sustainable aquaculture include the use of eco-friendly feed ingredients, waste management systems, and responsible water use. Sustainable practices also focus on reducing the carbon footprint of aquaculture operations and promoting biodiversity through integrated multi-trophic aquaculture (IMTA) or other ecosystem-based approaches.

Water Analysis: Water analysis in aquaculture involves the regular testing and monitoring of water quality parameters, such as pH, dissolved oxygen, ammonia, nitrite, nitrate, salinity, hardness, and temperature. These analyses are critical for maintaining optimal conditions for the cultured species and preventing problems such as disease outbreaks, poor growth, and mortality. Water analysis can be conducted using various methods, including chemical test kits, electronic meters, and laboratory analysis. Regular water analysis allows for timely adjustments to management practices, such as aeration, water exchange, or the application of water treatments, to ensure a healthy aquaculture environment.

Water Chemistry: Water chemistry in aquaculture involves the study and management of chemical parameters such as pH, alkalinity, hardness, ammonia, nitrite, nitrate, and dissolved oxygen in the aquatic environment. Optimal water chemistry is essential for the health and growth of aquaculture species. For instance, ammonia levels should be maintained below 0.02 mg/L, nitrite below 0.1 mg/L, and dissolved oxygen levels between 5-8 mg/L. Regular monitoring and adjustment of water chemistry are critical to prevent toxic conditions and ensure a conducive environment for aquaculture species.

Water Quality Management: Water quality management in aquaculture involves the continuous monitoring and regulation of key water parameters to maintain an optimal environment for the cultured species. Critical parameters include dissolved oxygen (typically above 5 mg/L), pH (6.5 to 8.5), temperature (species-dependent, generally between 20°C to 30°C), ammonia, nitrite, and nitrate levels, salinity, and turbidity. Effective water quality management ensures the health and growth of the cultured organisms by minimizing stress, preventing disease outbreaks, and optimizing feed efficiency. Techniques include aeration, water exchange, filtration, and the use of biofilters or chemical treatments to maintain water quality within acceptable ranges.

Water Quality Parameters: Water quality parameters in aquaculture include a range of physical and chemical factors that influence the health and productivity of cultured species. Key parameters include temperature (species-specific, e.g., 24°-28°C for tilapia), pH (6.5-8.5), dissolved oxygen (5-8 mg/L), ammonia (<0.02 mg/L), nitrite (<0.1 mg/L), and alkalinity (50-200 mg/L as $CaCO_3$). Regular monitoring and management of these parameters are critical to maintaining optimal conditions for growth, reproduction, and disease prevention. Water quality is influenced by factors such as stocking density, feeding practices, and system design.

Water Recirculation Systems: Water recirculation systems (RAS) in aquaculture are closed-loop systems that filter and reuse water, minimizing the need for water exchange and reducing environmental impact. These systems include mechanical, biological, and chemical filtration to remove solids, nitrogenous waste, and other contaminants. RAS allows for intensive farming of species at high densities with reduced water consumption and improved biosecurity. Optimal water quality parameters, such as ammonia levels below 0.02 mg/L and dissolved oxygen levels above 5 mg/L, are maintained through continuous monitoring and management.

Zero Water Exchange System: A Zero Water Exchange (ZWE) system in aquaculture is a closed or semi-closed system where water is recirculated and treated within the system, with minimal or no water exchange with the external environment. This system is designed to maintain water quality through filtration, biological treatment, and regular monitoring of parameters such as ammonia, nitrite, nitrate, and pH. ZWE systems are often used in high-density aquaculture operations, such as shrimp or tilapia farming, where water conservation and biosecurity are critical. These systems can reduce the environmental impact of aquaculture by minimizing effluent discharge and water usage, but they require advanced technology and management practices to prevent water quality deterioration.

CHAPTER 2

WATER AND SOIL QUALITY MANAGEMENT

Acidic Soil: Acidic soils have a pH below 6.5, which can affect nutrient availability and microbial activity. In aquaculture, acidic soils can lead to poor water quality, reduce the availability of essential nutrients, and increase the solubility of toxic metals like aluminum and iron. To mitigate the effects of acidic soils in aquaculture ponds, liming is commonly used to raise the pH to a more neutral range (6.5 to 8.0). This helps to stabilize the water environment, improve nutrient availability, and promote the growth of phytoplankton and other primary producers, ensuring a healthy environment for aquatic species.

Agricultural Soil Management: Agricultural soil management in aquaculture involves maintaining soil health through practices like crop rotation, organic amendments, and nutrient supplementation. Proper management ensures soil fertility, water retention, and reduced erosion, all of which are vital for integrated aquaculture systems where crops and fish are cultivated together.

Alkaline Soil: Alkaline soils have a pH above 8.0, often due to high concentrations of calcium carbonate or other alkaline salts. In aquaculture, alkaline soils can lead to high water pH levels, which can negatively impact the growth and survival of fish and other aquatic organisms. High pH can also reduce the availability of nutrients like phosphorus and cause imbalances in water chemistry. Managing alkaline soils in aquaculture ponds often involves the application of acidifying agents, such as sulfur or gypsum, to lower the pH and improve nutrient availability, creating a more suitable environment for aquatic life.

Alkalinity (Water Quality Parameter): Alkalinity measures the water's ability to neutralize acids, primarily due to the presence of bicarbonates (HCO_3^-), carbonates (CO_3^{2-}), and hydroxides (OH^-). It is expressed in milligrams per liter of calcium carbonate (mg/L as $CaCO_3$). In aquaculture, alkalinity acts as a buffer against pH fluctuations, ensuring a stable environment for aquatic organisms. Optimal alkalinity levels vary, with freshwater systems typically requiring 50 to 200 mg/L as $CaCO_3$, while brackish and marine systems may need higher levels. Proper alkalinity management helps prevent stress and improve fish health.

Alum Application: Alum (aluminum sulfate) is used in aquaculture to clarify water by precipitating suspended solids and reducing phosphorus levels, which helps control algal blooms. When added to water, alum reacts with phosphorus to form insoluble aluminum phosphate, which settles to the bottom, effectively reducing phosphorus availability in the water column. The application of alum is typically in the range of 15-30 mg/L, depending on the initial phosphorus concentration. However, care must be taken to avoid over-application, which can lower pH and negatively affect aquatic organisms, especially if the alkalinity of the water is below 20 mg/L as $CaCO_3$.

Amelioration of Problem Soils: Amelioration of problem soils involves the use of various techniques to improve the physical and chemical properties of soils that are unsuitable for aquaculture due to issues like acidity, alkalinity, salinity, or waterlogging. Common amelioration methods include liming to raise the pH of acidic soils, gypsum application to reduce alkalinity and sodium content, and organic matter addition to enhance soil structure and fertility. In aquaculture, amelioration of problem soils is crucial for optimizing pond conditions, ensuring proper water retention, and promoting the growth of aquatic organisms by improving the overall quality of the soil.

Ammonia (Water Quality Parameter): Ammonia (NH_3 and NH_4^+) is a nitrogenous waste product excreted by fish and other aquatic organisms. It exists in two forms: unionized ammonia (NH_3), which is toxic, and ionized ammonia (NH_4^+), which is less harmful. The relative concentration of each form depends on pH and temperature. In aquaculture, NH_3 levels should be kept below 0.05 mg/L to prevent toxicity, which can cause gill damage, stress, and mortality. Biological filtration and regular water exchange are key to managing ammonia levels in aquaculture systems.

Analytical Instruments: Analytical instruments in aquaculture refer to devices used for precise measurements of water and soil quality parameters, essential for maintaining optimal conditions in ponds and tanks. Instruments like spectrophotometers, ion-selective electrodes, and gas chromatographs are employed to measure nutrients (e.g., nitrate, phosphate), toxic compounds, and other water quality indicators. These instruments provide high sensitivity and accuracy, enabling detailed monitoring of parameters that influence aquatic species' health, such as nitrogen compounds (e.g., ammonia, which should be kept below 0.02 mg/L).

Available Nitrogen (Soil Quality Parameter): Available nitrogen refers to the forms of nitrogen in the soil that are accessible to plants and microorganisms, primarily as ammonium (NH_4^+) and nitrate (NO_3^-). In aquaculture, available nitrogen in pond soils plays a critical role in supporting primary productivity, which forms the base of the food chain for fish and other aquatic organisms. Ideal available nitrogen levels in aquaculture pond soils range from 50 to 150 mg/kg. Excessive nitrogen can lead to nutrient imbalances and algal blooms, while insufficient levels may reduce productivity, necessitating careful management of nitrogen inputs.

Available Nitrogen Analysis (Practical): Available nitrogen in soil, primarily in the form of ammonium (NH_4^+) and nitrate (NO_3^-), is measured using the alkaline permanganate method. Adequate nitrogen levels in pond soil (50–150 mg/kg) promote primary productivity but excessive nitrogen can cause eutrophication and algae blooms.

Available Phosphorus (Soil Quality Parameter): Available phosphorus represents the portion of phosphorus in the soil that is accessible to plants and aquatic organisms. Phosphorus is a vital nutrient for photosynthesis and growth, but its availability is often limited in soils due to fixation by minerals. In aquaculture, phosphorus levels in pond soils should be carefully managed to prevent deficiency or excess, which can lead to poor growth or eutrophication. Optimal available phosphorus levels in pond soils typically range from 5 to 15 mg/kg. Fertilization or soil amendments may be necessary to adjust phosphorus levels to support aquaculture productivity.

Available Phosphorus Analysis (Practical): Available phosphorus in soil is determined using the Olsen method, which involves extraction with sodium bicarbonate and colorimetric analysis. Phosphorus levels between 10 to 30 mg/kg are ideal for promoting primary productivity in aquaculture ponds. Higher levels may lead to algal blooms, which can impact water quality and fish health.

BOD (Biological Oxygen Demand): Biological Oxygen Demand (BOD) is a measure of the amount of oxygen consumed by microorganisms to decompose organic matter in water. It is determined using a 5-day incubation test at 20°C. High BOD levels, typically above 6 mg/L, indicate poor water quality and can lead to oxygen depletion, which stresses or kills aquatic organisms. In aquaculture, maintaining a BOD level below 3 mg/L is essential to ensure healthy water conditions.

Buffering Capacity of Soil: The buffering capacity of soil refers to its ability to resist changes in pH when acids or bases are added. This property is primarily influenced by the soil's carbonate content, clay minerals, and organic matter. In aquaculture, soils with good buffering capacity (pH 6.5 to 8.5) help stabilize water pH, preventing rapid fluctuations that can cause stress to aquatic organisms. Liming is often used to improve buffering capacity in acidic soils.

Chemical Fertilizers: Chemical fertilizers are inorganic compounds that provide essential nutrients to soil and water in a readily available form. In aquaculture, fertilizers such as urea (for nitrogen), superphosphate (for phosphorus), and potassium chloride (for potassium) are used to enhance the productivity of aquatic ecosystems. Fertilizers promote the growth of phytoplankton and aquatic plants, which serve as the primary food source for many cultured species. The use of chemical fertilizers in aquaculture should be balanced and carefully monitored, as excessive application can lead to nutrient imbalances, water pollution, and harmful algal blooms.

Chloride Measurement: Chloride ions (Cl^-) are a major component of salinity in both marine and brackish water. Chloride concentration can be measured using the argentometric titration method with silver nitrate. Chloride levels are critical for osmotic balance in aquatic organisms and must be monitored, especially in freshwater aquaculture systems where optimal levels typically range from 10 to 300 mg/L. Variations in chloride can affect fish physiology and water quality.

Chloride: Chloride (Cl^-) is a major anion found in both freshwater and marine environments, derived from the dissociation of salts such as sodium chloride (NaCl). It is a critical ion for osmoregulation in aquatic organisms and contributes to the overall salinity of water. In aquaculture, chloride levels must be monitored to ensure they fall within species-specific tolerances. For instance, freshwater fish generally thrive with chloride levels below 250 mg/L, while marine species can tolerate levels much higher due to the natural salinity of seawater.

Chlorination: Chlorination in aquaculture involves the use of chlorine compounds, such as sodium hypochlorite or calcium hypochlorite, to disinfect water, equipment, and facilities, effectively controlling pathogens and preventing disease outbreaks. Chlorine is a potent oxidizing agent that kills bacteria, viruses, and other microorganisms. However, it is toxic to aquatic life, so chlorinated water must be dechlorinated before being introduced into aquaculture systems. Typical chlorine concentrations for disinfection range from 2-10 mg/L, and residual chlorine levels should be reduced to less than 0.003 mg/L before reintroducing aquatic species.

Coastal Soil Characteristics: Coastal soils, often influenced by saline or brackish water, are characterized by higher salinity, sodicity, and unique mineral compositions. These soils require proper management in aquaculture to prevent salinization and nutrient imbalance, ensuring the health and productivity of ponds located in coastal areas.

COD (Chemical Oxygen Demand): Chemical Oxygen Demand (COD) measures the total quantity of oxygen required to chemically oxidize organic and inorganic matter in water. It is determined by oxidizing the sample with a strong oxidant, usually potassium dichromate. COD levels in aquaculture systems should ideally be below 100 mg/L to avoid water quality deterioration. High COD values often reflect excess organic waste from feed and fish excreta, leading to reduced dissolved oxygen.

Conductivity (Water Quality Parameter): Conductivity measures the ability of water to conduct electrical current, which depends on the concentration of dissolved ions, such as salts. It is usually expressed in microsiemens per centimeter (µS/cm). In aquaculture, conductivity is used to monitor salinity and ion levels, which are crucial for osmoregulation in fish and other aquatic species. Freshwater systems typically have conductivities between 50 and 1,500 µS/cm, while seawater has much higher conductivity (around 50,000 µS/cm). Monitoring conductivity helps maintain optimal ionic balance for aquatic life.

Deodorizers: Deodorizers in aquaculture are substances used to neutralize or mask unpleasant odors caused by the decomposition of organic matter, such as uneaten feed and fish waste. These odors are often associated with the production of hydrogen sulfide (H_2S) and ammonia, which are toxic to fish at concentrations above 0.01 mg/L and 0.02 mg/L, respectively. Deodorizers may include chemical agents that bind or neutralize these compounds or biological agents like probiotics that enhance the breakdown of organic matter and reduce odor emissions.

Dilute Solutions: Dilute solutions contain a relatively low concentration of solutes compared to the solvent. In water chemistry, dilute solutions are often used in titration and other analytical procedures to measure water quality parameters such as hardness, alkalinity, or nutrient concentrations. The behavior of solutes in dilute solutions follows ideal solution laws, making calculations predictable. In aquaculture, maintaining appropriate concentrations of dissolved nutrients and gases is critical for optimal growth and health of aquatic organisms.

Dissolved Oxygen Measurement: Dissolved oxygen (DO) is the concentration of oxygen dissolved in water, critical for respiration in aquatic organisms. It is measured using an oxygen meter or Winkler titration method. DO levels should be maintained above 5 mg/L for most aquaculture species. Low DO can lead to stress, reduced growth, and even mortality, while high levels of organic matter or excess feed can deplete DO through microbial activity.

Dissolved Oxygen: Dissolved oxygen (DO) refers to the amount of oxygen gas (O_2) dissolved in water, which is vital for the respiration of aquatic organisms. DO levels are typically measured in milligrams per liter (mg/L). In aquaculture, optimal DO levels vary by species but generally range from 5 to 7 mg/L for most fish. Low DO levels can lead to hypoxia, stressing or even killing fish and other aquatic organisms. Factors such as water temperature, salinity, and organic load affect DO levels, making regular monitoring essential.

Environmental Amelioratives: Environmental amelioratives are agents or practices used in aquaculture to improve water and soil quality, creating a healthier environment for aquatic species. These can include water conditioners, probiotics, organic amendments, and substances like zeolites or lime to neutralize toxins, enhance microbial activity, and stabilize pH. The application of environmental amelioratives is critical in maintaining optimal conditions such as dissolved oxygen levels above 5 mg/L, pH between 6.5 and 8.5, and reducing the buildup of harmful substances like ammonia, which should be kept below 0.02 mg/L.

Eutrophication: Eutrophication is the process where water bodies become enriched with excessive nutrients, primarily nitrogen and phosphorus, leading to rapid algal blooms. This can result in oxygen depletion, fish kills, and overall poor water quality. In aquaculture, eutrophication is often caused by overfeeding, improper waste management, and excess fertilizer application. Effective nutrient management and water aeration are crucial to prevent eutrophication in ponds.

Fertilizer Application: Fertilizer application in aquaculture ponds is aimed at promoting the growth of phytoplankton, which serves as a primary food source for many aquatic species. Fertilizers can be organic (e.g., manure) or inorganic (e.g., urea, superphosphate). The choice of fertilizer and its application rate depend on the nutrient content of the pond soil and water. For instance, inorganic fertilizers are often applied at a rate of 5-10 kg/ha for nitrogen and 2-4 kg/ha for phosphorus to achieve balanced nutrient levels that support phytoplankton blooms without causing excessive eutrophication.

Free Carbon Dioxide Measurement: Free carbon dioxide (CO_2) is a byproduct of respiration and decomposition in aquatic systems. It can be measured through titration with sodium hydroxide. High levels of CO_2, typically above 20 mg/L, can cause stress to fish by reducing oxygen uptake. Optimal levels are below 5 mg/L for most aquaculture species. Maintaining aeration and balanced stocking densities can help control CO_2 levels.

Free Carbon Dioxide: Free carbon dioxide (CO_2) in water is a byproduct of respiration and decomposition processes. In aquaculture, elevated CO_2 levels can lower water pH and interfere with fish respiration by reducing the blood's ability to carry oxygen. Ideally, CO_2 concentrations should remain below 10 mg/L in most aquaculture systems. CO_2 management is critical in recirculating aquaculture systems (RAS), where buildup can occur due to limited water exchange. Proper aeration and degassing methods are used to control CO_2 levels.

Ground Water Composition: Ground water is water that resides in subsurface aquifers and is generally characterized by higher mineral content compared to surface water. Its composition depends on the geology of the aquifer and can include dissolved minerals like calcium, magnesium, and iron, as well as bicarbonates. In some regions, groundwater may also contain contaminants like nitrates or heavy metals due to agricultural or industrial activities. Groundwater is often used in aquaculture systems where stable water temperature and mineral content are beneficial for certain species.

Groundwater Quality: Groundwater quality in aquaculture is essential for maintaining healthy water parameters. It is assessed by measuring parameters such as salinity, pH, hardness, and nutrient content. Groundwater should have salinity levels below 1 ppt and be free from heavy metals or pollutants to prevent negative impacts on aquatic life and pond ecosystems.

Gypsum Application: Gypsum (calcium sulfate) is used in aquaculture primarily to improve water hardness and clarity. It is beneficial in ponds with soft water, where calcium levels are insufficient for fish health and for promoting proper molting in crustaceans like shrimp. Gypsum also helps flocculate suspended clay particles, reducing turbidity. The recommended application rate for gypsum in aquaculture ponds ranges from 100 to 500 kg/ha, depending on the water's hardness and the specific requirements of the cultured species. The ideal water hardness for most freshwater aquaculture species is 20-150 mg/L as $CaCO_3$.

Hardness (Water Quality Parameter): Water hardness refers to the concentration of divalent cations, primarily calcium (Ca^{2+}) and magnesium (Mg^{2+}), in water. It is measured in milligrams per liter as calcium carbonate (mg/L as $CaCO_3$). In aquaculture, hardness influences osmoregulation in fish and shellfish. Soft water (hardness < 50 mg/L as $CaCO_3$) can stress organisms adapted to harder water, while very hard water (> 300 mg/L as $CaCO_3$) can lead to scaling and mineral buildup. Most freshwater species thrive in moderately hard water (75-200 mg/L as $CaCO_3$).

Heavy Metal Contamination in Soil: Heavy metal contamination refers to the accumulation of metals like lead (Pb), mercury (Hg), cadmium (Cd), and arsenic (As) in soil, often from industrial runoff or polluted water sources. These metals can accumulate in aquatic organisms, leading to toxic effects and health hazards for humans consuming contaminated seafood. In aquaculture, regular monitoring of soil and water for heavy metals is necessary to ensure food safety and environmental sustainability.

Hydrogen Bonding: Hydrogen bonding occurs when a hydrogen atom covalently bonded to an electronegative atom (such as oxygen in water) interacts with another electronegative atom nearby. In water, hydrogen bonds form between the hydrogen of one water molecule and the oxygen of another. These bonds are responsible for water's high boiling point, high specific heat, and the solid-liquid phase behavior. In aquaculture, hydrogen bonding contributes to the unique physical properties of water, which are critical for temperature regulation and nutrient transport in aquatic environments.

Laboratory Glassware: Laboratory glassware refers to a variety of tools made of glass used in scientific research, analysis, and experimentation. In aquaculture labs, glassware such as beakers, flasks, graduated cylinders, pipettes, and Petri dishes are essential for measuring, mixing, and observing chemical and biological samples. The glass used is typically borosilicate, which resists thermal shock and chemical corrosion. Proper handling and regular cleaning of glassware are crucial to ensure accuracy in analyses such as water quality testing (e.g., pH, salinity, dissolved oxygen).

Lime in Soil and Water: Lime is commonly used in aquaculture to adjust soil and water pH, increase the availability of essential nutrients, and reduce the toxicity of harmful substances. The application of agricultural lime (calcium carbonate) raises soil pH to the optimal range of 6.5-8.5, promoting healthy microbial activity and nutrient availability. In water, lime can increase alkalinity and hardness, buffering against pH fluctuations. The dosage of lime depends on the initial pH and alkalinity levels, with typical applications ranging from 1,000 to 2,000 kg/ha for pond soils.

Lime Requirement (Soil Quality Parameter): Lime requirement refers to the amount of lime (calcium carbonate or other alkaline substances) needed to raise the pH of acidic soils to a desired level. In aquaculture, lime is often applied to pond soils to neutralize acidity, improve water quality, and enhance the availability of nutrients like phosphorus. The lime requirement is determined by the initial soil pH, buffer capacity, and target pH. Typically, soils with pH values below 6.5 require liming, with application rates varying based on soil type and acidity level. Proper liming ensures optimal conditions for aquatic life by stabilizing pH and enhancing nutrient cycling.

Lime Requirement Determination (Practical): The lime requirement of soil is measured to adjust soil pH to optimal levels for aquaculture, typically using a buffer solution or titration. Lime is applied to raise soil pH, particularly in acidic ponds with pH below 6.5, where lime requirements can range from 1 to 5 tons/ha, depending on the initial pH and soil texture.

Manures: Manures are organic materials derived from the decomposition of animal feces, plant residues, and other organic matter. They are commonly used in aquaculture to enrich pond soils and water with nutrients such as nitrogen, phosphorus, and organic carbon. Manures support the growth of phytoplankton and other primary producers, which are crucial for sustaining fish and shrimp populations in pond-based systems. Common types of manures used in aquaculture include cattle dung, poultry manure, and compost. The application of manure should be carefully managed to prevent over-fertilization, which can lead to water quality issues like eutrophication.

Micronutrients in Soil and Water: Micronutrients such as zinc (Zn), copper (Cu), iron (Fe), and manganese (Mn) are essential for the growth and metabolism of aquatic plants and animals, albeit required in trace amounts. In aquaculture, the availability of these micronutrients is influenced by factors like pH, redox potential, and organic matter content. For example, iron is most bioavailable in slightly acidic conditions (pH 6-7), while zinc availability decreases in alkaline soils. Micronutrient deficiencies can impair aquatic plant growth and nutrient cycling, while excess levels may lead to toxicity.

Molality: Molality (m) is a concentration unit defined as the number of moles of solute per kilogram of solvent. Unlike molarity, molality is temperature-independent, making it useful in scenarios where temperature changes can affect solution volume. In water chemistry, molality is important in colligative property calculations, such as freezing point depression and boiling point elevation, which can influence aquatic systems. In aquaculture, molality may be used in the precise formulation of salt solutions for osmotic balance experiments.

Molarity: Molarity (M) is a unit of concentration that expresses the number of moles of a solute per liter of solution. In water quality management, molarity is often used to prepare solutions for titration or calibration in analytical procedures. For example, a 1 M solution of sodium chloride contains 1 mole of NaCl (58.44 g) dissolved in 1 liter of water. In aquaculture, molarity is crucial for preparing chemical treatments, such as pH buffers or nutrient solutions, to maintain optimal water conditions for aquatic organisms.

Nitrate Measurement: Nitrates (NO_3^-) are the final product of the nitrification process in the nitrogen cycle, measured via colorimetric methods or ion-selective electrodes. Elevated nitrate levels, typically above 50 mg/L, can be toxic to aquatic life and contribute to eutrophication. In aquaculture systems, nitrate levels should ideally be kept below 20 mg/L to avoid water quality degradation and fish stress.

Nitrates (Water Quality Parameter): Nitrates (NO_3^-) are the end products of nitrogenous waste decomposition and nitrification in aquatic systems. They are less toxic than ammonia or nitrites but can still be harmful at high concentrations, leading to algal blooms and eutrophication. In aquaculture, nitrate levels should ideally be kept below 100 mg/L, with optimal ranges varying based on species and system type. Regular monitoring and proper management, such as water exchange or biofiltration, help control nitrate levels and maintain water quality.

Normality: Normality (N) is a measure of concentration equivalent to the molarity of reactive species in a solution. It is defined as the number of equivalents of solute per liter of solution and is often used in acid-base titrations and redox reactions. For example, a 1 N solution of hydrochloric acid (HCl) contains 1 mole of H^+ ions per liter. In aquaculture, normality can be used to calculate the correct dosage of chemicals for water treatments, such as adjusting alkalinity or acidity.

Nutrient Cycling in Soil: Nutrient cycling in soil refers to the movement and transformation of essential elements like nitrogen, phosphorus, and carbon through biological, chemical, and physical processes. In aquaculture, proper nutrient cycling ensures a continuous supply of nutrients for primary productivity and the maintenance of water quality. Effective soil and pond management practices like organic matter control and liming help optimize nutrient cycling and prevent the buildup of harmful substances.

Organic Carbon (Soil Quality Parameter): Organic carbon is a key indicator of soil health and fertility, representing the amount of carbon stored in soil organic matter. It influences nutrient cycling, water retention, and soil structure. In aquaculture, organic carbon levels in pond soils affect the decomposition of organic material, nutrient availability, and overall water quality. Ideal organic carbon content in aquaculture ponds is between 0.5% and 2%, depending on the species being farmed and the pond management practices. Proper management of organic carbon helps support a healthy ecosystem in aquaculture systems.

Organic Carbon Analysis (Practical): Organic carbon in soil is measured using the Walkley-Black method, which involves oxidizing organic matter with potassium dichromate and sulfuric acid, followed by titration. Organic carbon levels indicate soil fertility and microbial activity. In aquaculture, soil organic carbon should ideally range from 0.5% to 2%, as excessive organic matter can lead to anoxic conditions.

pH (Water Quality Parameter): pH is a measure of the hydrogen ion concentration in water, indicating its acidity or alkalinity on a scale of 0 to 14. In aquaculture, pH levels directly affect fish physiology, the availability of nutrients, and the toxicity of compounds like ammonia. Freshwater species generally thrive in a pH range of 6.5 to 8.5, while marine species prefer a range of 7.5 to 8.5. Extreme pH levels can lead to stress, poor growth, and even mortality in aquatic organisms, making pH monitoring critical.

pH Measurement (Practical): pH measurement in aquaculture is vital for monitoring water quality, as pH levels influence the availability of nutrients and the toxicity of compounds like ammonia. The optimal pH range for most aquaculture systems is between 6.5 and 8.5. pH is measured using either a pH meter or pH paper, with the meter providing more accurate readings. Regular pH monitoring helps prevent stress in aquatic species, which can occur when pH fluctuates outside the optimal range.

Phosphate Measurement: Phosphates ($PO_4^{\ -}$) are nutrients that contribute to primary productivity, but excessive levels can lead to eutrophication. Phosphate levels are measured colorimetrically using molybdenum blue. In aquaculture, phosphate concentrations should be maintained below 0.05 mg/L to prevent algal blooms and poor water quality. Balanced feed management and effluent treatment can control phosphate levels.

Phosphates (Water Quality Parameter): Phosphates ($PO_4^{\ -}$) are essential nutrients for plant and algal growth, but excessive concentrations can lead to eutrophication and water quality degradation. In aquaculture, phosphates originate from feed, fertilizers, and organic waste. Optimal phosphate levels depend on the system; for example, levels below 0.1 mg/L are recommended for most freshwater systems to prevent algal blooms. In intensive systems, phosphate management is critical to prevent the overgrowth of algae, which can deplete oxygen and produce toxins harmful to fish.

Physical Properties of Soil: The physical properties of soil, including texture, structure, density, porosity, and water-holding capacity, play a crucial role in determining soil health and suitability for aquaculture pond construction. Soil texture, determined by the proportion of sand, silt, and clay particles, influences water retention and drainage. For aquaculture, clay loam soils with good water retention are preferred for pond construction, while sandy soils with high permeability may require additional lining. Understanding and managing soil physical properties ensures the stability and functionality of aquaculture ponds.

Polarity of Water: The polarity of water arises due to the difference in electronegativity between oxygen and hydrogen atoms. This creates a dipole moment with the oxygen end being slightly negative and the hydrogen end being slightly positive. The polar nature of water makes it an excellent solvent for ionic and polar substances, enabling the dissolution of essential nutrients, gases, and minerals in aquatic environments. This property is vital in aquaculture for maintaining water quality and ensuring the bioavailability of nutrients.

Pond Soil Management: Pond soil management focuses on maintaining the physical and chemical properties of pond bottoms to support aquaculture. Regular soil monitoring, liming, organic matter management, and proper fertilization ensure optimal nutrient availability, reduced acidification, and balanced microbial activity, thus promoting healthy aquatic environments.

Potassium (Soil Quality Parameter): Potassium (K) is an essential macronutrient in soil that plays a vital role in plant growth, particularly in water regulation, enzyme activation, and photosynthesis. In aquaculture, potassium influences the productivity of aquatic plants and algae, which are important for oxygen production and as food sources for cultured species. Optimal potassium levels in pond soils should be in the range of 60 to 150 mg/kg. Managing potassium levels is important for maintaining a balanced nutrient environment in aquaculture ponds, ensuring the health and productivity of the ecosystem.

Pure Water Properties: Pure water, with a pH of 7 at 25°C, is a neutral substance that exhibits a high specific heat capacity (4.18 J/g°C), surface tension (72.8 mN/m at 20°C), and dielectric constant. These properties allow water to regulate temperature and support life. Pure water has no dissolved minerals or impurities, making it an ideal medium for studying water chemistry. In aquaculture, deviations from the properties of pure water due to dissolved salts, gases, or pollutants can impact organism health and require careful monitoring.

Saline Water Intrusion: Saline water intrusion occurs when saltwater infiltrates freshwater systems or soils, often due to over-extraction of groundwater or rising sea levels. This process can lead to soil salinization, negatively impacting freshwater aquaculture by reducing water quality and affecting fish growth and reproduction. Coastal aquaculture farms are particularly vulnerable, requiring proper water management practices like constructing barriers or switching to salt-tolerant species.

Salinity: Salinity refers to the concentration of dissolved salts in water, typically measured in parts per thousand (ppt) or practical salinity units (PSU). It is a critical parameter in aquaculture, as different species have specific salinity requirements. For example, freshwater species like catfish thrive in salinities below 0.5 ppt, while marine species like shrimp prefer salinities around 20 to 35 ppt. Salinity affects osmoregulation, growth, and reproduction in aquatic organisms, making its management essential in both freshwater and marine aquaculture systems.

Sea Water Properties: Sea water is a complex solution containing dissolved salts, organic compounds, and gases, with an average salinity of 35 ppt (parts per thousand). The composition includes major ions such as chloride, sodium, magnesium, sulfate, and calcium. Sea water has a lower freezing point (-1.9°C) and higher density (1.025 g/cm at 4°C) than freshwater. The presence of dissolved salts affects osmotic balance in marine organisms, making sea water composition a critical factor in aquaculture, particularly for species with specific salinity requirements.

Seawater Composition: Seawater is primarily composed of water (about 96.5%) and salts (about 3.5%), with sodium chloride being the most abundant. It also contains minor components such as sulfate, magnesium, calcium, potassium, and trace elements like iodine and fluoride. The composition of seawater is relatively constant across the globe, making it suitable for marine aquaculture. However, variations in nutrient levels, pollutants, and dissolved oxygen can significantly impact marine organisms, making regular monitoring essential in aquaculture operations.

Sediment-Water Interface: The sediment-water interface is the boundary layer where pond or lake bottom sediments meet overlying water. It plays a crucial role in nutrient cycling, as nutrients like nitrogen and phosphorus can be released from the sediments into the water, influencing water quality and primary productivity. In aquaculture ponds, maintaining a healthy sediment-water interface is critical to prevent nutrient overload, oxygen depletion, and the buildup of harmful gases like hydrogen sulfide.

Soil Aeration: Soil aeration refers to the process of introducing air into the soil, which enhances oxygen availability for microbial activity and nutrient cycling. In aquaculture ponds, poor soil aeration can lead to anaerobic conditions, resulting in the production of harmful gases like methane and hydrogen sulfide. Regular aeration of pond bottoms and removal of excess organic matter are essential practices to maintain oxygenated soils and ensure a healthy aquatic environment.

Soil Amendments: Soil amendments in aquaculture refer to materials added to pond soils to improve physical, chemical, and biological properties. These amendments can help regulate pH, enhance nutrient availability, and reduce harmful compounds. Common amendments include lime, organic matter, and gypsum. Lime, for instance, is used to raise pH and increase calcium and magnesium availability, with an optimum pH range of 6.5-8.5 for aquaculture ponds. The type and quantity of soil amendments applied depend on the soil's texture, pH, organic matter content, and the specific requirements of the species being cultured.

Soil Classification: Soil classification is the process of categorizing soils based on their physical and chemical properties, such as texture, structure, and mineral content. Common classification systems include the USDA Soil Taxonomy and the World Reference Base for Soil Resources (WRB). In aquaculture, soil classification helps in determining the suitability of a site for pond construction and the necessary amendments. For example, clay soils are preferred for pond linings, while acidic soils may need liming to adjust pH. Accurate classification ensures that the soil management practices align with the requirements of the aquaculture species.

Soil Conductivity Analysis (Practical): Soil conductivity measures the ability of soil to conduct electrical current, which correlates with the concentration of soluble salts. It is measured using a conductivity meter, and in aquaculture, optimal soil conductivity values should be between 1 to 4 dS/m. High soil salinity can affect plant growth and the quality of water in ponds.

Soil Conductivity: Soil conductivity, or electrical conductivity (EC), measures the ability of soil to conduct electrical current, which is related to the concentration of soluble salts in the soil. High conductivity indicates elevated salinity levels, which can affect water quality in aquaculture ponds and influence the growth and health of aquatic organisms. In aquaculture, managing soil conductivity is important to prevent excessive salinity, which can stress or harm fish and other cultured species. Ideal soil conductivity for freshwater aquaculture should be below 4 dS/m (deciSiemens per meter) to avoid negative impacts on water quality.

Soil Distribution in Tamil Nadu: Tamil Nadu features diverse soil types due to its varied climate, topography, and geological formations. The major soil groups in the state include red soils, black soils, alluvial soils, lateritic soils, and saline/alkaline soils. Red soils, found in hilly and plateau regions, are suitable for rainfed crops, while black soils, known for their water retention capacity, are prominent in the river basins. Alluvial soils are found along the coastal plains, and lateritic soils are common in high rainfall areas. In aquaculture, understanding the local soil distribution helps in selecting suitable sites for ponds and managing soil quality for optimum water retention and fertility.

Soil Organic Matter: Soil organic matter (SOM) refers to decomposed plant and animal residues, including humus, in the soil. It is crucial for nutrient retention, water-holding capacity, and soil structure. In aquaculture, soils with 1–2% organic matter are considered optimal, as they provide nutrients for primary productivity while avoiding excessive decomposition, which can lead to oxygen depletion and the release of harmful gases in pond environments.

Soil pH: Soil pH is a measure of the acidity or alkalinity of soil, expressed on a scale of 0 to 14. It influences nutrient availability, microbial activity, and chemical processes in the soil. In aquaculture, soil pH affects water quality and the growth of aquatic organisms. Most aquaculture ponds require a neutral to slightly alkaline soil pH (6.5 to 8.5) to optimize nutrient availability and prevent the release of toxic substances like ammonia. Acidic soils may need liming to increase pH, while highly alkaline soils might require gypsum to lower pH.

Soil Probiotics: Soil probiotics in aquaculture are live microbial formulations applied to pond soils to improve soil health and nutrient cycling. These probiotics contain beneficial bacteria and fungi that enhance the decomposition of organic matter, reduce the buildup of harmful metabolites like hydrogen sulfide and ammonia, and promote the availability of nutrients such as nitrogen and phosphorus. The use of soil probiotics helps maintain a balanced microbial ecosystem, improving the overall productivity of the pond. Optimal application rates vary depending on the product but typically range from 1-2 kg/ha.

Soil Redox Potential: Soil redox potential (Eh) is a measure of the soil's tendency to gain or lose electrons, indicating the oxidation-reduction conditions. It is crucial for determining the availability of nutrients and the presence of toxic substances like hydrogen sulfide. In aquaculture ponds, maintaining positive redox potential (above 200 mV) is essential for healthy benthic environments. Negative redox conditions can lead to the release of toxic compounds that harm fish and other aquatic organisms.

Soil Sample Collection: Soil sample collection involves gathering representative soil samples from different locations and depths within an aquaculture site to assess the soil's physical and chemical properties. Proper collection techniques are essential to avoid contamination and ensure accurate results. In aquaculture, soil samples are typically collected from pond bottoms or potential pond construction sites. Samples should be taken from multiple spots to account for variability and mixed to create a composite sample. Properly collected soil samples provide crucial information for managing soil quality, improving water retention, and adjusting nutrient levels.

Soil Sample Preservation: Soil sample preservation refers to the proper handling and storage of soil samples to maintain their original physical and chemical characteristics until analysis. Samples should be air-dried or stored in sealed containers to prevent moisture loss or contamination. In aquaculture, preserving soil samples correctly is essential for accurate analysis of parameters like pH, organic matter, and nutrient content. For instance, air-drying helps stabilize the organic matter and nutrient content, which is crucial for assessing the fertility and amendment requirements of the soil for pond construction or management.

Soil Sample Processing: Soil sample processing involves collecting, drying, and sieving soil to remove debris and homogenize the sample for analysis.

Typically, soil is air-dried at room temperature and sieved through a 2 mm mesh. This ensures that the sample is representative of the pond or farm's soil conditions and can be used for nutrient, pH, and organic matter analysis.

Soil Texture Analysis: Soil texture refers to the relative proportion of sand, silt, and clay in a soil sample. The hydrometer method or sieve analysis is used to determine these proportions. Texture influences water retention, aeration, and nutrient availability in aquaculture ponds. A balanced texture with a moderate clay content (20–30%) is preferred for aquaculture ponds to maintain water quality and reduce seepage.

Soil Texture: Soil texture refers to the relative proportions of sand, silt, and clay particles in soil. It significantly influences water infiltration, retention, and drainage, as well as nutrient availability and root penetration. Soil texture classes include sandy, loamy, and clayey soils. In aquaculture, clay or loamy soils are ideal for constructing ponds due to their ability to hold water, while sandy soils, which drain quickly, may require lining to prevent water loss. Proper analysis of soil texture helps ensure the success of aquaculture operations by optimizing pond conditions for water management.

Soil Types: Soil types are classifications based on the composition, texture, structure, and other physical and chemical properties of soil. The major soil types include sandy soils, clayey soils, loamy soils, peaty soils, saline soils, and silty soils. Each soil type has unique characteristics that affect water retention, drainage, and fertility, which are critical considerations in aquaculture. For instance, clayey soils, with their fine particles and low permeability, are often preferred for aquaculture ponds because they help retain water, while sandy soils with high permeability may require sealing to minimize water loss.

Solution Preparation: Solution preparation involves creating mixtures of known concentration for use in various laboratory analyses in aquaculture. This process requires accurate weighing of solutes (e.g., salts, acids) and careful dilution with solvents, typically distilled water, to create solutions of specific molarity or normality. Properly prepared solutions are essential for titrations, standardizations, and other tests, such as determining the alkalinity or hardness of pond water. Maintaining consistency in solution preparation ensures the reliability of experimental results.

Standard Solutions: A standard solution is a solution with a precisely known concentration of a solute. In water quality analysis, standard solutions are used for calibration and to ensure accurate measurements of various parameters, such as pH, dissolved oxygen, and salinity. The preparation of standard solutions requires careful measurement and dilution, often using reagents with known purity. For example, a 0.1 M sodium chloride standard solution is commonly used in conductivity calibration.

Surface Water Composition: Surface water includes water in rivers, lakes, and reservoirs. Its composition is influenced by the geological characteristics of the surrounding area, runoff, and human activities. It typically contains dissolved organic matter, nutrients (e.g., nitrates, phosphates), and suspended solids. In aquaculture, surface water quality is important because it can carry pollutants, pathogens, and variable nutrient loads that affect aquatic organism health. Surface water is also subject to seasonal fluctuations in temperature and dissolved oxygen levels.

Temperature (Water Quality Parameter): Temperature is a fundamental water quality parameter that affects the solubility of gases (e.g., oxygen), metabolic rates of aquatic organisms, and chemical reaction rates. In aquaculture, the optimum temperature range varies by species. For example, tilapia thrives at temperatures between 25°C and 30°C, while salmon prefer cooler waters around 10°C to 15°C. Temperature fluctuations can lead to thermal stress, reduced oxygen levels, and increased susceptibility to disease, making temperature control vital in aquaculture systems.

Total Alkalinity Measurement: Total alkalinity is the water's ability to neutralize acids, primarily due to bicarbonates (HCO_3^-), carbonates (CO_3^{2-}), and hydroxides (OH^-). It is measured via titration with sulfuric acid. Alkalinity is crucial for pH stability, and in aquaculture, levels between 75 to 200 mg/L are optimal. Alkaline waters can buffer pH fluctuations, thus preventing stress on aquatic organisms.

Total Hardness Measurement: Total hardness refers to the concentration of calcium (Ca^{2+}) and magnesium (Mg^{2+}) ions in water, measured using EDTA titration. Hardness is important for shell formation and metabolic processes in fish and shellfish. Ideal hardness levels range from 75 to 300 mg/L in aquaculture, with calcium comprising at least 50 mg/L of this total. Extremely hard or soft waters can negatively affect growth and reproduction.

Total Nitrogen (Soil Quality Parameter): Total nitrogen refers to the sum of all forms of nitrogen present in the soil, including organic nitrogen, ammonium (NH_4^+), nitrate (NO_3^-), and nitrite (NO_2^-). It represents the soil's nitrogen reserve, which is critical for supporting plant and microbial growth in aquaculture ponds. Total nitrogen levels provide insight into the soil's long-term fertility and nutrient cycling capacity. In aquaculture, maintaining appropriate total nitrogen levels (around 0.1% to 0.3% by weight) is crucial for promoting the growth of phytoplankton, which forms the base of the food chain for fish and other aquatic species.

Total Nitrogen Analysis (Practical): Total nitrogen in soil, including organic and inorganic forms, is measured using the Kjeldahl method, which involves digestion with sulfuric acid and distillation. Total nitrogen levels give a comprehensive understanding of soil fertility. In aquaculture soils, maintaining nitrogen levels around 0.1–0.5% is essential for balanced nutrient cycling and productivity.

Turbidity Measurement: Turbidity measurement assesses the cloudiness or haziness of water, which is caused by suspended particles such as silt, algae, and organic matter. High turbidity can reduce light penetration, affecting photosynthesis and dissolved oxygen levels. In aquaculture, maintaining turbidity within a moderate range (e.g., 30-50 NTU) is critical for balancing phytoplankton growth and fish health. Turbidity is measured using a nephelometer or turbidity meter, which detects the scattering of light by particles in the water.

Turbidity: Turbidity is a measure of the cloudiness or haziness of water, caused by suspended particles such as silt, clay, organic matter, and microorganisms. It is usually measured in Nephelometric Turbidity Units (NTU). High turbidity can reduce light penetration, affecting photosynthesis in aquatic plants and decreasing dissolved oxygen levels. In aquaculture, maintaining low turbidity (below 25 NTU for most freshwater species) is important for ensuring optimal water quality and preventing clogging of fish gills or filter systems.

Water Amendments: Water amendments in aquaculture are substances added to improve water quality and create an optimal environment for aquatic species. They include chemicals and natural substances like lime, alum, and zeolites to regulate pH, reduce turbidity, and adsorb toxic substances like ammonia. The addition of lime, for instance, helps maintain a stable pH between 7.0 and 8.5, which is crucial for fish health. Alum is used to reduce phosphorus levels and control algal blooms, while zeolites are effective in adsorbing ammonia, which should be kept below 0.02 mg/L in most aquaculture systems.

Water Molecule: A water molecule (H_2O) consists of two hydrogen atoms covalently bonded to one oxygen atom. The molecular geometry is bent or V-shaped with an angle of about 104.5°, which results in a polar molecule with a partial negative charge near the oxygen atom and partial positive charges near the hydrogen atoms. This polarity allows water to form hydrogen bonds, which are critical for its solvent properties, high specific heat, and surface tension. The molecular weight of water is 18.015 g/mol, and it is essential for maintaining cellular processes in aquatic organisms.

Water Probiotics: Water probiotics are live microbial supplements introduced into aquaculture systems to enhance water quality by promoting beneficial microbial activity. These probiotics help break down organic waste, reduce ammonia and nitrate levels, and inhibit the growth of harmful pathogens. They are particularly useful in intensive aquaculture systems with high stocking densities. Water probiotics typically consist of bacteria like Bacillus and Nitrosomonas species, and are applied at rates ranging from 0.5-1 kg per 100,000 liters of water, depending on the specific system and water conditions.

Water Sample Collection: Water sample collection is a critical step in water quality management and involves obtaining representative samples from various aquatic environments (e.g., ponds, rivers, groundwater). Proper techniques ensure that the samples accurately reflect the water conditions without contamination. For aquaculture, water samples are typically collected from different depths and locations within a water body to assess parameters such as dissolved oxygen, temperature, pH, and nutrient levels. The use of clean, non-reactive containers and minimal handling helps preserve sample integrity.

Water Sample Preservation: Water sample preservation involves treating or storing water samples to prevent changes in chemical composition or biological activity before analysis. Preservation methods depend on the parameter being measured. For example, dissolved oxygen samples may be fixed with reagents, while samples for nutrient analysis (e.g., nitrates, phosphates) might be refrigerated or acidified. In aquaculture, timely preservation is crucial for accurate water quality assessment, which in turn affects the health and productivity of cultured species.

Waterlogged Soil: Waterlogged soils are soils that remain saturated with water for extended periods, often due to poor drainage, excessive irrigation, or flooding. In aquaculture, waterlogged conditions can negatively impact the oxygen availability in pond soils, leading to anaerobic conditions that promote the production of harmful gases like hydrogen sulfide (H_2S) and methane (CH_4). These conditions can stress or kill aquatic organisms. Managing waterlogged soils in aquaculture requires proper pond design, drainage systems, and soil aeration techniques to maintain an oxygenated environment that supports healthy aquatic life.

Water-Soil Interaction: Water-soil interaction refers to the dynamic exchange of nutrients, gases, and other substances between water and the underlying soil in aquatic systems. In aquaculture, this interaction plays a critical role in regulating water quality, nutrient availability, and the overall health of the pond ecosystem. For example, organic matter in the soil can release nutrients like nitrogen and phosphorus into the water, promoting the growth of algae and phytoplankton. Conversely, toxic substances like ammonia may be absorbed by pond soils. Effective management of water-soil interactions helps maintain a balanced environment, preventing issues such as nutrient depletion or toxic buildup.

Zeolites: Zeolites are microporous, aluminosilicate minerals used in aquaculture primarily for their ion-exchange properties, particularly in adsorbing ammonia from water. Ammonia toxicity, which can be lethal to fish at concentrations above 0.02 mg/L, is a common issue in aquaculture systems with high stocking densities. Natural zeolites, such as clinoptilolite, are highly effective at binding ammonium ions, thereby reducing ammonia concentrations and improving water quality. The dosage of zeolites in aquaculture ponds varies but typically ranges from 10 to 50 g/m^2 depending on the water's ammonia concentration and the system's size.

CHAPTER 3

ORNAMENTAL FISH CULTURE

Acclimatization: Acclimatization is the process of gradually adjusting fish to new environmental conditions, such as changes in water temperature, pH, or salinity. This is essential to prevent shock or stress when transferring fish from one habitat to another. Acclimatization can involve slowly mixing the water of the new environment with the original water or adjusting water chemistry over time.

Aeration Systems: Aeration systems introduce oxygen into the aquarium water, enhancing the oxygen exchange process at the water's surface. These systems use devices like air pumps, airstones, and diffusers to create air bubbles, promoting oxygen dissolution in water and aiding in the removal of carbon dioxide. Adequate aeration is vital for maintaining dissolved oxygen levels above 5 mg/L for fish health.

Algae Control: Algae control refers to managing unwanted algae growth in aquariums, which can occur due to excess nutrients, light, or poor water quality. Methods of control include reducing light exposure, maintaining proper nutrient levels, using algae-eating species (e.g., snails or shrimp), and applying chemical treatments when necessary.

Aquarium Accessories: Aquarium accessories include a wide range of tools and decorative items used to enhance the aquarium's aesthetics and functionality. These include thermometers, heaters, water pumps, substrates, and decorations like rocks, driftwood, and artificial plants. Accessories are chosen based on the type of species housed and their environmental requirements.

Aquarium Business: The aquarium business encompasses all commercial activities related to the sale, breeding, and maintenance of ornamental fish, plants, and related products. This includes aquarium equipment, live feed, fish tanks, and decorations. Successful businesses focus on market trends, quality of livestock, and adherence to regulations such as CITES for the trade of exotic species.

Aquarium Decorative: Aquarium decorative refers to the aesthetic elements placed inside an aquarium to simulate a natural environment and provide hiding spaces for fish. These include artificial plants, rocks, caves, and driftwood. Proper decoration improves fish behavior by reducing stress and providing territories, especially in species prone to aggression.

Aquarium Equipment: Aquarium equipment refers to the tools and devices essential for maintaining a healthy fish tank environment. This includes filters (mechanical, biological, and chemical), heaters, lighting systems, air pumps, thermometers, and water testing kits. Proper equipment ensures stable water quality, temperature, and oxygen levels, which are vital for fish health and plant growth.

Aquarium Fabrication: Aquarium fabrication refers to the process of designing and constructing tanks that hold water and sustain aquatic life. The materials used can be glass or acrylic, and the size depends on the type of aquatic organisms intended to be housed. Proper sealing, waterproofing, and structural support are critical to avoid leakage and ensure durability.

Aquarium Filters: Aquarium filters are essential for maintaining water quality by removing physical, chemical, and biological waste from the water. Filters can be biological (beneficial bacteria), mechanical (particulate removal), or chemical (activated carbon) and need to be cleaned and replaced regularly to function effectively.

Aquarium Lighting: Aquarium lighting is essential for maintaining the health of fish and plants. Proper lighting helps simulate day-night cycles, promotes plant growth through photosynthesis, and enhances the coloration of ornamental fish. Light intensity and spectrum depend on the aquarium type-freshwater plants require light in the range of 5000-7000K, while marine coral aquariums may need higher intensity light (10000K-14000K).

Aquarium Maintenance: Aquarium maintenance involves regular cleaning and upkeep tasks such as water changes, filter cleaning, substrate vacuuming, and glass scraping. Proper maintenance ensures that water quality remains stable and that fish and plants thrive without experiencing stress or disease outbreaks.

Aquarium Nets: Aquarium nets are mesh tools used for capturing and transferring fish or removing debris from the tank. They come in various sizes, with finer mesh used for delicate species. Nets should be soft and non-abrasive to prevent injury to fish, and their use should be minimized to avoid stressing the fish.

Aquarium Plants: Aquarium plants serve both aesthetic and functional purposes in fish tanks, providing shelter, oxygenating the water, and aiding in nutrient cycling. Some common species include Java fern, Anubias, and Vallisneria. Healthy plant growth depends on adequate lighting, nutrient availability, and CO_2 levels.

Aquatic Plant Nutrition: Aquatic plant nutrition involves the supply of essential macro and micronutrients, such as nitrogen, phosphorus, potassium, and iron. These nutrients can be absorbed from the substrate or water column, with the ideal concentration of nitrates between 10-20 ppm and phosphate levels of 0.5-1 ppm to support growth without causing algae blooms.

Aquatic Plant Propagation: Propagation of aquatic plants refers to the methods used to grow new plants from existing ones. Techniques include cutting, dividing, and using plantlets or runners. Each plant species has specific propagation requirements, such as light, substrate, and nutrient conditions, to ensure successful growth.

Bacterial Infections: Bacterial infections in fish are caused by pathogens like *Aeromonas* or *Pseudomonas*, leading to symptoms such as ulcers, fin rot, and septicemia. Poor water quality and stress can make fish more susceptible. Effective treatment includes antibiotics, water changes, and improving overall tank hygiene. It's essential to identify the specific bacterial strain for targeted treatment.

Biological Filtration: Biological filtration refers to the process of using beneficial bacteria to convert toxic ammonia, produced from fish waste, into less harmful substances like nitrites and nitrates. This process is known as the nitrogen cycle and is essential for maintaining a stable and healthy aquarium environment. The optimal functioning of biological filters depends on maintaining appropriate water conditions for bacterial colonization.

Brackish Water Fish Breeding: Brackish water fish breeding involves species that thrive in water with moderate salinity, such as mollies and scats. Successful breeding requires maintaining salinity levels between 5-15 ppt and monitoring water conditions to simulate the natural estuarine environment. These species often require gradual acclimatization to varying salinity levels for reproduction.

Broodstock Management: Broodstock management refers to the process of maintaining mature fish for breeding purposes. In ornamental aquaculture, this involves selecting healthy, genetically superior fish, providing optimal nutrition, and ensuring suitable environmental conditions like temperature (24°-28°C for most ornamental species) to induce spawning. Proper management results in high-quality offspring with desirable traits.

Broodstock Selection: Broodstock selection involves choosing mature, healthy fish with desirable traits (e.g., size, coloration, growth rate) for breeding. Proper selection ensures the production of high-quality offspring. Inbreeding is avoided by maintaining genetic diversity, and the broodstock should be free of diseases and parasites to prevent their transmission to the fry.

Chemical Filtration: Chemical filtration uses chemical media such as activated carbon or resins to remove dissolved impurities from the water, including toxins, odors, and discolorations. Activated carbon, for instance, adsorbs organic pollutants and helps maintain water clarity and quality. Regular replacement of chemical filter media is necessary to ensure efficiency.

CO_2 Injection: CO_2 injection involves adding carbon dioxide into the aquarium to boost the growth of aquatic plants, which use CO_2 for photosynthesis. This is especially important in high-light aquariums, where CO_2 deficiency can limit plant growth. Optimal CO_2 levels for planted tanks range from 20-30 ppm.

Commercial Fish Feed: Commercial fish feed refers to formulated diets available in various forms, such as pellets, flakes, or granules, designed to meet the nutritional needs of ornamental fish. These feeds are typically fortified with essential vitamins and minerals. High-quality commercial feeds are developed to enhance growth, immunity, and coloration while minimizing waste.

Common Ornamental Fish: Common ornamental fish include species widely kept in home aquariums due to their beauty, ease of care, and adaptability to captivity. Examples include bettas, goldfish, guppies, tetras, and angelfish. These species thrive in a range of water parameters, typically requiring temperatures of 24°-28°C and pH levels of 6.5-7.5.

Conditioning: Conditioning refers to the process of preparing ornamental fish for transportation, breeding, or stressful situations by gradually acclimating them to new conditions or enhancing their health. This is often done by adjusting water parameters, providing high-quality nutrition, and ensuring optimal environmental conditions for a period before the fish undergo changes or stress.

Cropping Methods: Cropping methods in aquaculture refer to the harvesting techniques used for ornamental plants or fish. For plants, this may involve periodic trimming, propagation, and maintaining optimal growth conditions. For fish, it includes selective breeding or removing excess individuals to maintain population density and aesthetics.

Disease Control: Disease control in ornamental fish involves preventive and curative measures to manage and treat diseases. Strategies include maintaining clean water, isolating infected fish, administering antibiotics or anti-parasitic treatments, and using disinfectants in the aquarium environment. Regular monitoring and quarantine of new stock are key preventive measures.

Disease Prevention: Disease prevention in aquariums focuses on maintaining optimal water quality, nutrition, and biosecurity measures to minimize the risk of disease outbreaks. Regular water changes, proper filtration, quarantine of new fish, and avoiding overcrowding are key practices. Immunostimulants, probiotics, and proper diet can also boost fish immunity and reduce susceptibility to pathogens.

Domestic Market: The domestic market in ornamental fish refers to the demand and trade of these species within a country's borders. Factors such as species preferences, availability, infrastructure for breeding, local regulations, and consumer awareness influence this market. In India, for instance, ornamental fish farming is expanding due to rising hobbyist interest and developing infrastructure for breeding both indigenous and exotic species.

Dry Feed: Dry feed refers to feed types like flakes, pellets, or granules that contain minimal moisture (usually less than 10%). These feeds are convenient for storage and feeding but may require soaking before use to prevent digestive issues in certain fish species. Dry feeds are typically formulated with a balanced nutrient profile suitable for ornamental fish.

Egg Hatching: Egg hatching refers to the process of fish embryos developing within eggs and emerging as fry. Conditions such as temperature, oxygen levels, and light are critical for successful hatching. Most tropical fish eggs hatch within 3-7 days, and water temperatures should be maintained around 26°C-28°C for optimal development. Proper aeration and antifungal treatments help ensure healthy hatch rates.

Egg-Laying Fish: Egg-laying fish deposit eggs either in the substrate, on plants, or in nests, depending on the species. Examples include tetras, barbs, and cichlids. The water conditions for breeding are critical, with soft, slightly acidic water (pH 6.5-7.0) and temperatures between 24°C and 28°C typically ideal for many egg-laying species.

Environmental Requirements: Environmental requirements refer to the ideal physical and chemical conditions necessary for the growth and health of ornamental fish and plants. Parameters such as water temperature, pH, light intensity, dissolved oxygen, and substrate type must be carefully managed. Different species have varying requirements that need to be tailored to their specific needs.

Exotic Species: Exotic species refer to non-native ornamental fish introduced from other regions for aquarium purposes. These species are often favored for their unique aesthetic value. Examples include species like discus, guppy, and neon tetra. However, careful regulation is needed to prevent the spread of diseases and invasiveness in local ecosystems when released into the wild.

Filtration Media: Filtration media refers to the materials used in aquarium filters to remove impurities and maintain water quality. There are three types: mechanical (e.g., sponge, filter floss) for trapping debris, biological (e.g., bio-balls, ceramic rings) for hosting beneficial bacteria, and chemical (e.g., activated carbon) for removing toxins. Biological filtration is the most critical for maintaining a balanced nitrogen cycle, with media typically requiring cleaning or replacement every 1-3 months.

Fish Breeding Tank: A fish breeding tank is a specialized setup designed to encourage the reproduction of ornamental fish. It typically includes soft substrate, plants or spawning mops for egg layers, and hiding places to reduce stress. The water parameters, such as temperature and pH, are adjusted to trigger spawning behaviors, and separate chambers may be used to protect eggs and fry from being eaten by adult fish.

Fish Diseases: Fish diseases refer to a variety of pathogenic conditions that affect ornamental fish, including bacterial, viral, fungal, and parasitic infections. Common symptoms include fin rot, white spots, and lethargy. Effective disease control relies on maintaining optimal water quality, regular monitoring, and timely treatment with medications or other remedies.

Fish Export Regulations: Fish export regulations are legal frameworks governing the trade and transportation of ornamental fish across international borders. These regulations often include health certifications, restrictions on endangered species, and adherence to CITES (Convention on International Trade in Endangered Species) guidelines. Exporters must comply with quarantine and biosecurity measures to prevent the spread of diseases.

Fish Farm Biosecurity: Fish farm biosecurity involves implementing practices that prevent the introduction and spread of diseases, parasites, and pathogens in aquaculture operations. Key measures include quarantine of new stock, disinfection of equipment, controlling water quality, and regular health checks. Biosecurity is essential for maintaining healthy fish populations and ensuring the success of farming operations.

Fish Hatchery Management: Fish hatchery management involves overseeing the breeding, hatching, and rearing of fish in controlled environments. This includes managing broodstock selection, maintaining optimal water quality, and providing appropriate nutrition at different life stages. Hatcheries must ensure conditions conducive to high survival rates, including temperatures around 26°C-28°C for many ornamental fish species.

Fish Hatchery Setup: Fish hatchery setup refers to the design and construction of facilities for breeding, hatching, and rearing fish in a controlled environment. A proper hatchery includes broodstock tanks, breeding tanks, fry tanks, filtration systems, and aeration. Water quality, temperature, and light are meticulously controlled, with optimal breeding conditions varying by species (e.g., 26-28°C for many tropical fish).

Fish Health Monitoring: Fish health monitoring involves regularly observing fish for signs of stress, illness, or abnormal behavior, such as loss of appetite, unusual swimming patterns, or discoloration. Early detection of health issues allows for timely intervention, reducing the risk of disease outbreaks. Regular testing of water parameters also plays a crucial role in maintaining fish health.

Fish Health: Fish health encompasses the overall well-being of ornamental fish, influenced by factors such as diet, water quality, and tank environment. Common health indicators include vibrant coloration, active swimming behavior, and normal feeding patterns. Poor health can result from stress, inadequate nutrition, or water quality issues, leading to disease outbreaks. Preventative care, proper nutrition, and regular monitoring are essential to maintaining fish health.

Fish Medication: Fish medication includes drugs used to treat diseases in ornamental fish, ranging from antibiotics and antiparasitics to antifungal and antiviral treatments. Medications can be administered in baths, orally, or directly into the water. It is crucial to follow proper dosage and treatment protocols to avoid harmful effects on fish and the aquarium ecosystem.

Fish Packing: Fish packing is the method of preparing fish for transport by placing them in oxygenated water-filled plastic bags or containers. Proper packing techniques ensure that fish experience minimal stress during transit. Factors like water temperature, oxygen saturation, and fish density in the packing medium are carefully controlled to prevent mortality.

Fish Parasites: Fish parasites are organisms that live on or inside fish, causing harm by feeding on their tissues, blood, or nutrients. Common parasites include protozoans (e.g., Ich), worms (e.g., nematodes), and crustaceans (e.g., lice). Symptoms may include scratching, clamped fins, weight loss, and visible external parasites. Treatment often involves antiparasitic medications or baths.

Fish Quarantine Tanks: Fish quarantine tanks are separate tanks used to isolate new or sick fish from the main aquarium to prevent the spread of diseases. Quarantine tanks are usually smaller and easier to clean, with similar water conditions to the main tank. Fish are monitored and treated for any health issues during a quarantine period of 2-4 weeks.

Fish Rearing: Fish rearing refers to the process of growing fish from the larval stage to adult size under controlled conditions. This involves managing water quality, providing proper nutrition, and ensuring suitable environmental parameters such as temperature and light. Rearing ornamental fish requires careful attention to feed, health, and habitat conditions to optimize growth and color development.

Fish Stress Management: Fish stress management involves strategies to reduce environmental stressors that can weaken fish and make them susceptible to diseases. Key factors include maintaining stable water quality, minimizing sudden changes in temperature or pH, providing adequate hiding places, and reducing overstocking. Proper handling, avoiding excessive netting, and providing a well-balanced diet also help minimize stress.

Fish Tank Decorations: Fish tank decorations include artificial or natural items like rocks, driftwood, caves, and plants that enhance the aesthetic appeal and provide hiding places for fish. These decorations contribute to creating a stress-free environment, helping fish feel secure. It's important to ensure that the materials are safe for aquatic environments and do not alter the water chemistry.

Fish Tank Heater: A fish tank heater maintains a stable water temperature, essential for the health of tropical fish. Heaters come in adjustable and preset models, with most tropical fish requiring temperatures between 24°C and 28°C (75°F-82°F). Proper heater sizing depends on the tank's volume, with a general rule of 5 watts per gallon. Consistent temperature prevents stress and disease in fish.

Fish Tank Quarantine: Fish tank quarantine is the practice of isolating new or sick fish in a separate tank to prevent the spread of diseases to healthy fish in the main aquarium. Quarantine tanks should be fully equipped with filtration, aeration, and temperature control, and fish are usually kept in quarantine for 2-4 weeks. This period allows for observation and treatment of any health issues before introducing the fish to the main tank.

Fish Tank Setup: Fish tank setup refers to the initial process of preparing an aquarium for ornamental fish, which includes choosing the appropriate tank size, installing equipment like filters, heaters, lighting, and adding substrate and decorations. The ideal tank size varies depending on the fish species, but generally, a 10-gallon tank is a minimum for small freshwater fish. Proper setup ensures stable water parameters and creates a suitable environment for fish health and well-being.

Fish Tank Substrate: Fish tank substrate refers to the material placed at the bottom of the aquarium, which can range from gravel, sand, to specialized plant substrates. The substrate plays a vital role in providing a surface for beneficial bacteria growth, anchoring plants, and enhancing the aesthetics of the tank. The ideal substrate type depends on the fish species and plants in the tank, with pH and mineral content varying accordingly.

Fish Transport Bags: Fish transport bags are specialized plastic bags used to safely move ornamental fish from one location to another, typically filled with a mix of air (or pure oxygen) and water. For long-distance transport, the ratio of air to water is generally 2:1. The bags are sealed to prevent leakage and often placed in insulated containers to maintain temperature during transit, reducing stress on the fish.

Fish Transport: Fish transport involves moving live ornamental fish over short or long distances, ensuring minimal stress and survival. This process includes using oxygenated bags, temperature control, and minimizing vibrations during travel. For international trade, transport regulations must comply with biosecurity measures to prevent disease transmission.

Formulated Feed: Formulated feed refers to commercially manufactured diets designed to meet the nutritional requirements of ornamental fish. These feeds are made from a combination of ingredients like fish meal, soy protein, vitamins, and minerals. Formulated feeds are available in various forms like pellets, flakes, and granules, and are designed to enhance fish health, coloration, and growth.

Freshwater Acclimatization: Freshwater acclimatization refers to the process of transitioning marine or brackish water fish into freshwater conditions. This is typically done for certain ornamental species that are tolerant of a wide range of salinities, like mollies. Slow salinity reduction helps the fish adjust without undergoing osmotic shock.

Freshwater Aquarium: A freshwater aquarium is a tank that simulates natural freshwater habitats such as rivers, lakes, and ponds. It contains species adapted to fresh water, such as guppies and goldfish. Water quality management, including maintaining pH between 6.5 and 7.5, temperature (20-28°C), and hardness, is essential for sustaining fish health.

Freshwater Ornamental Fish Breeding: This refers to the controlled breeding of ornamental fish species in freshwater environments. Species like guppies, tetras, and cichlids are commonly bred in freshwater aquariums. Key factors for successful breeding include maintaining appropriate water conditions (pH, temperature, hardness), creating breeding setups with hiding spaces or spawning surfaces, and ensuring optimal nutrition.

Fry Rearing: Fry rearing involves the care and feeding of newly hatched fish (fry) until they reach a size suitable for sale or release. This stage requires close monitoring of water quality, frequent feeding with appropriate diets like live feed or micro pellets, and protection from predators. Maintaining a stable temperature of 26°C-28°C and clean water with minimal ammonia levels is critical to fry survival.

Fungal Infections: Fungal infections in fish, typically caused by species like *Saprolegnia*, appear as cotton-like growths on the skin, gills, or fins. Fungal spores are common in aquatic environments and often infect fish with weakened immune systems or injuries. Treatment includes antifungal medications such as malachite green and improving water conditions.

Indigenous Fishes: Indigenous fishes are native species that occur naturally in a particular region. In ornamental fish culture, indigenous fishes are often bred for trade, such as the Indian Glass Fish (*Parambassis ranga*) and the Rosy Barb (*Pethia conchonius*). Indigenous fish are easier to maintain in local climates and ecosystems, reducing the risk of ecological imbalance.

International Fish Trade: International fish trade refers to the global exchange of live fish, fish products, and related commodities, with ornamental fish being a significant segment. This trade is regulated by international agreements, health certifications, and biosecurity measures to prevent the spread of diseases and invasive species. Major markets include the United States, Europe, and Japan.

International Market: The international market for ornamental fish involves the global trade of species across borders. This market is regulated by international trade laws, customs procedures, and agreements like CITES (Convention on International Trade in Endangered Species) to protect wild populations. Countries like Singapore and Thailand are major exporters, while Europe and North America dominate as large importers.

Lighting for Plants: Lighting for plants in aquariums needs to provide full-spectrum illumination, covering the PAR (Photosynthetically Active Radiation) range of 400-700 nm. The intensity and duration of light required depend on the plant species, with low-light plants needing around 20-40 $\mu mol/m^2/s$ and high-light plants requiring over 100 $\mu mol/m^2/s$ for 8-12 hours daily.

Live Bearers: Live bearers are fish species that give birth to live young, rather than laying eggs. Popular ornamental live bearers include guppies, mollies, and swordtails. They are known for their ease of breeding in captivity, as they do not require special setups for spawning. Optimum breeding conditions include a temperature range of 24°-28°C and a pH of 7-8.

Live Feed: Live feed includes small, live organisms such as R 6otifers, Artemia (brine shrimp), Daphnia, and micro-worms, which are fed to ornamental fish, especially in larval or juvenile stages. These feeds are rich in nutrients and stimulate natural hunting behaviour in fish. Live feed is often used during critical life stages when high-protein diets are necessary for growth and survival.

Livebearer Breeding: Livebearer breeding involves species that give birth to free-swimming fry instead of laying eggs, such as guppies, mollies, and platies. Livebearers reproduce easily in captivity, and the breeding tank should provide hiding spots for fry, such as dense plants. These species often breed at temperatures between 24°C and 28°C.

Marine Aquarium: A marine aquarium houses saltwater species and is designed to mimic ocean or sea environments. Key parameters for marine aquariums include salinity levels of 30-35 ppt (parts per thousand), pH levels of 8.1-8.4, and calcium and alkalinity balance for coral health. Marine aquariums support a wide range of species, including fish, corals, and invertebrates.

Marine Ornamental Fish Breeding: Marine ornamental fish breeding is a complex process due to the specific environmental needs of marine species. Breeding species like clownfish or angelfish requires maintaining precise salinity (30-35 ppt), water temperature (24°-28°C), and pH (8.1-8.4). Larval rearing is particularly challenging, often requiring live feeds such as rotifers or copepods for the survival of hatchlings.

Marine Ornamental Fish: Marine ornamental fish are species that live in saltwater environments and are bred or captured for aquarium trade. Examples include clownfish, tangs, and damselfish. These species require specific water parameters, including high salinity (30-35 ppt), stable pH (8.1-8.4), and controlled temperatures (24°-28°C), to thrive in captivity.

Mechanical Filtration: Mechanical filtration is the physical removal of particulate matter, such as uneaten food, fish waste, and debris, from aquarium water. This is achieved through the use of filter media, such as sponges, filter floss, or sand, which trap solids. Mechanical filtration helps keep water clear and reduces the accumulation of organic matter that could decay and harm water quality.

Moist Feed: Moist feed is intermediate between dry and wet feeds, containing 10-30% moisture. This type of feed provides higher palatability and nutritional value than dry feed, making it suitable for fast-growing fish or species with specific dietary needs. Moist feeds must be handled properly to prevent spoilage.

Nutrient Requirements: Nutrient requirements for ornamental fish and plants include essential macro and micronutrients. For fish, nutrients include proteins, fats, carbohydrates, vitamins, and minerals. For aquatic plants, essential nutrients include nitrogen, phosphorus, potassium, and trace elements like iron and magnesium, which promote growth, color development, and root stability.

Ornamental Aquatic Plants: Ornamental aquatic plants are species grown for aesthetic purposes in aquariums, ponds, or water gardens. These plants provide shelter, oxygenate the water, and help maintain water quality. Examples include species like Java fern, Amazon sword, and Anubias. Proper lighting and nutrient management are essential for their growth.

Ornamental Fish Export: Ornamental fish export is the international trade of aquarium fish species, typically bred in captivity or collected from wild populations. Export processes include proper fish conditioning, packaging, and adherence to health certifications and international trade regulations. Leading exporting countries include Singapore, Thailand, and Indonesia, known for supplying a wide variety of ornamental species.

Ornamental Fish Farm: An ornamental fish farm is a facility where ornamental fish species are bred, reared, and maintained for commercial purposes. These farms require specialized systems for water quality management, broodstock conditioning, breeding setups, and disease control. They may focus on either freshwater or marine species and include operations for both local and international markets.

Ornamental Fish Nutrition: Ornamental fish nutrition focuses on providing balanced diets tailored to the species' needs, promoting growth, color vibrancy, and disease resistance. Diets should include a mix of proteins, fats, carbohydrates, vitamins, and minerals, with specific requirements varying by species. For example, carnivorous fish require higher protein content (around 45%), while herbivorous fish thrive on fiber-rich diets.

Ornamental Fish Trade: The ornamental fish trade is a global industry involving the breeding, rearing, and sale of ornamental fish species. This trade includes both captive-bred and wild-caught species and requires adherence to regulations that ensure sustainable practices and prevent the spread of diseases. Key markets include North America, Europe, and Asia.

Ornamental Fish: Ornamental fish are species kept in aquariums or water gardens for decorative purposes due to their vibrant colors, unique patterns, and aesthetic appeal. These fish can be freshwater, brackish water, or marine species, bred primarily for the aquarium trade. Common examples include goldfish, guppies, and bettas. They are subject to specific breeding, rearing, and health management practices to maintain their desirable traits and ensure their survival in artificial environments.

pH Regulation: pH regulation refers to maintaining the aquarium water's acidity or alkalinity at optimal levels for the species being kept. The pH scale ranges from 0 to 14, with 7 being neutral. Freshwater fish typically thrive in pH levels between 6.5 and 7.5, while marine fish prefer a pH of 8.1 to 8.4. Adjustments can be made using buffers, chemical additives, or natural methods like driftwood or crushed coral.

Plant Lighting: Plant lighting is critical for the growth of live aquatic plants in the aquarium, providing the energy needed for photosynthesis. The intensity, spectrum, and duration of light influence plant health. Full-spectrum lighting or LED grow lights that mimic natural sunlight (with wavelengths in the range of 400-700 nm) are ideal, with a light duration of 8-10 hours per day for most plants.

Plant Packing: Plant packing involves preparing aquatic plants for transport by ensuring they are packaged in a way that retains moisture and protects their structure. Plants are usually wrapped in wet paper towels or plastic bags, ensuring they stay hydrated and free from physical damage during transport.

Plant Substrates: Plant substrates are nutrient-rich materials, such as specialized aquarium soils or fertilized gravel, designed to support the root growth and nutrition of aquatic plants. These substrates often contain minerals like iron and other nutrients that are slowly released into the water column, promoting healthy plant development.

Plant Transport: Transporting ornamental aquatic plants requires maintaining appropriate moisture levels and avoiding temperature fluctuations. Plants are often packed in water-resistant bags or boxes with air holes to ensure airflow. Ideal transport conditions prevent physical damage and stress, ensuring that plants remain healthy upon arrival.

Protozoan Infections: Protozoan infections, such as *Ichthyophthirius multifiliis* (Ich) or *Costia*, cause white spots or velvet-like growth on fish skin and gills. These infections can lead to respiratory distress, lethargy, and death if untreated. Treatments include raising water temperature, applying copper-based medications, or using formalin-based treatments to eradicate the protozoans.

Quarantine Methods: Quarantine methods are procedures used to isolate and observe newly acquired fish before introducing them into established aquariums or farms. The quarantine period (usually 2-4 weeks) helps detect and treat diseases or parasites without affecting the existing stock. During quarantine, the fish are treated with medications and monitored for signs of stress or infection.

Spawning Techniques: Spawning techniques are methods used to encourage fish reproduction in captivity. Techniques vary by species and may include temperature manipulation, adjusting pH levels, or creating spawning habitats (e.g., spawning mops, caves). Some fish may require hormonal injections to trigger spawning. For example, induced breeding with hormones is common in egg-laying species like carps.

Substrate Management: Substrate management involves maintaining the material at the bottom of the aquarium, such as gravel, sand, or soil. Proper substrate management ensures healthy plant growth, provides a habitat for beneficial bacteria, and prevents the buildup of toxic substances like ammonia or hydrogen sulfide.

Tank Aeration: Tank aeration refers to the process of increasing oxygen levels in the water through the use of air stones, bubble curtains, or surface agitation created by filters. Adequate oxygenation is crucial for fish respiration and beneficial bacterial function in the filter media. Ideally, dissolved oxygen levels should remain above 6 ppm for optimal fish health.

Tank Cycling: Tank cycling is the process of establishing beneficial bacterial colonies in the aquarium's filtration system, allowing for the conversion of toxic ammonia produced by fish waste into less harmful nitrite, and finally into nitrate. This process typically takes 4-6 weeks, and the goal is to reach stable ammonia and nitrite levels of 0 ppm and nitrate levels under 40 ppm. Proper cycling is crucial for the long-term health of the fish.

Tank Lighting Control: Tank lighting control involves adjusting the light intensity and duration to suit the needs of the aquarium inhabitants. Fish, plants, and corals all have specific lighting requirements, with fish needing 8-10 hours of light daily. Overexposure can cause stress or algae overgrowth, so timers and dimmers are often used to regulate the lighting cycles.

Tank Stocking Density: Tank stocking density refers to the number of fish that can be kept in an aquarium without compromising their health and well-being. Overstocking can lead to poor water quality, increased stress, and disease outbreaks. A general rule of thumb is one inch of fish per gallon of water, but this can vary depending on species, tank filtration, and aeration.

Tank Water Testing: Tank water testing involves regularly measuring key water parameters such as pH, ammonia, nitrite, nitrate, and hardness to ensure the aquatic environment remains stable and safe for fish. Test kits or electronic meters are used for monitoring, and optimal ranges depend on the fish species, but for most freshwater tanks, ammonia and nitrite should be 0 ppm, and nitrate below 20 ppm.

Trade Regulations: Trade regulations in ornamental fish refer to the rules and laws governing the import, export, and sale of species across borders. These regulations often include health certifications, permits, and adherence to international treaties like CITES, which protect endangered species from exploitation.

Water Conditioning: Water conditioning involves treating tap water before adding it to an aquarium, ensuring it is free from harmful chemicals like chlorine, chloramines, and heavy metals. Conditioners neutralize these toxins, making the water safe for fish. Additionally, conditioners may contain substances to promote a protective slime coat on fish, further reducing stress during water changes.

Water Quality Management: Water quality management in aquaculture involves maintaining optimal chemical, physical, and biological conditions in an aquarium. Critical parameters include temperature (20°-30°C), pH (6.5-8.5 depending on the species), dissolved oxygen (>5 mg/L), ammonia, nitrate, and phosphate levels. Regular monitoring and adjustments ensure the healthy growth of fish and plants.

Wet Feed: Wet feed contains a higher moisture content (30-70%) and includes freshly prepared diets like minced fish, shrimp, or specially formulated gels. Wet feed is often used for species that require more natural diets or for conditioning broodstock to induce spawning. Care must be taken to avoid contamination and maintain freshness.

Wildlife Act: The Wildlife Act refers to national or international laws that regulate the capture, trade, and protection of wildlife, including ornamental fish. In India, for example, the Wildlife Protection Act (1972) helps control the exploitation of native fish species and ensures sustainable management of biodiversity.

CHAPTER 4

FINFISH NUTRITION AND FEEDING MANAGEMENT

Acid-Insoluble Ash: Acid-insoluble ash (AIA) is a measure of indigestible material in feed, particularly silica and other contaminants that cannot be broken down by acid in the stomach. AIA is used as an indicator of feed quality and digestibility. High levels of AIA suggest the presence of impurities or poor-quality ingredients, leading to reduced feed efficiency. Monitoring AIA levels helps ensure that feed is free from excessive contaminants that could affect the health and growth of aquatic species.

Adult Fish Nutrition: Adult fish nutrition focuses on maintaining health, reproductive capability, and moderate growth, with reduced nutrient demands compared to earlier life stages. Protein requirements are lower (25-35%), while lipids remain around 5-10% to provide energy and maintain tissue integrity. Carbohydrate content may increase to around 20-40%, as adults can metabolize carbohydrates more efficiently. Essential micronutrients such as vitamins A, D, and E, as well as minerals like zinc and magnesium, are critical for reproductive success and overall metabolic functions.

Amino Acid Profile: The amino acid profile of aquaculture feed refers to the composition and ratio of essential and non-essential amino acids present in the feed. Essential amino acids like lysine, methionine, and tryptophan must be provided through the diet as aquatic species cannot synthesize them. An optimal amino acid profile is crucial for protein synthesis, tissue repair, and growth. For instance, lysine requirements for species like Nile tilapia range from 4.5-5.5% of total protein.

Anti-Nutritional Factors: Anti-nutritional factors (ANFs) are naturally occurring compounds in plant-based feed ingredients that reduce nutrient availability and digestibility in aquaculture species. Common ANFs include phytates, tannins, trypsin inhibitors, and saponins. These compounds can interfere with protein digestion, mineral absorption, and overall feed efficiency. For example, phytates bind to phosphorus, making it unavailable to fish and shrimp. Detoxification methods or the use of enzymes such as phytase can mitigate the impact of ANFs in aquafeeds.

Antioxidants: Antioxidants are compounds added to aquafeeds to prevent the oxidation of fats and oils, which can lead to rancidity and the degradation of essential nutrients like vitamins and fatty acids. Common antioxidants include synthetic additives like ethoxyquin and butylated hydroxytoluene (BHT), as well as natural sources like vitamin E and ascorbic acid (Vitamin C). Antioxidants protect feed quality during storage and improve the health of fish and shrimp by reducing oxidative stress, which can damage cells and impair immune function.

Automatic Feeders: Automatic feeders are devices used to dispense feed at predetermined intervals and quantities, allowing for precise and consistent feeding without the need for manual labor. These feeders are commonly used in intensive aquaculture systems to optimize feed management and reduce labor costs. They can be programmed to deliver feed multiple times per day, matching the species' natural feeding behavior. Automatic feeders help maintain feeding schedules, reduce feed waste, and improve feed efficiency, particularly in large-scale or commercial operations.

Balanced Feed: Balanced feed provides the necessary nutrients in the correct proportions to meet the specific dietary needs of fish or shrimp at various life stages. A well-balanced feed contains the right amounts of proteins, lipids, carbohydrates, vitamins, and minerals, ensuring proper growth, health, and reproductive success. A balanced diet minimizes nutrient deficiencies or excesses, which can lead to poor growth performance or health problems. In aquaculture, balanced feed is crucial for maintaining high feed efficiency and supporting sustainable production.

Binders: Binders are substances added to aquafeeds to improve pellet integrity and water stability, preventing the feed from disintegrating too quickly in water. Common binders include natural materials like guar gum, alginates, and starch, as well as synthetic binders like lignosulfonates. Strong pellet binding ensures that nutrients remain available to fish and shrimp without leaching into the water, reducing feed waste and improving water quality. Binders are especially important in shrimp feeds, which are typically exposed to water for longer periods.

Bioenergetics: Bioenergetics in aquaculture refers to the study of energy flow and transformation in aquatic organisms, particularly how they obtain, store, and use energy for growth, reproduction, and maintenance. Understanding bioenergetics helps optimize feeding strategies, ensuring that energy intake from feed meets the metabolic demands of the organism. Energy from proteins, lipids, and carbohydrates is measured as gross energy (GE), with fish and shrimp converting a portion of this energy into growth and activity while the rest is lost as waste or heat.

Biological Value (BV): Biological value (BV) is a measure of the proportion of absorbed protein that is retained in the body for growth and maintenance. It reflects the quality of protein based on its amino acid composition, with higher BV values indicating that the protein contains a more complete or balanced set of essential amino acids. Protein sources like fish meal typically have a high BV (80-90%), while plant-based proteins may have lower BV due to limiting amino acids. BV is crucial for evaluating protein sources in feed formulations.

Broadcast Feeding: Broadcast feeding is a manual or mechanical method of distributing feed across the surface of the water, typically used in pond or cage systems. This method allows for even distribution of feed, ensuring that fish have equal access to the feed. Broadcast feeding is commonly used for species like tilapia and catfish, which feed near the water surface. Properly executed broadcast feeding helps reduce feed waste and ensures that all individuals have access to the feed.

Carbohydrate Requirement: Carbohydrates serve as a readily available energy source in aquaculture diets, though fish generally have a lower capacity to digest carbohydrates compared to terrestrial animals. Typically, 20-40% carbohydrates are included in the diet, depending on species and life stage. Herbivorous fish like tilapia can utilize carbohydrates more efficiently than carnivorous species like salmon. Insufficient carbohydrates in the diet lead to protein being used for energy, which can affect growth, while excessive carbohydrates can lead to digestive issues and fatty liver.

Conditioning: Conditioning in feed production involves adding heat, moisture, and sometimes steam to feed ingredients before pelletization or extrusion. This process softens the ingredients, improving their plasticity and binding properties, which enhances pellet durability and water stability. Conditioning also improves the digestibility of starches and proteins, making nutrients more bioavailable to the cultured species. Proper conditioning is critical to producing high-quality pellets that maintain their structure and nutritional integrity during storage and feeding.

Conventional Feed Ingredients: Conventional feed ingredients are commonly used raw materials in aquafeeds, such as fish meal, soybean meal, wheat, corn, and rice bran. These ingredients are widely available, well-studied, and provide essential nutrients like proteins, lipids, carbohydrates, vitamins, and minerals. For example, fish meal is a rich source of high-quality protein and essential fatty acids, making it a staple in many fish and shrimp diets. The use of conventional ingredients ensures predictable nutritional profiles and feed performance.

Corn Gluten Meal: Corn gluten meal is a by-product of corn processing and contains around 60-70% protein, making it a valuable plant-based protein source for aquafeeds. It is rich in amino acids such as methionine but is deficient in lysine. Corn gluten meal also provides carotenoids, which can enhance the coloration of some species. Due to its high protein content and digestibility, it is often used as a partial replacement for fish meal in aquaculture diets.

Crude Fiber: Crude fiber refers to the indigestible portion of plant material in feed, primarily cellulose and lignin. While fiber is not a significant energy source for most aquatic species, it plays a role in promoting healthy digestion and gut function. Fiber levels are typically kept low in aquafeeds, usually between 2-8%, as excessive fiber can reduce nutrient digestibility and energy availability. Fiber content is especially important in diets for herbivorous species, where moderate levels are required to mimic natural feeding habits.

Crude Lipid: Crude lipid represents the total fat content in a feed, which provides a concentrated source of energy and essential fatty acids for fish and shrimp. Lipid levels in aquafeeds typically range from 5-20%, depending on the species and life stage. Lipids are vital for growth, reproduction, and maintaining cell membrane integrity. The inclusion of high-quality lipid sources, such as fish oil or plant oils, is important for ensuring the proper balance of omega-3 and omega-6 fatty acids in aquaculture diets.

Crude Protein: Crude protein refers to the total amount of protein in a feed, calculated based on nitrogen content. It is a critical component of aquafeeds, as protein is essential for growth, tissue repair, and enzyme production in aquatic species. The protein requirement varies by species and life stage; for example, carnivorous fish may require 40-50% crude protein, while herbivorous species may require 25-35%. The quality of protein, including its digestibility and amino acid profile, is as important as the crude protein level.

Crumble Feeds: Crumble feeds are a type of feed that is produced by crushing pellets into smaller particles. This form is particularly suitable for juvenile fish and shrimp, which have small mouth sizes and require easily digestible feed particles. Crumble feeds offer better water stability and nutrient retention compared to powder feeds. They provide a balanced nutrient profile in a size-appropriate form, promoting efficient growth and reducing feed waste during the early developmental stages.

Demand Feeding: Demand feeding involves the use of a feeding system that allows fish or shrimp to access feed as needed, typically by triggering a sensor or lever. This method is beneficial in reducing feed waste, as it allows the animals to control their own feeding based on hunger. Demand feeders are often used in cage or recirculating systems for species that exhibit clear feeding behaviors. However, demand feeding requires careful monitoring to prevent overfeeding and ensure all individuals have access to feed.

Detoxification Methods: Detoxification methods are processes used to reduce or eliminate the harmful effects of anti-nutritional factors and toxins in feed ingredients. Common methods include heat treatment (to inactivate trypsin inhibitors), fermentation (to break down phytates and tannins), and the use of enzymes like phytase. Chemical treatments, such as ammoniation, can also detoxify certain toxins like aflatoxins. These methods improve the digestibility and nutritional value of feed ingredients, making them safer and more effective for aquaculture use.

Digestibility Trial: A digestibility trial is an experimental process used to determine the proportion of nutrients from feed that are absorbed by the aquatic organism. It typically involves feeding a test diet containing a known marker (such as chromic oxide), collecting fecal samples, and analyzing nutrient content to calculate apparent digestibility coefficients. These trials help in formulating more efficient feeds by improving nutrient uptake.

Digestibility: Digestibility refers to the proportion of nutrients in the feed that are absorbed by the digestive system of the aquatic species. It is a key factor in determining feed efficiency and growth performance. Digestibility is influenced by the feed's ingredient composition, processing methods, and the presence of anti-nutritional factors. High digestibility is essential for minimizing feed waste and maximizing nutrient absorption. Digestibility trials typically measure apparent digestibility coefficients (ADCs) for protein, lipids, and carbohydrates in aquafeeds.

Drying: Drying is the process of removing moisture from feed pellets after pelletization or extrusion to improve shelf life and prevent mold growth. Excess moisture in feed can lead to spoilage and the growth of harmful microorganisms. Drying methods include air drying, oven drying, or the use of specialized drying equipment. Optimal moisture levels for stored feed are typically between 10-12%. Proper drying ensures feed stability, reduces the risk of contamination, and maintains the nutritional quality of the feed.

Eco-Friendly Feeds: Eco-friendly feeds are formulated to minimize environmental impacts by using sustainable ingredients, reducing nutrient waste, and optimizing feed efficiency. These feeds may include alternative protein sources like insect meal, algae, or plant-based ingredients to reduce reliance on fish meal. Eco-friendly feeds aim to improve nutrient retention and reduce phosphorus and nitrogen emissions, which can contribute to eutrophication. In addition, they are often free from harmful additives and contaminants, promoting sustainable aquaculture practices.

Energy Budget Equation: The energy budget equation in aquaculture is a mathematical expression used to describe the balance between energy intake (from feed) and energy expenditure. It can be represented as: **E (intake) = E (maintenance) + E (growth) + E (reproduction) + E (activity) + E (waste).** This equation helps in understanding how feed energy is allocated and optimizing feeding strategies to ensure efficient energy use for growth and reproduction. The goal is to maximize the energy available for production by minimizing waste energy.

Energy Partitioning: Energy partitioning refers to the division of dietary energy into different physiological functions: maintenance, growth, reproduction, and activity. In aquaculture, understanding energy partitioning helps in designing feeds that provide sufficient energy for all these functions. Energy not used for maintenance is allocated to growth and reproduction. Optimizing energy partitioning ensures that more energy is directed toward growth rather than maintenance or waste, improving feed conversion efficiency and overall farm productivity.

Essential Fatty Acids: Essential fatty acids (EFAs), particularly long-chain polyunsaturated fatty acids like omega-3 (EPA, DHA) and omega-6 (ARA), are critical for cell membrane structure, growth, and immune function in fish and shrimp. EFAs must be provided through the diet as aquatic organisms cannot synthesize them. Deficiencies in EFAs can lead to poor growth, reduced fertility, and weakened immune systems. Marine fish generally require higher levels of omega-3 fatty acids, while freshwater species may have balanced requirements for both omega-3 and omega-6.

Exogenous Enzymes: Exogenous enzymes are externally sourced enzymes added to aquafeeds to improve the digestibility of nutrients, particularly in plant-based ingredients that contain anti-nutritional factors. Examples include phytase, which breaks down phytate to release phosphorus, and proteases, which aid in protein digestion. The inclusion of these enzymes enhances nutrient absorption, reduces feed costs, and minimizes environmental pollution from undigested nutrients. The use of exogenous enzymes is especially important in species with limited natural enzyme production.

Extrusion: Extrusion is a high-temperature, high-pressure process used to produce floating or sinking aquafeeds. Feed ingredients are mixed and forced through a die, where the heat and pressure cause the starches to gelatinize, improving digestibility. Extruded feeds have better water stability and can be tailored to float or sink, depending on the feeding behavior of the species. Floating feeds are often used for surface-feeding species, while sinking feeds are preferred for bottom-dwellers like shrimp.

Farm-Made Feeds: Farm-made feeds are feeds produced on-site using locally available ingredients and simple processing methods. These feeds are often formulated to meet the specific nutritional needs of the cultured species and can be more cost-effective than commercial feeds. However, farm-made feeds may have variable nutrient composition and lower digestibility compared to factory-produced feeds. Ensuring proper formulation, ingredient selection, and processing methods are critical for farm-made feeds to be nutritionally adequate and cost-efficient.

Feed Additives: Feed additives are substances added to aquafeeds in small quantities to enhance growth, health, or feed quality. These include vitamins, minerals, probiotics, antioxidants, preservatives, growth promoters, and flavor enhancers. Additives improve feed efficiency, enhance nutrient absorption, and support immune function. For example, Vitamins A, C, and E are commonly added as antioxidants to protect against oxidative stress. The selection and dosage of feed additives must be optimized to meet the specific needs of the species being cultured.

Feed Attractants: Feed attractants are substances added to aquafeeds to enhance palatability and encourage feed intake, especially in species with low feeding responses in captivity. These attractants include amino acids, peptides, nucleotides, and flavors like fish hydrolysates, krill meal, and squid meal. Attractants improve feed consumption and reduce feed waste, leading to better growth performance and feed efficiency. The effectiveness of attractants depends on species-specific preferences and sensory abilities, and they are commonly used in feeds for carnivorous fish and shrimp.

Feed Conversion Ratio (FCR): FCR is a measure of feed efficiency in aquaculture, calculated as the ratio of feed intake (in kilograms) to weight gain (in kilograms) by the cultured species. Lower FCR values indicate better feed efficiency, meaning less feed is required to produce a given amount of biomass. Optimal FCR values vary by species but are typically around 1.0 to 1.5 for fish like tilapia and salmon. FCR is influenced by feed quality, feeding management, and environmental conditions such as water quality and temperature.

Feed Efficiency Ratio (FER): FER is a measure of the efficiency with which feed is converted into body mass, calculated as weight gain divided by the amount of feed consumed. It is similar to FCR but expressed as a positive value. Higher FER values indicate better feed utilization. FER is influenced by the nutrient composition of the feed, the species' ability to digest and metabolize the feed, and feeding practices. Optimizing FER is critical for reducing feed costs and improving production sustainability.

Feed Formulation: Feed formulation is the process of combining various feed ingredients in precise proportions to meet the nutritional requirements of aquatic species. The goal is to create a balanced diet that optimizes growth, health, and feed efficiency while minimizing costs. Feed formulation involves selecting appropriate protein, lipid, carbohydrate, vitamin, and mineral sources based on the species' needs, life stage, and farming conditions. Advanced software and nutritional models are often used to ensure the correct nutrient composition in formulated feeds.

Feed Intake: Feed intake refers to the quantity of feed consumed by fish or shrimp over a certain period, typically measured in grams per day. It is influenced by several factors, including water temperature, species, metabolic rate, and feed composition. Proper feed intake ensures adequate nutrient absorption and optimal growth. For many species, feed intake is around 1-3% of body weight per day, depending on the growth stage and environmental conditions.

Feed Manufacturing Equipment: Feed manufacturing equipment refers to the machinery used in the production of aquafeeds, including grinders, mixers, pelletizers, extruders, dryers, and coolers. These machines are essential for converting raw ingredients into pellets or crumbles with the desired nutrient profile and physical properties. High-quality equipment ensures uniform mixing, optimal particle size, proper pelletization, and water stability. For example, extruders produce floating feeds by gelatinizing starches under high pressure, while pellet mills compact ingredients into sinking pellets. Regular maintenance of equipment is crucial for efficient feed production.

Feed Palatability: Feed palatability refers to the acceptability of feed to fish or shrimp, influencing their willingness to consume it. It is determined by factors such as flavor, texture, and odor, which are enhanced by feed additives like attractants (e.g., krill meal, fish oil). Palatable feeds encourage higher feed intake, leading to improved growth performance. Proper palatability can also reduce feed waste and FCR.

Feed Processing: Feed processing refers to the various mechanical and thermal treatments applied to raw ingredients during feed production, including grinding, mixing, conditioning, pelletization, and extrusion. Processing improves the physical properties of feed, such as water stability, digestibility, and pellet integrity, while preserving or enhancing nutrient availability. Proper feed processing ensures that aquafeeds meet the nutritional needs of the species and maintain their quality during storage and feeding. Techniques like extrusion also allow for the production of specialized feeds, such as floating or sinking pellets.

Feed Storage: Feed storage refers to the conditions and practices used to preserve the quality of aquafeeds over time. Proper storage involves maintaining feeds in cool, dry, and well-ventilated areas to prevent spoilage from moisture, heat, or pests. Storage duration should be minimized to avoid nutrient degradation, particularly of sensitive compounds like vitamins and fats. The use of preservatives and antioxidants can extend the shelf life of feed, but feeds should ideally be used within 6 months to maintain optimal quality.

Feed Trial Evaluation: Feed trial evaluation is a systematic process of testing and analyzing the performance of a formulated feed under controlled or semi-controlled aquaculture conditions. These trials typically measure parameters such as growth rate, feed conversion ratio (FCR), survival rate, nutrient digestibility, and overall health of the cultured species. Feed trial evaluations help determine the efficacy of different feed formulations, ingredient inclusions, and feeding regimes, ensuring the development of cost-effective and nutritionally balanced diets

Feeding Frequency: Feeding frequency refers to how often feed is provided to aquatic species, which can range from multiple times per day to less frequent intervals. The appropriate feeding frequency depends on species, life stage, and digestive physiology. Frequent feeding ensures that smaller amounts of feed are consumed more efficiently, particularly in juvenile fish and shrimp with high metabolic rates. Reduced feeding frequencies may be used for older or larger fish with slower growth rates. Optimizing feeding frequency can improve feed conversion and reduce waste.

Feeding Management: Feeding management involves the strategies and practices used to optimize feed delivery, feed intake, and feed utilization in aquaculture systems. It includes determining the appropriate feed type, feeding rate, frequency, and method based on species requirements, life stage, and environmental conditions. Effective feeding management improves growth performance, feed efficiency, and water quality while minimizing feed waste and production costs. Regular monitoring and adjustment of feeding protocols are essential for maintaining optimal production outcomes.

Feeding Methods: Feeding methods refer to the techniques used to deliver feed to cultured species in aquaculture. Common methods include hand-feeding, broadcast feeding, automatic feeders, and demand feeders. The choice of method depends on the species, size of the farming operation, and type of feed (e.g., floating or sinking). Effective feeding methods ensure that all animals have access to the feed, reduce feed waste, and maintain good water quality. Automation can improve feeding precision and labor efficiency in large-scale operations.

Feeding Rate: Feeding rate refers to the amount of feed provided to the cultured species over a specific period, typically expressed as a percentage of the biomass per day. It is a key factor in optimizing growth, feed efficiency, and water quality. Overfeeding can lead to feed waste and deteriorating water conditions, while underfeeding can result in suboptimal growth and health issues. The optimal feeding rate depends on species, life stage, water temperature, and environmental conditions. For many fish species, the feeding rate ranges from 2-5% of body weight per day.

Feeding Strategy: Feeding strategy refers to the overall approach to feeding aquatic species, including the selection of feeds, feeding rates, frequencies, and methods used to achieve specific production goals. Strategies may vary depending on the species, production system, and economic considerations. A well-designed feeding strategy ensures that the nutritional needs of the species are met while maximizing feed efficiency, minimizing waste, and maintaining water quality. Adjusting the feeding strategy to account for seasonal changes, growth stages, and environmental conditions is crucial for optimal production.

Fish Meal: Fish meal is a high-protein feed ingredient made from whole fish or fish processing by-products. It typically contains 60-70% protein, along with essential amino acids, omega-3 fatty acids, vitamins, and minerals. Fish meal is highly digestible and provides a balanced nutrient profile, making it a key component in aquaculture feeds, especially for carnivorous species. However, concerns over sustainability and cost have driven efforts to find alternative protein sources.

Floating Feeds: Floating feeds are produced by extrusion technology, which allows the feed to remain buoyant on the water surface. These feeds are ideal for surface-feeding fish species such as tilapia and catfish. Floating feeds offer the advantage of easy monitoring of feed consumption, reducing feed waste and improving water quality. They typically have a lower density and are designed to remain intact for extended periods without leaching nutrients. The inclusion of specific ingredients and processing techniques ensures the correct buoyancy.

Freshwater Fish Nutrition: Freshwater fish nutrition focuses on providing a balanced diet that meets the specific energy and nutrient needs of various freshwater species. A typical diet includes 30-40% protein, 5-10% lipid, and 20-30% carbohydrates. Essential amino acids, fatty acids (e.g., omega-3), vitamins, and minerals must also be included. The balance of these nutrients supports optimal growth, reproductive success, immune function, and overall health. Feeds are usually formulated with ingredients like fish meal, soybean meal, and corn gluten meal to ensure the availability of these nutrients.

Grinding: Grinding is a mechanical process used to reduce the particle size of raw materials in feed production. The finer the particle size, the better the digestibility and nutrient absorption in aquatic species. Grinding also ensures uniform mixing of ingredients, improving the consistency and quality of the final feed product. However, excessively fine grinding can reduce pellet durability and increase dust formation. The optimal grinding process balances feed digestibility and physical quality for efficient aquaculture production.

Gross Energy: Gross energy (GE) is the total amount of energy present in a feed ingredient or diet, measured in kilojoules or kilocalories per gram. It includes all the energy contained in proteins, lipids, and carbohydrates. However, not all gross energy is available for growth or metabolism, as some is lost through waste (feces) and heat. For aquatic organisms, the energy content of feed is critical for determining how much feed is required to meet the energy needs for growth and activity.

Growth Performance: Growth performance in aquaculture refers to the overall development and weight gain of aquatic organisms, which is a direct indicator of the effectiveness of feeding regimes and environmental conditions. It is often measured by growth metrics such as specific growth rate (SGR), feed conversion ratio (FCR), and weight gain percentage. Optimal growth performance is achieved when the diet provides the right balance of nutrients, such as proteins, lipids, and essential vitamins, with a typical SGR range of 1-2% for many finfish species under ideal conditions.

Growth Stimulants: Growth stimulants are feed additives that promote faster growth rates and improved feed efficiency in aquaculture species. These may include hormones, growth factors, or bioactive compounds such as nucleotides and peptides. Some growth stimulants, like -glucans or immunostimulants, enhance the immune system, indirectly supporting better growth by reducing disease susceptibility. Regulatory standards often limit the use of synthetic hormones, and natural alternatives are increasingly being explored to ensure food safety and sustainability in aquaculture.

Heat Increment of Feeding: The heat increment of feeding (HIF) refers to the increase in metabolic heat production following food intake, which is a result of digestion, absorption, and assimilation of nutrients. HIF is an important consideration in aquaculture, as the energy expended during these processes reduces the amount of energy available for growth. It varies with the type of feed, nutrient composition, and species. Reducing HIF through optimized feed formulation can improve feed efficiency by making more energy available for growth and other productive activities.

Ingredient Selection: Ingredient selection in feed formulation involves choosing raw materials that provide the required nutrients in the most cost-effective and sustainable way. Ingredients like fish meal, soybean meal, corn gluten, and wheat are common in aquafeeds due to their high digestibility and nutrient content. However, alternative ingredients, such as insect meal or algae, are increasingly being explored to reduce the environmental impact. The selected ingredients must be free from anti-nutritional factors, toxins, and contaminants to ensure feed quality and animal health.

Juvenile Nutrition: Juvenile fish nutrition is tailored to promote continued growth, development, and the establishment of a strong immune system. Protein levels are typically high (35-45%), while lipids (8-12%) provide essential energy and fatty acids. The carbohydrate content is lower, as juveniles rely more on protein and lipid metabolism. The inclusion of vitamins like vitamin C and minerals such as calcium and phosphorus is essential for bone development and disease resistance. High-quality feeds improve feed conversion ratios (FCR) and growth rates during this critical growth phase.

Larval Nutrition: Larval nutrition is crucial for the early development stages of fish and shrimp, where nutrient requirements are significantly higher due to rapid growth and incomplete digestive system development. Larvae require easily digestible proteins (40-60%) and lipids (10-20%) rich in essential fatty acids (EPA and DHA). Larval feeds often include live prey such as rotifers and Artemia, along with micro diets that deliver necessary vitamins and minerals. The critical balance of nutrients supports survival, development of organs, and skeletal growth in the early stages of life.

Lipid Requirement: Lipids are an essential source of energy and fatty acids in aquaculture nutrition, comprising 5-15% of the diet. Lipids provide essential fatty acids (EFAs) such as omega-3 and omega-6, which are necessary for cell membrane integrity, growth, and reproduction. Lipid requirements vary by species and life stage, with carnivorous species generally requiring more. Excessive lipids can lead to fatty liver disease, while deficiencies can result in poor growth, reduced immunity, and impaired reproduction.

Maintenance Energy: Maintenance energy is the minimum amount of energy an aquatic organism requires to sustain vital physiological functions such as respiration, circulation, and osmoregulation without growth or reproduction. This energy demand is met through basal metabolism and routine activities. Maintenance energy must be provided before any additional energy can be used for growth or production. It varies depending on species, water temperature, activity level, and body size, typically accounting for 30-50% of total energy intake.

Mineral Requirement: Minerals are vital for the structural and metabolic functions in aquatic organisms, including bone formation, osmoregulation, and enzyme activation. Key minerals include calcium, phosphorus, magnesium, and trace elements like zinc, selenium, and copper. The requirement for calcium and phosphorus, especially, is high during periods of growth and development.

Mixing: Mixing is the process of combining various feed ingredients uniformly to ensure that each pellet or portion of feed contains the correct nutrient composition. Proper mixing is essential to prevent nutrient imbalances and ensure consistent feed quality. In aquafeed production, specialized mixers, such as ribbon or paddle mixers, are used to achieve homogeneity. Inadequate mixing can result in nutrient segregation, leading to uneven growth and health issues in fish or shrimp populations.

Moisture Content: Moisture content refers to the amount of water present in feed, usually expressed as a percentage of the total weight. Proper moisture levels in feed are essential for preventing spoilage and microbial growth. For most pelleted and extruded aquafeeds, the ideal moisture content is between 10-12%. Excess moisture can lead to mold growth and nutrient degradation, while overly dry feeds may crumble easily and lose physical integrity. Controlling moisture content is critical for maintaining feed quality during storage.

Net Energy: Net energy is the portion of feed energy that remains available to the organism after accounting for energy losses in feces, urine, and heat increment of feeding. It represents the actual energy that can be used for maintenance, growth, reproduction, and activity. In aquaculture, net energy is a key factor in formulating diets that maximize growth and minimize waste. Accurate estimation of net energy allows for efficient feed utilization, reducing feed costs and improving growth performance.

Net Protein Utilization (NPU): NPU is a measure of the efficiency with which ingested protein is used for tissue synthesis, calculated as the percentage of ingested protein that is retained by the body. It combines protein digestibility and biological value (BV). NPU provides insight into the quality of dietary protein and its ability to support growth. High NPU values indicate that the protein is well-absorbed and efficiently used by the organism. Optimal NPU values depend on species and feed composition but typically range between 50-70%.

Non-Conventional Feed Ingredients: Non-conventional feed ingredients refer to alternative raw materials that are used to reduce dependency on conventional ingredients like fish meal or soybean meal. These may include insect meal, algae, duckweed, and agricultural by-products like palm kernel cake or cassava meal. While often more sustainable and cost-effective, non-conventional ingredients must be carefully evaluated for nutrient content, digestibility, and the presence of anti-nutritional factors. They offer potential for reducing the environmental footprint of aquaculture feeds while maintaining nutritional quality.

Non-Demand Feeding: Non-demand feeding refers to feeding practices where the feed is provided at regular intervals without the animals directly controlling the feeding process. This method includes hand-feeding, automatic feeding, or timed feeding systems. Non-demand feeding requires accurate determination of the species' feeding needs to avoid overfeeding or underfeeding. Proper management of non-demand feeding systems ensures that the animals receive the appropriate amount of feed for optimal growth and health.

Nutrient Composition: Nutrient composition refers to the specific levels of essential nutrients such as proteins, lipids, carbohydrates, vitamins, and minerals in aquafeeds. The composition is carefully designed to meet the physiological needs of the species being cultured. For example, carnivorous fish require diets with higher protein (35-50%) and lipid (10-20%) content, while herbivorous species may require more carbohydrates. Proper nutrient composition ensures optimal growth, health, and feed conversion ratios (FCR) while reducing nutrient waste.

Nutrient Digestibility: Nutrient digestibility refers to the extent to which aquatic species can break down and absorb nutrients from their feed. It is a key factor determining feed efficiency, growth, and overall health. Digestibility coefficients are used to quantify the proportion of nutrients, such as protein or fat, that is absorbed versus excreted, with digestibility rates of 85-95% being desirable for protein in well-formulated feeds.

Nutrient Leaching: Nutrient leaching refers to the loss of soluble nutrients from feed pellets into the water after they are introduced into the culture system. This can result in reduced nutrient availability to the organism and degraded water quality due to increased organic matter and nutrient loading. Minimizing leaching involves improving pellet durability and using water-stable binders in feed formulations, with leaching rates ideally less than 10% over a 2-hour period in aquatic systems.

Nutrient Stability: Nutrient stability refers to the ability of a feed to retain its nutritional value during storage, processing, and after being introduced into the water. Nutrients like vitamins (e.g., Vitamin C) and lipids are prone to degradation due to oxidation, heat, and light exposure. Ensuring nutrient stability requires the use of stabilizers, antioxidants, and proper storage conditions to maintain feed quality and prevent nutrient loss.

Nutritional Bioavailability: Nutritional bioavailability in aquaculture refers to the proportion of ingested nutrients that are effectively absorbed and utilized by the organism for growth, reproduction, and maintenance. Factors affecting bioavailability include feed processing, the form of nutrients (e.g., organic vs. inorganic), and interactions with other feed components. For example, phosphorous bioavailability can range between 30-50% in plant-based feeds due to the presence of anti-nutritional factors like phytates.

Optimal Ration: Optimal ration is the ideal amount of feed provided to an aquatic species to meet its nutritional needs without causing overfeeding or underfeeding. It is determined based on the species' life stage, size, activity level, and environmental factors such as water temperature. Providing the optimal ration ensures efficient feed utilization, promotes growth, and minimizes feed waste. For example, juvenile fish may require higher feeding rates than adults due to their faster growth rates.

Organic Acids: Organic acids, such as citric acid, formic acid, and propionic acid, are used in aquafeeds as acidifiers to enhance nutrient digestion and reduce the growth of pathogenic bacteria in the gut. These acids lower the pH in the digestive tract, improving the solubility of minerals like phosphorus and calcium and promoting a favorable gut microflora. Organic acids also act as natural preservatives, preventing spoilage during feed storage. Their inclusion in diets can lead to better growth performance and feed utilization.

Overfeeding: Overfeeding in aquaculture occurs when fish or shrimp are provided with more feed than they can consume within a specific timeframe, often leading to excess uneaten feed accumulating in the culture environment. This can cause deterioration of water quality due to nutrient leaching, increased ammonia and nitrate levels, and oxygen depletion, which can negatively affect animal health and growth. An optimal feeding strategy should balance the feed supply with the species' appetite and growth stage, with feed conversion ratios (FCR) of 1.2 to 1.8 considered efficient for many aquaculture species.

Oxalates: Oxalates are naturally occurring organic compounds found in certain plant-based feed ingredients, such as spinach, beetroot, and rhubarb, that can form insoluble salts with calcium, reducing its bioavailability. In aquaculture, high levels of oxalates can lead to calcium deficiency, impaired bone formation, and reduced growth. Diets with high oxalate levels should be avoided or balanced with calcium-rich ingredients to prevent nutrient imbalances. Methods such as fermentation or soaking can reduce oxalate content in feed ingredients.

Pellet Binders: Pellet binders are specific types of binders used during the feed pelletization process to enhance the cohesion and durability of feed pellets. They ensure that the pellets hold their shape during storage, transportation, and feeding, while also resisting breakdown when exposed to water. Common pellet binders include wheat gluten, lignosulfonates, and bentonite clay. Pellet binders are crucial for producing high-quality feeds that maintain their structural integrity, thereby reducing nutrient loss and improving feed efficiency.

Pellet Durability: Pellet durability refers to the physical strength of feed pellets, ensuring that they do not crumble or disintegrate during handling, storage, and water immersion. Durable pellets maintain their shape and nutrient content when exposed to water, reducing feed waste and nutrient leaching. Optimal pellet durability can be achieved by controlling moisture content, extrusion pressure, and the inclusion of binders like starch.

Pelleted Feeds: Pelleted feeds are compressed feed particles produced through a pelletizer. These feeds are commonly used in aquaculture for their durability, ease of handling, and improved water stability. Pellets can be designed to float or sink depending on the species' feeding habits. Pelleted feeds provide uniform nutrient distribution, reduce waste, and improve feed efficiency. The size and composition of pelleted feeds can be tailored to the species and life stage, ensuring that nutritional needs are met.

Pelletization: Pelletization is the process of compressing feed ingredients into compact, uniform pellets using mechanical pressure and heat. This process enhances feed stability, reduces dust, and improves handling and storage. Pelletized feeds also maintain their structure in water for longer periods, reducing nutrient leaching and feed waste. The size and density of the pellets can be adjusted based on the species being cultured, with larger, denser pellets used for larger fish and smaller, less dense pellets for juveniles or shrimp.

Permissible Levels: Permissible levels refer to the maximum allowable concentrations of certain substances, including anti-nutritional factors, toxins, or contaminants in aquafeeds. These levels are established to ensure the safety and health of the cultured species and the final product for human consumption. For example, the permissible level of aflatoxins (toxins from mold) in aquafeeds is typically less than 20 ppb (parts per billion). Exceeding permissible levels can lead to health issues in aquatic animals, poor growth, or accumulation of harmful substances in fish tissues.

Phytate: Phytate is an anti-nutritional compound found in plant-based feed ingredients like soybean meal, which binds to minerals such as phosphorus, calcium, and zinc, making them unavailable to aquatic species. Fish and shrimp lack the enzyme phytase to break down phytate, leading to reduced mineral absorption and potential deficiencies. To counter this, phytase is often added to aquafeeds, or ingredients are pre-treated to reduce phytate levels. Phytate management improves nutrient bioavailability and reduces environmental pollution from undigested phosphorus.

Preservatives: Preservatives are substances added to aquafeeds to inhibit the growth of microorganisms, such as bacteria, fungi, and mold, thereby extending the shelf life of feed products. Common preservatives include organic acids like propionic and sorbic acids, which prevent spoilage and maintain feed quality. In aquaculture, maintaining proper levels of preservatives in feeds is crucial for preventing the growth of harmful pathogens and ensuring the nutritional integrity of the feed over time, particularly in moist or farm-made feeds.

Probiotics: Probiotics are live microorganisms, such as beneficial bacteria or yeast, added to aquafeeds to improve gut health, enhance digestion, and boost the immune system of aquatic species. Common probiotics include strains of *Lactobacillus*, *Bacillus*, and *Saccharomyces*. Probiotics promote a healthy balance of gut microflora, reduce the prevalence of pathogenic bacteria, and enhance feed utilization, leading to improved growth performance and disease resistance. The dosage and specific strains must be carefully selected to match the species and farming conditions.

Protein Efficiency Ratio (PER): PER is a measure of how effectively dietary protein is utilized for growth, calculated as the weight gain per unit of protein consumed. A higher PER indicates more efficient protein utilization. PER is influenced by the protein quality, digestibility, and amino acid composition of the feed. For example, feeds with high-quality protein sources such as fish meal or soybean meal tend to have higher PER values. In carnivorous fish species, a typical PER ranges from 1.5 to 3.0.

Protein Requirement: Protein requirement in aquaculture refers to the need for high-quality, digestible proteins for growth, maintenance, and tissue repair. Fish and shrimp typically require 30-50% protein in their diet, depending on the species and life stage. Proteins must provide essential amino acids such as lysine and methionine, which cannot be synthesized by the animals. Insufficient protein can lead to poor growth, weakened immune systems, and higher feed conversion ratios (FCR), while excess protein is metabolized as energy, increasing nitrogenous waste.

Protein Sparing: Protein sparing refers to the ability to reduce the proportion of dietary protein by substituting it with non-protein energy sources (like lipids or carbohydrates) without compromising growth performance. This concept is important in aquaculture as protein is the most expensive component of feed. A protein-sparing effect can be achieved when dietary lipids are provided in sufficient quantities, reducing protein catabolism for energy and allowing it to be used for growth and tissue repair.

Rice Bran: Rice bran is the outer layer of the rice grain and is rich in carbohydrates, lipids, and moderate levels of protein (12-15%). It is used as an energy source in aquaculture diets, particularly for herbivorous and omnivorous species. Rice bran can be a cost-effective ingredient, though its high fat content may lead to rancidity if not stored properly. Its nutrient content varies, so it is often combined with protein-rich ingredients to create a balanced feed.

Sinking Feeds: Sinking feeds are designed to submerge in water, catering to bottom-feeding species like shrimp and some fish species. These feeds are typically denser and can be produced through pelletization or low-temperature extrusion processes. Sinking feeds reduce feed competition among surface feeders and provide nutrients directly to species that prefer feeding near the substrate. Ensuring proper water stability and minimal nutrient leaching is important to prevent water quality degradation.

Sinking Rate: Sinking rate is the speed at which a feed pellet descends through the water column, typically measured in centimeters per second. It is an important parameter for matching feed to the feeding habits of aquatic species. For example, bottom-feeding species like shrimp require slow-sinking or sinking feeds, while surface-feeding species like tilapia prefer floating feeds. Adjusting the density and composition of the feed through extrusion or pelletization processes allows control over the sinking rate, optimizing feed availability and reducing waste.

Soybean Meal: Soybean meal is a plant-based protein source commonly used in aquafeeds, with a protein content of around 44-48%. It is highly digestible and contains essential amino acids, though it is deficient in methionine and may contain anti-nutritional factors like trypsin inhibitors and phytates. Despite these limitations, soybean meal is widely used as a substitute for fish meal due to its availability, lower cost, and environmental sustainability. Processing techniques such as fermentation can reduce its anti-nutritional properties.

Specific Dynamic Action: Specific dynamic action (SDA) is the metabolic cost associated with the processing of ingested food, including digestion, absorption, and nutrient assimilation. SDA is typically expressed as a percentage of the energy consumed and can account for 5-15% of the total energy intake. In aquaculture, understanding SDA is crucial for optimizing feed efficiency, as energy lost to SDA is not available for growth or reproduction. Feeds with lower SDA values, such as those with high digestibility, improve growth rates and reduce waste.

Tannins: Tannins are polyphenolic compounds present in many plant-based feed ingredients, such as legumes and grains. While they have some antioxidant properties, tannins are generally considered anti-nutritional factors because they can bind to proteins and inhibit digestive enzymes, reducing feed intake, protein digestibility, and growth in aquatic species. Processing methods such as fermentation, soaking, or heat treatment can reduce tannin content in feed ingredients, improving their nutritional quality for use in aquaculture.

Total Ash: Total ash represents the inorganic mineral content of a feed, which includes essential macro- and micronutrients like calcium, phosphorus, magnesium, and trace elements. Ash levels in feed are typically around 5-15%, depending on the formulation. High ash content may indicate the presence of excessive fillers or low-quality ingredients, while low ash content may suggest a lack of essential minerals. A balance of minerals is crucial for bone development, metabolism, and overall health in aquatic species.

Trypsin Inhibitors: Trypsin inhibitors are anti-nutritional factors commonly found in raw soybeans and other legumes that inhibit the activity of trypsin, an essential enzyme for protein digestion in fish and shrimp. These inhibitors reduce the availability of dietary proteins, leading to poor growth and feed efficiency. Heat treatment, such as toasting or extrusion, is commonly used to deactivate trypsin inhibitors in feed ingredients. The allowable levels of trypsin inhibitors in fish and shrimp diets are minimized to avoid negative impacts on nutrient utilization and health.

Underfeeding: Underfeeding refers to providing less feed than required for optimal growth and development, often leading to poor growth rates, lower biomass yield, and compromised health. Fish or shrimp deprived of adequate nutrition may exhibit reduced immune response and delayed sexual maturation. In practical terms, underfeeding can lower feed costs, but it also results in suboptimal growth performance and higher FCR, indicating inefficient feed utilization. The feed should be adjusted to meet daily nutritional needs based on biomass, water temperature, and species-specific metabolic rates.

Vitamin Requirement: Vitamins are essential micronutrients in aquaculture, required for various physiological processes such as metabolism, immunity, and reproduction. Water-soluble Vitamins (e.g., B-complex and C) and fat-soluble Vitamins (A, D, E, K) must be provided through diet. Deficiencies in vitamins like Vitamin C can lead to poor collagen formation, skeletal deformities, and reduced immune response. Vitamin E and C also act as antioxidants, protecting against oxidative stress. Proper vitamin supplementation ensures optimal growth, health, and stress resistance.

Water Quality Impact: Water quality impact in aquaculture refers to the influence of feed and feeding practices on the water parameters, such as ammonia, nitrite, nitrate, dissolved oxygen, and pH. Overfeeding, poor feed quality, and high nutrient leaching can deteriorate water quality, leading to hypoxia, toxic buildup of nitrogenous wastes, and algal blooms. Maintaining water quality is critical for the health and growth of cultured species, with ammonia levels typically kept below 0.1 mg/L.

Water Stability: Water stability refers to the ability of feed pellets to maintain their structural integrity when submerged in water without disintegrating or leaching nutrients. High water stability is essential in aquaculture feeds to ensure that nutrients are available for consumption and do not leach into the water, which can deteriorate water quality. Good water stability depends on the use of binders, the pelletizing or extrusion process, and ingredient composition. Shrimp feeds typically require water stability for at least 2-4 hours, while fish feeds require 30-60 minutes.

Wheat Bran: Wheat bran is a by-product of wheat milling and is used in aquaculture diets primarily as a carbohydrate and fiber source. It contains around 12-16% protein and 50-60% carbohydrates. Wheat bran helps bind pellets together and provides energy to herbivorous and omnivorous species. However, its high fiber content can limit digestibility, so it is usually included in moderate amounts. Wheat bran is often combined with other ingredients to balance the nutrient profile of the feed.

CHAPTER 5

FINFISH DISEASES AND HEALTH MANAGEMENT

Acanthocephalans: Acanthocephalans, or thorny-headed worms, are parasitic worms that infect the intestines of finfish. These parasites attach to the intestinal wall using a spiny proboscis, causing tissue damage, inflammation, and reduced nutrient absorption. Infections are often linked to the presence of intermediate hosts, such as crustaceans. Control measures include managing the intermediate host populations and using anthelmintic treatments when necessary.

Aeromonas: Aeromonas is a genus of Gram-negative bacteria commonly found in freshwater and brackish environments. Aeromonas species, such as *A. hydrophila*, are opportunistic pathogens that cause diseases like hemorrhagic septicemia and ulcerative conditions in finfish. These bacteria thrive in warm water temperatures (25°-30°C) and can cause rapid outbreaks in aquaculture systems, especially under stress conditions like poor water quality and high stocking densities. Control involves maintaining good water quality and using antibiotics when necessary.

Aetiology: Aetiology refers to the study of the causes or origins of diseases. In aquaculture, understanding the aetiology of diseases involves identifying the pathogens, host factors, and environmental conditions that contribute to disease outbreaks. This knowledge is essential for developing effective prevention and control strategies, including improving water quality, managing stress, and implementing biosecurity measures.

Antibiotics: Antibiotics are drugs used to treat bacterial infections in aquatic animals by inhibiting bacterial growth or killing bacteria. Common antibiotics in aquaculture include oxytetracycline, florfenicol, and enrofloxacin. Optimal use involves correct dosing, adherence to withdrawal times, and avoiding unnecessary or prophylactic use to prevent the development of antibiotic resistance.

Antibody: An antibody is a protein produced by the immune system in response to the presence of an antigen. Antibodies specifically bind to antigens, neutralizing pathogens, marking them for destruction, or preventing their attachment to host cells. In aquaculture, antibody levels are often measured to assess the immune status of animals or the effectiveness of vaccination programs.

Antifungals: Antifungals are used to treat fungal infections in aquatic animals, often caused by water molds like *Saprolegnia*. Common antifungal agents include formalin, potassium permanganate, and copper sulfate. Treatment protocols depend on the species, water conditions, and severity of the infection. Proper use helps prevent secondary infections and maintain fish health.

Antigen: An antigen is a substance, often a protein or polysaccharide, that is recognized by the immune system as foreign, prompting an immune response. In aquaculture, antigens include components of pathogens (e.g., bacteria, viruses, parasites) that trigger the production of antibodies or the activation of immune cells. Antigens are used in vaccines to stimulate protective immunity in aquatic animals.

Antiparasitics: Antiparasitics are chemicals or drugs used to treat parasitic infections in aquatic animals, such as protozoa, helminths, and crustaceans. Examples include formalin, praziquantel, and hydrogen peroxide. Effective use involves correct dosing, treatment duration, and consideration of environmental factors to ensure parasite elimination and minimize stress to the host.

Antivirals: Antivirals are agents used to prevent or treat viral infections in aquatic animals by inhibiting viral replication. Though less commonly used than antibiotics, some antiviral drugs or compounds, like ribavirin or natural extracts, are being explored for controlling viral diseases in aquaculture. The effectiveness of antivirals depends on the specific virus, stage of infection, and delivery method.

Aquaculture: Aquaculture is the farming of aquatic organisms, including fish, shellfish, crustaceans, and aquatic plants, in controlled environments. This practice involves breeding, rearing, and harvesting under optimized conditions to enhance productivity and sustainability. Aquaculture plays a critical role in global food security, providing nearly half of the seafood consumed worldwide. It requires careful management of water quality parameters such as temperature, dissolved oxygen, ammonia, and pH to ensure optimal growth, health, and survival of cultured species.

Aquatic Environment: The aquatic environment includes the physical, chemical, and biological components of water habitats where aquatic animals live. Key parameters include temperature, pH, dissolved oxygen, salinity, ammonia, nitrites, and nitrates. Maintaining optimal conditions is essential for the survival, growth, and health of aquatic organisms, as poor environmental quality can lead to stress, disease outbreaks, and mortality.

Bacteria: Bacteria are single-celled, prokaryotic microorganisms that can exist as free-living entities or as pathogens in aquatic environments. They play various roles, from nutrient cycling and biodegradation to causing diseases in aquatic animals, such as vibriosis and columnaris. Optimal water quality parameters for limiting pathogenic bacteria include maintaining low ammonia (<0.05 mg/L), nitrite (<0.1 mg/L), and controlling environmental stressors.

Bacterial Diseases: Bacterial diseases in finfish are caused by pathogenic bacteria such as *Aeromonas, Vibrio,* and *Flavobacterium* species. These diseases can result in symptoms like ulcers, hemorrhages, fin rot, and systemic infections, often leading to high mortality rates. Bacterial infections are influenced by environmental stressors, including poor water quality and high stocking densities. Management involves good husbandry practices, regular monitoring, and the use of antibiotics or other antibacterial treatments when necessary.

Biochemical Methods: Biochemical methods in aquaculture are used to analyze the metabolic and enzymatic activities of pathogens or the physiological responses of finfish to disease. Techniques include enzyme assays, protein electrophoresis, and tests for specific metabolites or biomarkers associated with disease. These methods help in identifying pathogenic strains, understanding disease mechanisms, and assessing the health status of fish, contributing to more targeted and effective treatments.

Biochemical Tests: Biochemical tests are laboratory procedures used to identify microorganisms based on their biochemical properties, such as enzyme activities, metabolic pathways, and chemical reactions. Examples include catalase test, oxidase test, carbohydrate fermentation, and API test strips. These tests help in differentiating species, assessing pathogenicity, and guiding appropriate treatment strategies in aquatic health management.

Biological Tests: Biological tests involve using live organisms or biological systems to evaluate the presence or activity of pathogens, toxins, or immune responses. Examples include bioassays, where test organisms are exposed to samples to observe effects, and in vivo tests, such as injecting suspected pathogens into healthy animals to confirm disease causality. Biological tests are crucial for understanding pathogen-host interactions and the efficacy of treatments or vaccines.

Bioremediation: Bioremediation in aquaculture involves the use of biological agents such as bacteria, fungi, or plants to degrade or detoxify pollutants in water, sediments, and aquaculture systems. This process helps in reducing harmful substances like ammonia, nitrites, and organic wastes, thereby maintaining water quality and supporting a healthy environment for aquatic organisms. Optimum parameters for bioremediation include pH 6.5-8.5, temperature 20°-30°C, and dissolved oxygen levels above 5 mg/L.

Biosecurity Measures: Biosecurity measures in aquaculture are practices designed to prevent the introduction, spread, and establishment of pathogens in aquatic systems. These include quarantine, disinfection protocols, control of water sources, use of pathogen-free stock, and monitoring for early detection of diseases. Biosecurity is critical for maintaining healthy aquaculture operations and reducing the risk of disease outbreaks.

Biosecurity: Biosecurity in aquaculture refers to practices designed to prevent the introduction, spread, and impact of pathogens within and between aquaculture facilities. It includes measures such as disinfection of equipment, use of pathogen-free stock, control of visitor access, and implementation of quarantine protocols. Effective biosecurity reduces the risk of disease outbreaks, protects the health of cultured finfish, and ensures the sustainability of aquaculture production.

Branchial Mycosis: Branchial mycosis involves fungal infections of the gills in finfish, leading to gill damage, impaired respiration, and increased mortality, especially in young or stressed fish. Fungi such as *Saprolegnia* can proliferate in gill tissues under suboptimal water conditions. Control measures focus on maintaining optimal water quality, minimizing stressors, and treating affected fish with appropriate antifungal agents like potassium permanganate.

Cestodes: Cestodes, or tapeworms, are parasitic flatworms that infect the intestines of finfish, where they absorb nutrients and can cause growth retardation, emaciation, and mortality. Common cestodes in aquaculture include species of the genera *Bothriocephalus* and *Ligula*. Management of cestode infections involves controlling intermediate hosts, such as copepods, and using antiparasitic drugs like praziquantel.

Chemicals: Chemicals used in aquaculture include disinfectants, water conditioners, anesthetics, and treatments for controlling pathogens or improving water quality. Common chemicals include chlorine, formalin, potassium permanganate, and copper sulfate. The use of chemicals requires careful management to ensure the safety of finfish, minimize environmental impact, and comply with regulatory standards for residues in fish products.

Control: Control measures in aquaculture aim to prevent, reduce, or eliminate the occurrence of diseases. These measures include biosecurity protocols, vaccination, water quality management, quarantine of new stocks, and regular health monitoring. Control strategies are essential for minimizing the impact of diseases on fish production, improving survival rates, and ensuring the sustainability of aquaculture operations.

Conventional Techniques: Conventional diagnostic techniques in aquaculture include culture methods, Gram staining, biochemical testing, and microscopic examination to identify bacterial, viral, fungal, and parasitic pathogens. These methods are often time-consuming but provide reliable results for identifying common pathogens. They are widely used in routine disease investigations, though they may be complemented by more advanced, rapid diagnostic methods for quicker results.

Crustacean Parasites: Crustacean parasites, such as sea lice (*Lepeophtheirus salmonis*) and fish lice (*Argulus* spp.), are ectoparasites that infest the skin, gills, and fins of finfish. These parasites feed on mucus, blood, and skin, causing irritation, tissue damage, and increased susceptibility to secondary infections. Control of crustacean parasites involves regular monitoring, biological controls (e.g., cleaner fish), and chemical treatments with antiparasitic agents like hydrogen peroxide or emamectin benzoate.

Dermal Mycosis: Dermal mycosis refers to fungal infections affecting the skin of finfish, often caused by fungi like Saprolegnia spp. It presents as white, cotton-like growths on the skin and can lead to ulcers, secondary bacterial infections, and impaired osmoregulation. Dermal mycosis is typically associated with environmental stressors such as poor water quality or injuries. Management includes improving water conditions, reducing handling stress, and applying antifungal treatments.

Diagnosis: Diagnosis in aquaculture involves identifying diseases in finfish through clinical examination, laboratory testing, and analysis of environmental conditions. Effective diagnosis relies on accurate detection of pathogens, assessment of fish health, and understanding of potential stressors. Diagnostic methods include microscopy, culture techniques, histopathology, molecular diagnostics, and serology. Timely and accurate diagnosis is essential for effective treatment and management of diseases in aquaculture.

Diagnostic Techniques: Diagnostic techniques in aquaculture encompass a range of methods used to detect and identify pathogens causing disease in finfish. These techniques include conventional methods such as culture and sensitivity tests, advanced molecular methods like PCR and ELISA, and imaging techniques such as microscopy and histopathology. Effective diagnostics are crucial for early detection, accurate identification of pathogens, and the formulation of appropriate treatment and control measures.

Disease Development: Disease development in finfish is a multifactorial process involving the interaction of the host, pathogen, and environment. It typically begins with exposure to a pathogen, followed by colonization, invasion, and multiplication of the pathogen within the host. The host's immune response, environmental stressors, and pathogen virulence factors play crucial roles in the progression and severity of the disease. Early detection and management are key to controlling disease outbreaks in aquaculture systems.

Disease Investigation: Disease investigation in aquaculture is a systematic process that involves identifying the cause of morbidity or mortality in aquatic organisms. It includes gathering background information, conducting clinical examinations, performing necropsies, collecting and analyzing samples (e.g., water quality, histopathology, pathogen testing), and reviewing management practices. The goal is to determine the etiology of the disease, contributing factors, and provide recommendations for treatment or management interventions.

Disease Prevention: Disease prevention in aquaculture encompasses strategies and practices aimed at minimizing the occurrence of diseases. This includes maintaining optimal water quality, proper nutrition, biosecurity measures, regular health monitoring, and reducing stress. Preventative measures help in maintaining high productivity, reducing losses, and minimizing the use of medications.

Disease Transmission: Disease transmission in aquaculture involves the spread of pathogens between individuals, populations, or species, either directly (e.g., contact, waterborne) or indirectly (e.g., vectors, fomites). Understanding transmission routes is critical for implementing biosecurity measures, such as disinfection, quarantine, and controlled stocking densities, to prevent and control disease outbreaks.

Disease: Disease in aquatic organisms is a state of abnormality that disrupts normal physiological functions, often caused by pathogens, environmental stressors, or genetic factors. Diseases in aquaculture can lead to significant economic losses due to mortality, reduced growth rates, and decreased feed efficiency. Common types of diseases include bacterial, viral, fungal, and parasitic infections, which can be exacerbated by poor water quality, overstocking, and inadequate nutrition.

Disinfection: Disinfection is the process of eliminating or reducing harmful microorganisms from surfaces, equipment, and water in aquaculture systems. Common disinfectants include chlorine, iodine, hydrogen peroxide, and UV light. Effective disinfection requires proper selection of agents, concentration, contact time, and application methods to ensure the safety and health of aquatic animals without harmful residues.

Drugs: Drugs in aquaculture refer to pharmaceutical agents used to treat or prevent diseases in finfish, including antibiotics, antiparasitics, antifungals, and antivirals. The selection of appropriate drugs is based on the type of pathogen, sensitivity of the pathogen to specific drugs, and regulatory guidelines. Responsible use of drugs is essential to prevent the development of drug resistance, ensure the safety of fish products, and protect the environment.

Edwardsiella: Edwardsiella is a genus of Gram-negative bacteria that includes pathogens like *Edwardsiella* tarda, which cause diseases such as Edwardsiellosis in finfish. This disease can result in systemic infections, with symptoms including hemorrhages, ascites, and necrosis of internal organs. Edwardsiella infections are more common in warm water conditions and can be exacerbated by poor water quality and high stocking densities. Management involves improving water quality, reducing stress, and administering effective antibiotics.

Environmental Stress: Environmental stress refers to adverse conditions in the aquatic environment, such as suboptimal water quality (e.g., low oxygen, high ammonia), extreme temperatures, or pollution, that negatively impact the health of aquatic animals. Environmental stress can weaken the immune system, increase susceptibility to diseases, and affect growth and reproduction. Proper management includes monitoring and adjusting key water parameters to maintain a stable and supportive environment.

Epidemiology: Epidemiology is the study of the distribution, patterns, and determinants of diseases within populations. In aquaculture, epidemiology involves analyzing disease outbreaks, identifying risk factors, and developing strategies to prevent and control the spread of infections. Key epidemiological factors include host density, water quality, pathogen virulence, and environmental conditions, all of which influence disease dynamics in aquaculture systems.

Finfish: Finfish are aquatic vertebrates characterized by the presence of fins, gills, and a streamlined body adapted for swimming. They belong to the group Osteichthyes (bony fish) or Chondrichthyes (cartilaginous fish), and are an important group in aquaculture due to their economic value. Finfish species commonly cultured include salmon, tilapia, catfish, and carp. These species have specific environmental requirements such as optimal temperature ranges (e.g., 20°-28°C for tilapia), dissolved oxygen levels (usually above 5 mg/L), and pH levels (typically between 6.5 and 8.5) for optimal growth and health.

Fish Health: Fish health refers to the overall well-being of fish, encompassing their physiological, immunological, and behavioral status. It is influenced by factors such as water quality, nutrition, stocking density, stress levels, and exposure to pathogens. Maintaining optimal fish health involves monitoring and managing these factors to prevent disease, promote growth, and enhance productivity in aquaculture systems.

Flavobacterium: Flavobacterium is a genus of Gram-negative bacteria that includes pathogens such as Flavobacterium columnare, which causes columnaris disease in finfish. This disease is characterized by lesions on the skin, gills, and fins, leading to significant mortality in aquaculture. Flavobacterium thrives in warm water conditions (20°-30°C) and can rapidly spread in crowded or poorly managed systems. Control measures include reducing water temperature, improving hygiene, and using antibiotics or other therapeutic agents.

Fungal Diseases: Fungal diseases in finfish are primarily caused by pathogenic fungi such as *Saprolegnia*, which can infect the skin, gills, and eggs of fish. Fungal infections, or mycoses, are often secondary to other stressors or injuries and can lead to high mortality rates if not managed properly. Control of fungal diseases involves improving water quality, reducing stress, and using antifungal treatments, such as formalin or potassium permanganate, in severe cases.

Fungi: Fungi in aquaculture include molds, yeasts, and water molds (e.g., Saprolegnia) that can act as pathogens or opportunistic invaders, especially in stressed or immunocompromised aquatic animals. Fungal infections are characterized by lesions, cotton-like growths, or systemic infections. Optimum conditions to prevent fungal outbreaks include maintaining low organic matter, adequate water flow, and a stable environment with pH 6.5-8.5 and dissolved oxygen above 5 mg/L.

Genetic Resistance: Genetic resistance is the inherent ability of certain aquatic animals to resist or tolerate infections or environmental stressors due to their genetic makeup. Selective breeding programs in aquaculture often aim to enhance genetic resistance to specific pathogens or adverse conditions, improving overall health, productivity, and survival rates in cultured species.

Haematopoietic Necrosis: Haematopoietic Necrosis is a viral disease affecting the blood-forming tissues of finfish, leading to severe anemia, immune suppression, and high mortality rates. It is caused by viruses such as Infectious Hematopoietic Necrosis Virus (IHNV), which is highly pathogenic to species like salmonids. The disease spreads rapidly in water temperatures between 8°-15°C and can result in significant losses in aquaculture. Preventive measures include strict biosecurity, vaccination, and avoidance of stressors.

Health Management: Health management in aquaculture involves the implementation of practices and protocols to maintain the health and well-being of cultured finfish. This includes regular monitoring of water quality, nutrition management, disease prevention through vaccination and biosecurity measures, and prompt intervention when diseases occur. Effective health management reduces the risk of disease outbreaks, improves fish growth and survival rates, and enhances the overall productivity of aquaculture systems.

Health Monitoring: Health monitoring involves regular observation, sampling, and analysis of aquatic animals and their environment to detect signs of disease, stress, or suboptimal conditions. This includes checking for clinical signs, conducting water quality tests, and performing routine diagnostics. Health monitoring helps in early detection of problems, enabling timely interventions and reducing losses due to disease.

Histology: Histology is the study of the microscopic structure of tissues, often used in diagnosing diseases in aquatic animals by examining tissue samples under a microscope. Techniques include fixation, sectioning, and staining (e.g., H&E stain) of tissues to identify pathological changes such as inflammation, necrosis, or neoplastic lesions. Histology provides insights into disease mechanisms and the extent of tissue damage.

Histopathology: Histopathology involves the microscopic examination of tissues to study the effects of diseases on the cells and organs of finfish. Tissue samples are fixed, sectioned, stained, and analyzed to identify cellular changes, such as inflammation, necrosis, or presence of pathogens. Histopathology provides insights into the pathogenesis of diseases and helps in diagnosing conditions that are not detectable through other methods, making it a valuable tool in fish health management.

Host Defense Mechanism: Host defense mechanisms in aquatic animals involve a complex interplay of physical, chemical, and biological processes that protect against pathogen invasion. These mechanisms include barriers like skin and mucous, cellular responses such as phagocytosis, and humoral responses involving antimicrobial peptides and proteins. The effectiveness of these defenses is influenced by factors such as stress, nutrition, and environmental conditions, which can alter the host's susceptibility to disease.

Host-Pathogen Interaction: Host-pathogen interaction refers to the dynamic relationship between a host organism (e.g., fish) and a pathogen (e.g., bacteria, virus, parasite). This interaction determines the outcome of infection, influenced by the host's immune response, pathogen virulence factors, and environmental conditions. Understanding these interactions is critical for developing strategies for disease prevention, treatment, and management in aquaculture.

Immune Modulation: Immune modulation involves altering the immune response to enhance or suppress immunity, often through the use of immunostimulants, vaccines, or immunosuppressive agents. In aquaculture, immune modulation aims to enhance disease resistance, improve vaccine efficacy, or manage autoimmune conditions. Understanding the mechanisms of immune modulation helps in developing strategies for health management.

Immune Response: The immune response in aquatic animals is the biological process by which the immune system detects and defends against pathogens. It involves innate responses, such as phagocytosis and the production of antimicrobial peptides, and adaptive responses, including antibody production and memory cell formation. The effectiveness of the immune response is influenced by factors like genetics, nutrition, stress, and environmental conditions.

Immune System: The immune system of aquatic animals, including finfish, comprises specific and non-specific defense mechanisms that protect against pathogens. It includes physical barriers (skin, scales, mucous), cellular responses (phagocytes, lymphocytes), and humoral components (antibodies, complement proteins). The effectiveness of the immune system can be influenced by environmental conditions such as water temperature, salinity, and the presence of pollutants, which can either enhance or suppress immune function.

Immunological Techniques: Immunological techniques are methods used to study and manipulate the immune response in aquatic animals, including ELISA, immunohistochemistry, western blotting, and flow cytometry. These techniques help in detecting pathogens, assessing immune function, and evaluating vaccine efficacy by measuring antibodies, antigens, or cellular immune responses.

Immunostimulants: Immunostimulants are substances that enhance the innate and adaptive immune responses of aquatic animals, making them more resistant to infections. Common immunostimulants include beta-glucans, peptidoglycans, and nucleotides, administered through feed, water, or injection. Optimum doses vary but are generally in the range of 0.1-1% of the diet for feed-based delivery. They work by activating immune cells, increasing the production of antibodies, and enhancing overall immune surveillance.

Infection: Infection occurs when a pathogen invades and multiplies within a host, causing disease or triggering an immune response. In aquaculture, infections can be caused by bacteria, viruses, fungi, or parasites, and can range from subclinical to severe. Managing infections involves early detection, appropriate treatment, and preventive measures to reduce transmission and impact on the population.

Infectious Pancreatic Necrosis (IPN): Infectious Pancreatic Necrosis (IPN) is a viral disease primarily affecting young salmonids, such as trout and salmon, but can also affect other species. It is caused by the IPN virus (IPNV), leading to necrosis of the pancreas, digestive organs, and high mortality in juvenile fish. IPN is particularly problematic in hatcheries and can persist in water temperatures of 10°-15°C. Control measures include biosecurity protocols, breeding for resistance, and vaccination.

Inflammation: Inflammation is the immune system's response to infection, injury, or irritants, characterized by redness, swelling, heat, and pain. In aquatic animals, inflammation can be observed in tissues through histopathology, and it plays a critical role in containing infections and initiating healing processes. Chronic or excessive inflammation, however, can lead to tissue damage and impaired function.

Live Examination: Live examination is a non-invasive diagnostic approach where the external appearance and behavior of live aquatic animals are observed for signs of disease or stress. This includes checking for abnormalities such as lesions, discolorations, fin erosion, abnormal swimming, or respiratory distress. Live examinations help in the initial assessment of health status and are often followed by more specific diagnostic procedures.

Lymphocystis: Lymphocystis is a viral disease caused by Lymphocystis disease virus (LCDV), affecting a broad range of finfish species. It is characterized by the formation of nodular lesions on the skin, fins, and gills due to the hypertrophy of infected cells. While the disease is generally non-lethal, it can cause aesthetic damage and secondary infections, affecting the marketability of fish. Control primarily involves maintaining good water quality and reducing stress, as there is no specific treatment for the virus.

Microbiological Methods: Microbiological methods in aquaculture involve the isolation, culture, and identification of microorganisms, including bacteria and fungi, from diseased fish or their environment. Techniques include agar plate cultures, broth cultures, and biochemical tests to identify specific pathogens. These methods are crucial for diagnosing infections, understanding pathogen dynamics, and developing appropriate treatment strategies, including selecting effective antimicrobial agents.

Microscopy: Microscopy is a fundamental diagnostic tool in aquaculture used to examine tissues, cells, and microorganisms at high magnification. It is essential for identifying pathogens such as bacteria, fungi, and parasites in finfish. Techniques include light microscopy, electron microscopy, and fluorescent microscopy, each providing different levels of detail. Microscopy is critical for assessing the presence of pathogens, evaluating tissue damage, and guiding treatment decisions.

Mode of Action: The mode of action refers to the mechanism by which a drug or chemical exerts its effects on pathogens or the host. In aquaculture, understanding the mode of action of treatments helps in selecting the most effective agents for controlling specific diseases. For example, antibiotics inhibit bacterial growth by targeting cell wall synthesis or protein production, while antiparasitic drugs may disrupt the nervous system of parasites.

Molecular Diagnostic Techniques: Molecular diagnostic techniques involve the use of molecular biology methods, such as PCR, qPCR, DNA sequencing, and in situ hybridization, to detect and identify pathogens at the genetic level. These techniques offer high sensitivity and specificity, allowing for early and accurate diagnosis of infections, even when pathogen levels are low or in complex samples.

Morphological Tests: Morphological tests involve the examination of the physical and structural characteristics of organisms, such as bacteria, parasites, or tissues, using techniques like light microscopy, scanning electron microscopy, and staining methods (e.g., Gram stain, Giemsa stain). These tests help in identifying and classifying organisms based on their size, shape, and structural features, and are essential in diagnosing diseases and differentiating between species.

Mycosis: Mycosis refers to fungal infections in finfish, caused by fungi such as *Saprolegnia* and *Aphanomyces*. Mycoses can affect the skin, gills, and internal organs, leading to symptoms like white cottony growths, lesions, and impaired breathing. These infections are often secondary to environmental stress or injury. Control involves improving water quality, minimizing handling stress, and using antifungal treatments, including formalin or iodine-based solutions.

Non-Specific Immunity: Non-specific immunity, or innate immunity, provides the first line of defense against infections in finfish and other aquatic organisms. It includes physical barriers (skin, mucous membranes), cellular defenses (phagocytes, natural killer cells), and soluble factors (lysozymes, interferons). This form of immunity responds rapidly and does not rely on prior exposure to pathogens, making it crucial for immediate defense against infections.

Organ Systems: Organ systems in aquatic animals, such as the respiratory, circulatory, digestive, immune, and nervous systems, work together to maintain homeostasis and overall health. Each system has specific functions and can be affected by pathogens, environmental factors, or management practices. Understanding the structure and function of these systems is essential for diagnosing diseases and developing effective treatments.

Outbreak: An outbreak is the sudden occurrence of a disease in a population of aquatic animals, often characterized by increased morbidity and mortality. Outbreaks can result from factors like pathogen introduction, environmental stress, or lapses in biosecurity. Effective management involves rapid identification, containment measures, and addressing underlying causes to minimize impact and prevent recurrence.

Parasite Identification: Parasite identification in aquaculture involves the detection and classification of parasitic organisms, such as protozoans, helminths, and crustaceans, that infect aquatic species. Techniques include microscopy for morphological analysis, staining methods, and molecular diagnostics like PCR. Accurate identification is essential for understanding the life cycle, pathogenicity, and transmission dynamics, enabling targeted control measures.

Parasitic Diseases: Parasitic diseases in finfish are caused by a wide range of parasites, including protozoans, trematodes, cestodes, and crustacean parasites like lice. These parasites can cause external and internal infections, leading to symptoms such as lesions, anemia, and impaired growth. Parasitic infections are often associated with poor water quality and overcrowding. Control measures include regular health monitoring, treatment with antiparasitic agents, and maintaining optimal environmental conditions.

Pathogen Identification: Pathogen identification involves detecting and characterizing the microorganisms (bacteria, viruses, fungi, parasites) responsible for diseases in aquatic animals. Methods include culturing on selective media, microscopy, biochemical assays, and molecular techniques like PCR and sequencing. Accurate identification is crucial for effective disease management and treatment, and it typically requires confirming the pathogen's presence, virulence, and impact on the host.

Pathogen: A pathogen is any microorganism-such as bacteria, viruses, fungi, or parasites-that can cause disease in its host. In aquaculture, common pathogens include bacteria like Aeromonas and Pseudomonas, viruses such as Viral Haemorrhagic Septicaemia Virus, and parasites like trematodes and crustaceans. Pathogens can spread rapidly in aquaculture settings due to high stocking densities and shared water systems, necessitating rigorous health management and biosecurity measures.

Pathogenesis: Pathogenesis is the process by which a disease develops, including the mechanisms through which a pathogen invades, colonizes, and damages the host. In aquaculture, pathogenesis studies focus on how pathogens like bacteria, viruses, and parasites interact with finfish, overcoming immune defenses and causing clinical signs of disease. Understanding pathogenesis is crucial for developing targeted treatments and preventive measures.

Pathological Changes: Pathological changes refer to structural and functional abnormalities in tissues or organs caused by disease, injury, or stress. In aquaculture, these changes can include cellular alterations, tissue necrosis, inflammation, fibrosis, and organ dysfunction. Identifying pathological changes is crucial for diagnosing diseases, understanding disease progression, and evaluating the impact of environmental stressors or toxicants.

Post Mortem Examination: Post mortem examination, or necropsy, involves the dissection and examination of deceased aquatic animals to determine the cause of death. It includes external inspection, opening the body cavity, and examining internal organs for signs of disease, parasites, or abnormalities. Key parameters assessed include organ color, size, texture, and presence of lesions. Tissue samples may be collected for further histological, microbiological, or molecular analysis.

Preventive Strategies: Preventive strategies in aquaculture aim to reduce the risk of disease outbreaks through proactive measures such as biosecurity, vaccination, optimal nutrition, and environmental management. These strategies include quarantine of new stock, regular health monitoring, water quality management, and reducing stress through appropriate stocking densities and handling practices. Preventive measures are more cost-effective and sustainable than reactive treatments, contributing to the overall health and productivity of aquaculture operations.

Probiotics: Probiotics are live microorganisms, typically bacteria or yeast, that are administered to aquatic animals to improve their intestinal microbial balance, enhance digestion, boost immune response, and inhibit pathogenic bacteria. Optimal concentrations in aquaculture water range from 10 to 10 CFU/mL. Probiotics work by competing with pathogens for nutrients and space, producing antimicrobial substances, and stimulating the host's immune system, thereby contributing to overall health and disease resistance.

Prophylaxis: Prophylaxis refers to preventive measures taken to reduce the risk of disease in aquaculture, including vaccination, biosecurity practices, use of probiotics, and water quality management. The aim is to enhance the overall health and resilience of aquatic animals, minimizing the impact of potential pathogens and reducing the need for therapeutic interventions.

Pseudomonas: *Pseudomonas* is a genus of Gram-negative, rod-shaped bacteria that are ubiquitous in aquatic environments. *Pseudomonas* species, particularly *P. fluorescens*, are known to cause diseases in finfish, including fin rot and septicemia. These bacteria are opportunistic and can thrive in a range of environmental conditions, often taking advantage of stressed or immunocompromised hosts. Management of *Pseudomonas* infections includes improving water quality, reducing stress, and using appropriate antimicrobial treatments.

Quarantine: Quarantine is the practice of isolating new or sick aquatic animals from the main population to prevent the spread of diseases. It involves monitoring and treating animals for specific pathogens, acclimating them to the new environment, and observing biosecurity protocols. Quarantine periods typically range from 2 to 4 weeks, depending on the species and potential disease risks.

Rapid Diagnostic Techniques: Rapid diagnostic techniques in aquaculture involve advanced methods such as polymerase chain reaction (PCR), loop-mediated isothermal amplification (LAMP), enzyme-linked immunosorbent assay (ELISA), and lateral flow immunoassays. These techniques provide quicker, more specific results than conventional methods, enabling early detection and intervention in disease outbreaks. Rapid diagnostics are essential for effective health management and biosecurity in aquaculture operations.

Resistance: Resistance refers to the ability of pathogens (bacteria, viruses, fungi, parasites) or hosts (aquatic animals) to withstand the effects of drugs, environmental stressors, or infections. Pathogen resistance, such as antibiotic resistance, can complicate disease management and necessitate alternative treatments. Host resistance involves genetic or immune factors that enable aquatic animals to resist infections or environmental challenges.

Sampling Procedures: Sampling procedures in aquaculture involve systematically collecting water, sediment, and biological samples (e.g., tissue, blood) from aquaculture systems or individual animals for health assessment and diagnosis. Proper sampling includes maintaining aseptic conditions, selecting appropriate sample sizes (e.g., 5-10% of a population), and timing (e.g., before feeding or during peak stress times). The procedures are critical for disease monitoring, environmental assessment, and ensuring representative data for analysis.

Sanitation: Sanitation in aquaculture involves regular cleaning and maintenance of facilities, equipment, and water to reduce the load of pathogens and organic matter. It includes practices such as removing uneaten feed, cleaning tanks and filters, and controlling biofilm formation. Proper sanitation helps in preventing disease outbreaks, maintaining water quality, and ensuring a healthy environment for aquatic animals.

Specific Immunity: Specific immunity, also known as adaptive immunity, involves a targeted response against specific pathogens. It is mediated by lymphocytes (T-cells and B-cells) that recognize antigens, leading to the production of antibodies and the development of immunological memory. This type of immunity is slower to develop compared to non-specific immunity but provides long-lasting protection. Vaccination is a key tool in aquaculture to enhance specific immunity in finfish against common pathogens.

Stress: Stress in aquatic organisms is a physiological response to external stimuli that disrupt homeostasis, such as poor water quality, handling, overcrowding, or sudden changes in temperature. Stress can weaken the immune system, making fish more susceptible to diseases. Chronic stress can impair growth, reproduction, and survival, highlighting the importance of maintaining optimal environmental conditions, including stable temperatures, appropriate oxygen levels, and minimal disturbance.

Stressors: Stressors are factors that disrupt the normal physiological balance of aquatic animals, leading to stress responses that can affect growth, reproduction, and disease resistance. Common stressors in aquaculture include poor water quality, overcrowding, handling, transportation, and changes in temperature or salinity. Managing stress involves optimizing environmental conditions and minimizing disruptions to support animal welfare and health.

Susceptibility: Susceptibility is the likelihood of an aquatic animal to be affected by a pathogen or disease, influenced by factors such as immune status, genetics, environmental conditions, and stress levels. Susceptibility testing for pathogens, such as antimicrobial susceptibility testing, helps guide appropriate treatment choices and understand disease dynamics.

Systemic Mycosis: Systemic mycosis refers to fungal infections that spread beyond localized areas to affect multiple organs within the body of finfish. Such infections can cause severe systemic symptoms, including organ failure and high mortality. Pathogenic fungi like *Aphanomyces* invadans can penetrate deeply into tissues, making treatment challenging. Management involves early detection, improved water quality, and the use of systemic antifungal treatments, though options are often limited.

Therapeutics: Therapeutics in aquaculture refer to the use of drugs, chemicals, or other interventions to treat diseases or enhance the health of aquatic animals. This includes antibiotics, antiparasitics, antivirals, antifungals, and other medications. Therapeutics must be used judiciously to avoid resistance, adhere to withdrawal times, and minimize impacts on the environment.

Treatment: Treatment in aquaculture refers to the medical management of diseases in finfish, including the use of antibiotics, antivirals, antifungals, and antiparasitic agents. Treatment strategies are chosen based on the type of pathogen, disease severity, and sensitivity of the pathogen to specific drugs. Effective treatment requires accurate diagnosis, appropriate drug selection, and adherence to recommended dosages and withdrawal times to ensure fish health and consumer safety.

Trematodes: Trematodes, or flukes, are parasitic flatworms that can infect various organs of finfish, including the gills, skin, and internal organs. Common trematode infections in aquaculture include those caused by species like *Dactylogyrus* and *Gyrodactylus*. These parasites can cause significant tissue damage, impaired respiration, and increased susceptibility to secondary infections. Control involves regular monitoring, reducing snail populations (intermediate hosts), and using antiparasitic treatments.

Vaccination Methods: Vaccination methods in aquaculture involve administering antigens to aquatic animals to stimulate an immune response and provide protection against specific pathogens. Common methods include immersion, oral, and injection vaccines. The choice of method depends on the type of vaccine, species, and the disease being targeted. Effective vaccination requires proper handling, dosing, and timing to ensure optimal protection.

Vaccination: Vaccination in aquaculture is a preventive measure used to enhance the specific immunity of finfish against infectious diseases. Vaccines can be administered through injection, immersion, or oral delivery, depending on the type of pathogen and the species of fish. Vaccination stimulates the immune system to recognize and respond more effectively to pathogens, reducing the incidence and severity of disease outbreaks. Vaccines are a key tool in disease management, promoting fish health and reducing the reliance on antibiotics.

Viral Diseases: Viral diseases in finfish are caused by various viruses, including IPN, VHS, and Koi Herpesvirus (KHV). These diseases can lead to severe outbreaks with high mortality rates, particularly in intensive aquaculture systems. Viral infections often have no direct treatments, making prevention through biosecurity measures, vaccination, and optimal management practices crucial. Controlling viral diseases involves limiting the introduction of infected stock and reducing stressors that can exacerbate infections.

Viral Haemorrhagic Septicaemia (VHS): Viral Haemorrhagic Septicaemia (VHS) is a highly contagious viral disease affecting a wide range of finfish species, including trout, salmon, and pike. It is caused by the VHS virus, which induces hemorrhages, anemia, and high mortality, particularly in cold water (4°-14°C). VHS can be transmitted through water, contact with infected fish, and contaminated equipment. Control measures include biosecurity protocols, monitoring, and avoiding the introduction of infected stocks into aquaculture facilities.

Virus: Viruses are microscopic infectious agents composed of genetic material (DNA or RNA) encased in a protein coat. They require a host cell to replicate and can cause severe diseases in aquatic animals, such as viral hemorrhagic septicemia and infectious pancreatic necrosis. Viral infections are often managed through biosecurity measures, vaccination, and maintaining optimal environmental conditions to reduce stress and susceptibility.

Water Quality Management: Water quality management involves monitoring and maintaining key water parameters such as temperature, pH, dissolved oxygen, ammonia, nitrites, nitrates, and salinity to ensure a suitable environment for aquatic organisms. Optimal ranges depend on the species but typically include dissolved oxygen >5 mg/L, ammonia <0.05 mg/L, and pH 6.5-8.5. Good water quality management reduces stress, enhances growth, and prevents disease outbreaks.

CHAPTER 6

SHRIMP FARMING

Acclimatization: Acclimatization is the gradual adjustment of shrimp to new environmental conditions, such as temperature, salinity, and pH, before stocking. This process reduces stress and improves survival rates. Acclimatization typically involves slowly mixing pond water with transport water over several hours to allow shrimp to adapt.

Aeration Devices: Aeration devices enhance oxygen levels in shrimp ponds, essential for maintaining water quality and supporting shrimp metabolism. Common aeration devices include paddle wheel aerators, diffused aerators, and airlifts. They help mix water layers, reduce stratification, and promote the breakdown of organic matter, reducing harmful ammonia and nitrite levels.

Aeration: Aeration is the process of increasing the oxygen content of water in shrimp ponds, essential for maintaining dissolved oxygen levels above 4-5 mg/L, which is critical for shrimp survival and growth. Aeration can be achieved using paddle wheel aerators, air diffusers, or other mechanical devices that agitate the water, enhancing oxygen transfer from the atmosphere. Proper aeration also helps in the breakdown of organic waste and maintaining overall water quality.

Aerobic Treatment: Aerobic treatment involves using oxygen and aerobic bacteria to break down organic matter in shrimp farming effluents. This process takes place in aerated ponds or tanks, reducing biochemical oxygen demand (BOD) and improving water quality. Aerobic treatment is an effective method for managing waste and maintaining environmental sustainability.

Anaerobic Treatment: Anaerobic treatment uses bacteria that thrive in oxygen-free environments to decompose organic matter, producing methane and carbon dioxide as byproducts. This process is used in sludge digesters or anaerobic ponds, helping reduce waste volume and stabilize organic content before final disposal or use as fertilizer.

Antibiotic Use: Antibiotic use in shrimp farming is employed to treat bacterial infections, but should be carefully managed due to the risk of antibiotic resistance and residue concerns. Antibiotics must be used according to veterinary guidance and regulations, with emphasis on alternatives like probiotics and biosecurity for disease prevention.

Biofloc Technology: Biofloc technology (BFT) is an aquaculture system that promotes the growth of beneficial microorganisms in the water to recycle nutrients and improve water quality. In this system, waste products such as ammonia are converted into microbial biomass that serves as an additional feed source for shrimp. BFT systems require high aeration, careful monitoring of parameters like carbon-to-nitrogen ratio, and can reduce the need for water exchange, enhancing sustainability and reducing costs.

Biomass Estimation: Biomass estimation in shrimp farming involves calculating the total live weight of shrimp in a pond to guide feeding, water management, and harvesting decisions. Methods include sampling with cast nets, seine nets, or feeding trays, followed by extrapolation based on average individual weights and estimated population size. Accurate biomass estimation is crucial for adjusting feed rates, avoiding overfeeding, and scheduling harvests. A typical target biomass for semi-intensive systems is 5,000-8,000 kg/ha, while intensive systems can reach up to 10,000-15,000 kg/ha, depending on pond management and water quality.

Biomass Estimation: Biomass estimation involves calculating the total weight of shrimp in a pond, used to adjust feeding rates, stocking density, and harvest timing. Methods include sampling and extrapolation, based on average individual weights and population counts. Accurate biomass estimation helps optimize production efficiency and economic returns.

Biosecurity: Biosecurity in shrimp farming involves practices designed to prevent the introduction and spread of diseases within and between farms. Measures include quarantine of new stock, disinfection of equipment, controlling visitor access, and regular health monitoring. Biosecurity protocols are essential to maintain healthy shrimp populations and prevent economic losses.

CAA Guidelines: CAA (Coastal Aquaculture Authority) Guidelines refer to the regulations and standards established by the Coastal Aquaculture Authority in countries like India, which govern the establishment and operation of aquaculture farms. These guidelines cover aspects such as site selection, farm design, species selection, water quality management, and environmental impact, aiming to promote sustainable and responsible aquaculture practices.

Capital Costs: Capital costs in shrimp farming are the initial investments required to establish the farm, including the construction of ponds, purchase of equipment (aerators, pumps, feed storage), and infrastructure development (e.g., water supply and drainage systems). These costs are typically one-time expenditures that form the foundation for the operational phase of the farm.

Carapace Hardening: Carapace hardening is the process following molting where the new exoskeleton of shrimp calcifies and strengthens. This process typically takes a few hours to a day and requires sufficient calcium availability in the water. Ensuring a stable environment during this vulnerable period is crucial to reduce stress and predation risks among shrimp.

Carapace Hardening: Carapace hardening, or mineralization, occurs immediately after molting, where the new soft exoskeleton absorbs calcium and other minerals from the surrounding water to become rigid and protective. This process is essential for shrimp survival, providing a barrier against pathogens and physical damage. Carapace hardening usually takes 24-48 hours and requires sufficient levels of calcium (50-100 mg/L) and alkalinity (80-120 mg/L). During this vulnerable period, shrimp are susceptible to predation and environmental stress, so maintaining optimal water quality and reducing handling are critical.

Cold Chain Management: Cold chain management in shrimp farming involves maintaining the shrimp at optimal temperatures throughout the supply chain, from harvest to final market. Proper cold chain management includes immediate chilling after harvest, refrigerated transport, and storage at temperatures below 4°C to preserve freshness, quality, and shelf life.

Cold Chain Management: Cold chain management in shrimp farming is the process of maintaining the shrimp at optimal temperatures throughout the supply chain, from harvest to final delivery. This includes immediate chilling after harvest, refrigerated transport, and storage at temperatures below 4°C to prevent spoilage and preserve quality. Effective cold chain management reduces the risk of bacterial growth, maintains freshness, and extends shelf life, ensuring that shrimp reach consumers in the best possible condition.

Cost of Feed: The cost of feed is one of the most significant variable costs in shrimp farming, often accounting for 50-60% of total production costs. Feed costs depend on the type, quality, and price of the feed used, and optimizing feed management, including minimizing feed waste and improving feed conversion ratios, is critical to reducing overall production costs and enhancing profitability.

Cultivable Shrimps: Cultivable shrimps are species of shrimp that are suitable for aquaculture due to their fast growth rates, tolerance to a range of environmental conditions, and market demand. Commonly farmed species include the Pacific white shrimp (*Penaeus vannamei*), black tiger shrimp (*Penaeus monodon*), and giant freshwater prawn (*Macrobrachium rosenbergii*). These species are selected based on factors such as growth performance, resistance to disease, and adaptability to farming systems.

Diffused Aeration: Diffused aeration systems use air blowers and diffusers placed at the pond bottom to release fine air bubbles, which rise and mix with water, enhancing oxygen transfer. These systems are efficient for deep ponds and provide uniform oxygen distribution. Diffused aeration also promotes the breakdown of organic matter and reduces the buildup of harmful gases.

Disease Management: Disease management in shrimp farming includes the prevention, diagnosis, and treatment of diseases. Strategies involve maintaining optimal water quality, biosecurity, regular health checks, and the use of probiotics, vaccines, or medications as needed. Effective disease management reduces mortality, maintains shrimp growth, and ensures sustainable production.

Disease Prevention: Disease prevention in shrimp farming involves a combination of biosecurity measures, proper pond management, vaccination (where applicable), and maintaining optimal water quality and nutrition. It includes practices such as the use of pathogen-free broodstock, disinfection of equipment, regular health monitoring, and the application of probiotics or immunostimulants to enhance shrimp immunity. Preventing diseases is crucial for sustaining high production levels and ensuring the economic viability of shrimp farming operations.

Disinfection: Disinfection in shrimp farming involves the use of chemicals or physical methods to kill or inactivate pathogens on equipment, pond surfaces, and water. Common disinfectants include chlorine, formalin, and iodine-based compounds. Proper disinfection protocols help prevent disease outbreaks and maintain biosecurity.

Drainage Systems: Effective drainage systems are crucial for managing water quality in shrimp ponds. These systems include drainpipes, sluices, and canals that allow for the efficient removal of water, sludge, and waste. Proper drainage helps control pond water levels, prevents flooding, and facilitates water exchange, contributing to a healthier growing environment for shrimp.

Effluent Treatment Systems: Effluent treatment systems in shrimp farming are designed to manage and treat wastewaters generated from the ponds, including excess feed, feces, and other organic matter. These systems may include sedimentation ponds, biofilters, or constructed wetlands that help reduce nutrient loads, suspended solids, and other pollutants before discharge. Proper effluent treatment is crucial to prevent environmental degradation and comply with regulatory standards.

Environmental Concerns: Environmental concerns in shrimp farming include the impact of effluent discharge on surrounding ecosystems, habitat destruction (e.g., mangroves), and the spread of diseases and invasive species. Practices such as responsible site selection, implementation of effluent treatment systems, reduction of chemical use, and adherence to sustainable farming guidelines are critical to minimizing the environmental footprint of shrimp aquaculture.

Export Standards: Export standards for shrimp include international regulations and quality requirements that must be met to access global markets. These standards cover aspects such as food safety (e.g., HACCP), residue limits for antibiotics and chemicals, traceability, labeling, and packaging. Compliance with export standards ensures that shrimp products are safe, high-quality, and suitable for international trade, protecting public health and maintaining the reputation of the exporting country

Export Standards: Export standards for shrimp include international regulations and quality requirements that must be met for global market access. These standards cover aspects like food safety, residue limits, labeling, and packaging. Compliance with export standards ensures that shrimp products are safe, high-quality, and suitable for international trade.

Extensive Farming: Extensive farming refers to a low-input, low-output shrimp farming system that relies on natural food availability and minimal human intervention. Ponds are usually large (greater than 5 hectares) with low stocking densities (1-5 shrimp/m^2). Water exchange is minimal or non-existent, and supplemental feeding and aeration are not commonly used. This system is environmentally friendly but has lower productivity and is more susceptible to external environmental fluctuations.

Feed Conversion Ratio (FCR): The Feed Conversion Ratio (FCR) measures the efficiency of feed utilization in shrimp farming, calculated as the amount of feed required to produce one unit of shrimp biomass. Lower FCR values (1.2-1.8) indicate more efficient feed use, which is desirable for cost-effective and sustainable shrimp production.

Feed Management: Feed management involves strategies to optimize feed delivery, minimize waste, and ensure shrimp receive adequate nutrition. Key practices include adjusting feed rates based on shrimp biomass and environmental conditions, using feeding trays to monitor consumption, and distributing feed evenly. Proper feed management helps improve growth rates and reduce feed costs.

Feed Supplements: Feed supplements in shrimp farming include additives like vitamins, minerals, enzymes, and immune-stimulants that enhance shrimp growth, health, and feed efficiency. Supplements may also include binders, antioxidants, and preservatives to improve feed stability and shelf life. They help address specific nutritional deficiencies and promote overall shrimp well-being.

Feeding Frequency: Feeding frequency in shrimp farming refers to how often feed is administered to shrimp, typically ranging from 3 to 5 times per day. Higher frequencies ensure constant availability of food, reduce competition, and improve feed intake. Feeding frequency may be adjusted based on shrimp size, water temperature, and feeding response.

Feeding Management: Feeding management in shrimp farming involves the strategic provision of feed to ensure optimal growth and health of shrimp. It includes selecting the appropriate type of feed, determining the correct feeding rates, and scheduling feed times based on shrimp size, water temperature, and other environmental factors. Proper feeding management helps maximize feed conversion efficiency, minimize waste, and reduce the risk of water quality deterioration.

Fertilization: Fertilization in shrimp farming involves adding nutrients to the pond to promote the growth of natural food sources such as plankton, which are beneficial for shrimp, especially in the early stages of growth. Fertilization can be organic or inorganic, depending on the nutrients needed. It helps maintain the balance of the pond ecosystem, supporting better shrimp health and growth.

Fixed Costs: Fixed costs in shrimp farming are expenses that do not change with the level of production, such as lease payments, salaries of permanent staff, depreciation of equipment, and insurance. These costs must be covered regardless of the farm's production output, and effective cost management can help improve overall profitability.

Freshwater Shrimp Farming: Freshwater shrimp farming involves the cultivation of shrimp species that thrive in freshwater environments, such as the giant freshwater prawn (*Macrobrachium rosenbergii*). This type of farming requires careful control of water quality parameters, including maintaining a pH of 7-8.5, temperature of 25°-30°C, and dissolved oxygen levels above 5 mg/L. Freshwater shrimp farming can be done in ponds, tanks, or recirculating systems, and often involves polyculture with other fish species.

Grow-Out Phase: The grow-out phase is the stage of shrimp farming where post-larvae are raised to market size in ponds or tanks. This phase focuses on optimizing growth through effective feeding, water quality management, and health monitoring. The grow-out phase typically lasts 3-4 months, depending on species, with shrimp reaching harvest size of 20-40 grams.

Growth Stages: Shrimp growth stages include embryonic, larval, post-larval, juvenile, sub-adult, and adult phases, each with distinct developmental and nutritional requirements. Growth rates are influenced by water temperature, which is optimal between 28°-32°C, feed quality, and management practices. The transition between stages involves several molts, with increased frequency in early stages and gradually lengthening as shrimp mature. Understanding and managing these stages is crucial for optimizing feeding strategies, health management, and overall production efficiency in shrimp farming.

Growth Stages: Shrimp growth stages include embryonic, larval, post-larval, juvenile, and adult phases. Each stage has specific environmental and nutritional requirements, with growth rates influenced by factors like temperature, feed quality, and water conditions. Understanding these stages is essential for effective management and achieving optimal production outcomes.

Harvesting Techniques: Harvesting techniques in shrimp farming include draining the pond completely, using cast nets or seines, or employing mechanical harvesters. Timing is critical to avoid overgrowth and maximize market value, typically targeting shrimp sizes of 20-40 grams. Partial harvesting can also be employed to selectively remove market-size shrimp while allowing others to continue growing. Proper handling during harvest is essential to minimize stress and damage to the shrimp, preserving quality for market sales.

Harvesting Techniques: Harvesting techniques in shrimp farming include draining ponds, using nets or traps, and mechanical harvesters. Timing is critical to avoid overgrowth and maximize market value. Techniques vary based on pond design and shrimp size, with careful handling required to minimize stress and physical damage to the shrimp.

Harvesting: Harvesting is the process of collecting shrimp from ponds at the end of the production cycle, typically when they reach marketable size. Methods include draining the pond and collecting shrimp manually or using nets, or partially harvesting using traps. Timing and method of harvesting are critical to minimizing stress and ensuring the quality of the shrimp. Proper post-harvest handling, including immediate cooling, is essential to preserve shrimp quality.

Inorganic Fertilizers: Inorganic fertilizers are synthetic compounds containing specific nutrients, such as nitrogen, phosphorus, and potassium, used to enhance the growth of phytoplankton in shrimp ponds. Common inorganic fertilizers include urea, ammonium phosphate, and superphosphate. These fertilizers provide rapid nutrient availability, promoting a quick response in plankton growth, which serves as a natural food source for shrimp.

Intensive Farming: Intensive farming is a high-input, high-output system characterized by high stocking densities (above 25 shrimp/m^2) and advanced management practices, including frequent feeding, continuous aeration, and regular water exchange or recirculation. Ponds are usually small (0.1-1 hectare), and production can exceed 10,000 kg/ha per cycle. This system maximizes yield but requires significant investment in infrastructure, technology, and ongoing management, including strict water quality control.

Larval Rearing: Larval rearing is the process of growing shrimp from eggs to post-larvae in controlled hatchery conditions. It involves providing optimal water quality, temperature (27°-30°C), salinity (28-35 ppt), and feeding regimes, including algae, Artemia, and formulated feeds. Successful larval rearing produces healthy post-larvae ready for stocking in grow-out ponds.

Liming: Liming is the application of lime (calcium carbonate, calcium hydroxide, or calcium oxide) to aquaculture ponds to stabilize pH, reduce acidity, and improve water hardness and alkalinity. The optimal pH range for shrimp ponds is 7.5-8.5, and liming helps maintain this by buffering against pH fluctuations. Liming also enhances the availability of nutrients and promotes the growth of beneficial algae, improving the overall pond environment.

Low Exchange Systems: Low exchange systems in shrimp farming are designed to minimize water exchange, reducing the impact on the environment and decreasing the risk of disease introduction from external water sources. Instead, water quality is maintained through aeration, biofloc systems, or biological filtration. These systems are particularly beneficial in areas with limited access to clean water or where water discharge regulations are stringent.

Market Specifications: Market specifications for shrimp include criteria such as size (count per kilogram), appearance (color and uniformity), texture, and absence of defects or diseases. Specifications vary by market segment and end use (e.g., retail, food service, or processing). Meeting these specifications is crucial for achieving optimal prices and consumer satisfaction, requiring careful management of shrimp farming practices, harvesting, and post-harvest handling.

Market Specifications: Market specifications for shrimp include size, weight, appearance, and quality standards required by buyers. These specifications vary by market segment and can include criteria like color, shell hardness, and the absence of defects. Meeting market specifications is crucial for maximizing the value and acceptability of shrimp products.

Molting: Molting is a critical physiological process in shrimp growth, where they shed their old exoskeleton to form a new, larger one, allowing for growth. This process occurs in multiple stages: pre-molt, molt, post-molt, and inter-molt. The frequency of molting depends on factors such as age, environmental conditions (like temperature and salinity), and nutritional status. Optimal water conditions for molting include a calcium level of 50-100 mg/L and a magnesium level of 30-50 mg/L, which are necessary for proper exoskeleton formation. Stress during molting can lead to molting-related mortality, so stable and favorable environmental conditions are crucial.

Molting: Molting is the process by which shrimp shed their exoskeleton to grow. It is a critical physiological function that occurs every few weeks, influenced by factors such as temperature, nutrition, and water quality. Proper management of molting, including ensuring sufficient calcium and magnesium levels in water, is essential to prevent mortality and promote growth.

Nutrient Recycling: Nutrient recycling in shrimp farming involves reusing nutrients from waste or effluents to support the growth of beneficial organisms like algae, biofloc, or periphyton. This process improves water quality, reduces feed costs, and enhances the sustainability of shrimp farming systems by creating a more self-sufficient production cycle.

Nutritional Requirements: The nutritional requirements of shrimp include essential amino acids, fatty acids, vitamins, minerals, and energy to support growth, reproduction, and immune function. Feed formulations must be balanced to provide optimal nutrition, with protein levels typically ranging from 30-40% for grow-out feeds. Proper nutrition is crucial for maximizing growth rates, feed efficiency, and resistance to disease.

Organic Fertilizers: Organic fertilizers are derived from natural sources, such as animal manure, compost, or plant materials, and are used to promote the growth of natural food organisms in shrimp ponds. They release nutrients slowly, improving soil and water quality over time. Organic fertilizers support the growth of phytoplankton and zooplankton, which serve as supplementary food for shrimp, enhancing their nutrition and overall pond productivity.

Paddle Wheel Aerators: Paddle wheel aerators are widely used in shrimp farming to oxygenate pond water and improve circulation. These devices work by splashing water into the air, increasing the surface area for oxygen exchange. Typically, 1 HP paddle wheel aerators can oxygenate 1,000-1,500 square meters of pond area, helping maintain dissolved oxygen levels above 5 mg/L.

Pathogen Identification: Pathogen identification is the process of detecting and diagnosing disease-causing organisms in shrimp farming. Techniques include microscopy, histopathology, PCR, and other molecular methods. Accurate identification of pathogens is essential for effective treatment and management of diseases.

Pellet Size: Pellet size in shrimp feed is critical for proper ingestion and digestion. Feed pellets should be appropriately sized for the shrimp's developmental stage, with smaller pellets for post-larvae and larger pellets for adult shrimp. Proper pellet size ensures efficient feeding, reduces waste, and improves growth performance.

Periphyton-Based Systems: Periphyton-based systems utilize natural substrates (e.g., bamboo, nets, or plant material) placed in aquaculture ponds to support the growth of periphyton-a complex mixture of algae, bacteria, and detritus. This periphyton provides a supplemental food source for shrimp, enhances water quality by absorbing nutrients, and increases the pond's carrying capacity. This system mimics natural habitats and can improve shrimp growth and survival rates.

Polyculture: Polyculture is the practice of raising multiple species, such as shrimp and fish, in the same aquaculture system. The goal is to utilize different trophic levels to enhance system productivity, reduce waste, and improve water quality through natural biological interactions. Common combinations include shrimp with tilapia or carp, where fish help control algae and detritus, improving overall pond conditions.

Pond Construction: Pond construction in shrimp farming involves designing and building ponds to create an optimal environment for shrimp growth. Key factors include site topography, soil type, and water supply. Ponds are typically rectangular with a depth of 1.0-1.5 meters, designed for easy water exchange and sludge removal. Proper dike construction and compaction prevent seepage, while inlet and outlet structures manage water flow and prevent the entry of predators.

Pond Lining: Pond lining involves using materials like HDPE (high-density polyethylene) or clay to cover pond bottoms and dikes, preventing seepage and erosion. Liners improve water retention, facilitate easier cleaning, and help maintain stable water quality parameters. HDPE liners are durable and resistant to UV radiation, offering a lifespan of 10-20 years.

Pond Preparation: Pond preparation is a crucial step in shrimp farming that involves cleaning, drying, and treating the pond before stocking shrimp. This process includes the removal of sludge, repair of pond structures, application of lime to adjust pH, and disinfection to eliminate pathogens and unwanted organisms. Proper pond preparation enhances water quality and reduces the risk of disease outbreaks during the culture period.

Post-Harvest Handling: Post-harvest handling refers to the practices undertaken immediately after shrimp harvesting to maintain product quality and minimize losses. This includes washing, sorting, chilling, and storing the shrimp under optimal conditions to prevent spoilage and contamination. Effective post-harvest handling is crucial for extending shelf life and ensuring that the shrimp meet food safety and quality standards required by the market.

Post-Harvest Management: Post-harvest management involves the steps taken after harvesting shrimp to ensure quality control, proper storage, transportation, and processing. It includes cooling, packaging, and implementing measures to prevent spoilage and contamination. Proper post-harvest management is essential for maintaining the marketability of shrimp, complying with food safety standards, and maximizing the economic returns from shrimp farming.

Post-Larvae: Post-larvae are the stage of shrimp development following the larval phase, typically 10-20 days old, and ready for stocking in grow-out ponds. They have fully developed swimming and feeding structures, making them more resilient and capable of rapid growth in pond conditions. Quality post-larvae are crucial for successful shrimp farming.

Probiotics: Probiotics are beneficial microorganisms added to shrimp diets or pond water to enhance digestion, inhibit pathogens, and improve overall health. Common probiotic strains include *Bacillus*, *Lactobacillus*, and *Saccharomyces*. Probiotics help maintain a balanced gut microbiome, reduce disease incidence, and improve feed conversion efficiency.

Quarantine: Quarantine in shrimp farming is the isolation of new or diseased shrimp to prevent the spread of pathogens to healthy populations. Quarantine protocols include separate holding facilities, monitoring for signs of disease, and treatment as necessary before introducing shrimp to main production ponds.

Sedimentation Ponds: Sedimentation ponds are used to remove suspended solids and organic matter from effluents in shrimp farming. They slow water flow, allowing particles to settle at the bottom. These ponds improve water quality before discharge or recirculation, helping to comply with environmental regulations and reduce pollution.

Seed Quality Evaluation: Seed quality evaluation involves assessing the health, vigor, and potential of shrimp post-larvae before stocking. This includes checking for size uniformity, active swimming, response to stimuli, and absence of pathogens. Quality evaluation helps predict survival rates and performance in the grow-out phase.

Seed Selection: Seed selection in shrimp farming refers to choosing high-quality post-larvae for stocking, which exhibit good health, uniform size, and active behaviour. Selection criteria include origin, genetic strain, and absence of deformities or signs of disease. Proper seed selection is critical for ensuring high survival rates and optimal growth.

Semi-Intensive Farming: Semi-intensive farming involves moderate stocking densities (10-25 shrimp/m^2) and enhanced management practices compared to extensive systems. This approach includes supplemental feeding, occasional water exchange, and basic aeration to improve shrimp growth and survival rates. Ponds are typically smaller (1-5 hectares), and production yields range between 1,000-5,000 kg/ha per cycle. The system balances increased productivity with manageable operational costs.

Shrimp Diseases: Shrimp diseases are caused by a variety of pathogens including viruses (e.g., White Spot Syndrome Virus), bacteria (e.g., *Vibrio* species), fungi, and parasites. These diseases can lead to significant mortalities and economic losses in shrimp farming. Effective management includes disease diagnosis, monitoring, prevention through biosecurity measures, and sometimes the application of therapeutics, although the use of antibiotics is strictly regulated due to environmental and food safety concerns.

Shrimp Farming: Shrimp farming is the practice of raising shrimp in controlled aquatic environments such as ponds, tanks, or raceways for commercial purposes. It involves the cultivation of shrimp species, primarily from the family Penaeidae, in both coastal and inland areas. Optimal farming requires careful management of water quality, feed, stocking densities, and environmental conditions to ensure high yields and minimize disease outbreaks. Typical water quality parameters include maintaining salinity levels between 15-25 ppt, pH between 7.5-8.5, temperature between 28°-32°C, and dissolved oxygen above 4 mg/L.

Shrimp Feed Formulation: Shrimp feed formulation involves creating balanced diets that meet the nutritional needs of shrimp, including protein, lipids, carbohydrates, vitamins, and minerals. Optimal protein levels range from 30-50%, depending on the growth stage. Formulations include fishmeal, soybean meal, fish oil, and other ingredients to ensure rapid growth and high feed conversion efficiency.

Shrimp Feed: Shrimp feed is a formulated diet designed to meet the nutritional requirements of shrimp at various growth stages. It typically includes proteins, lipids, carbohydrates, vitamins, and minerals, with formulations varying depending on the species, age, and farming system. High-quality feed improves growth rates, feed conversion efficiency, and overall health of the shrimp, contributing significantly to the success of shrimp farming operations.

Shrimp Grading: Shrimp grading involves categorizing shrimp into different sizes and quality grades based on market specifications. Grades are typically defined by count per kilogram and overall appearance. Grading ensures consistency and standardization, making it easier to meet buyer requirements and optimize pricing.

Shrimp Grading: Shrimp grading involves classifying shrimp into different size and quality categories based on market standards, such as count per kilogram (e.g., 30-40 count). Grading ensures uniformity in shrimp batches, which is important for meeting buyer specifications and optimizing pricing. Grades can also include considerations for shell hardness, color, and absence of defects. Proper grading enhances product presentation and consumer satisfaction, impacting the overall marketability of the shrimp.

Shrimp Growth Monitoring: Shrimp growth monitoring involves regular measurement of shrimp size, weight, and overall health to track growth rates and adjust management practices accordingly. Monitoring typically includes sampling shrimp at regular intervals to assess average body weight (ABW), feed conversion ratio (FCR), and survival rates. Effective growth monitoring helps optimize feeding, stocking densities, and overall farm performance.

Shrimp Health Management: Shrimp health management encompasses all practices aimed at maintaining optimal health and preventing diseases in shrimp farms. This includes monitoring water quality, biosecurity measures, disease diagnosis, and the use of probiotics and vaccines. Effective health management reduces mortality rates and improves overall farm productivity.

Shrimp Market Analysis: Shrimp market analysis involves assessing the demand, supply, price trends, and competition within the shrimp industry. It includes evaluating factors such as global production, export markets, consumer preferences, and trade regulations. Understanding market dynamics helps farmers make informed decisions on production planning, marketing strategies, and identifying opportunities for expansion or diversification.

Shrimp Packaging: Shrimp packaging involves protecting the harvested shrimp during transport and storage, maintaining product quality and extending shelf life. Common methods include using ice, vacuum sealing, or modified atmosphere packaging to reduce exposure to air and prevent spoilage. Packaging must comply with food safety standards, protect shrimp from physical damage, and be labeled accurately to meet regulatory and market requirements.

Shrimp Packaging: Shrimp packaging involves protecting the harvested shrimp from physical damage and contamination while maintaining product quality during transport and storage. Common packaging methods include using ice, vacuum sealing, or modified atmosphere packaging. Proper packaging is essential for extending shelf life and meeting market standards.

Shrimp Production: Shrimp production refers to the total yield of shrimp harvested from an aquaculture system over a given period, usually measured in kilograms or tons per hectare. Production levels depend on factors such as stocking density, management practices, feed quality, and environmental conditions. Optimizing production involves balancing inputs and costs with outputs, ensuring high growth rates, survival, and minimizing the risk of disease.

Shrimp Sorting: Shrimp sorting is the process of categorizing harvested shrimp based on size, quality, and market specifications. Sorting is typically done manually or with mechanical graders immediately after harvest to separate shrimp into uniform groups, enhancing market value and meeting buyer requirements. Sorting also helps in processing efficiency and packaging, as well-sorted batches are easier to manage and command better prices.

Shrimp Sorting: Shrimp sorting is the process of separating harvested shrimp based on size and quality. Sorting improves market value and ensures uniformity in batches, facilitating better pricing. Manual sorting or automated grading machines can be used, with attention to handling practices to maintain shrimp quality.

Shrimp Yield: Shrimp yield refers to the total production of shrimp, usually expressed in kilograms per hectare per crop cycle. Yield depends on factors such as stocking density, feed quality, water management, and disease control. Typical yields in semi-intensive farming range from 3,000 to 8,000 kg/ha/crop, while intensive systems can achieve 10,000-20,000 kg/ha/crop. Maximizing yield involves optimizing all aspects of farm management, including feed conversion ratios (FCR), water quality maintenance, and biosecurity measures to prevent disease outbreaks.

Shrimp Yield: Shrimp yield refers to the total quantity of shrimp produced per unit area or volume, typically measured in kilograms per hectare or tonnage. Yield depends on factors such as stocking density, feed management, and water quality. High yields are achieved through intensive management practices and optimal environmental conditions.

Site Selection: Site selection is a critical step in shrimp farming, determining the overall success of the operation. Ideal sites have access to a reliable water source, suitable soil with low permeability, and are protected from natural disasters like floods and storms. Proximity to resources, infrastructure, and markets also plays a role. Avoidance of industrial areas helps minimize pollution risks.

Sludge Management: Sludge management in shrimp farming involves the removal, treatment, and disposal of accumulated organic waste from pond bottoms. Effective sludge management prevents water quality deterioration, reduces the risk of disease, and minimizes environmental impact. Methods include periodic sludge removal, drying, and use as fertilizer.

Soil Quality: Soil quality affects the water retention and overall productivity of shrimp ponds. Ideal pond soils are clay or sandy clay with low permeability, which helps maintain water levels and prevents excessive seepage. The pH should be neutral to slightly alkaline (6.5-8.5), and organic matter should be minimal to avoid oxygen depletion from excessive decomposition.

Stocking Density: Stocking density refers to the number of shrimp stocked per unit area or volume in an aquaculture system. It is a critical parameter that influences shrimp growth, survival, and overall farm productivity. Optimum stocking densities vary based on farming systems: 1-5 shrimp/m^2 for extensive, 10-25 shrimp/m^2 for semi-intensive, and above 25 shrimp/m^2 for intensive systems. High stocking densities require enhanced management practices, including increased feeding, aeration, and water quality control.

Stocking Procedures: Stocking procedures in shrimp farming involve the introduction of shrimp post-larvae into ponds at appropriate densities to optimize growth and survival. Key steps include acclimatization to pond conditions, monitoring water quality, and ensuring uniform distribution. Proper stocking ensures optimal use of resources and reduces stress on shrimp.

Un-drainable Ponds: Un-drainable ponds are aquaculture systems that lack the infrastructure to be fully drained between production cycles. These ponds often rely on partial harvesting and are typically lined with clay or other materials to retain water. Management of these ponds requires careful water quality control, including regular monitoring of parameters like pH, dissolved oxygen, and ammonia levels, as the inability to fully drain can lead to the buildup of organic matter and waste.

Variable Costs: Variable costs in shrimp farming are ongoing expenses that vary with the level of production. They include feed, seed (post-larvae), labor, electricity, fuel, water treatment chemicals, and other inputs that are consumed during the production cycle. Effective management of variable costs is essential to maintain profitability and competitiveness in the market.

Water Exchange: Water exchange in shrimp farming refers to the periodic replacement of pond water to remove waste, maintain water quality, and regulate temperature and salinity. Exchange rates vary depending on the intensity of farming, with intensive systems requiring more frequent exchanges. Optimal exchange helps sustain shrimp health and growth.

Water Filling: Water filling refers to the process of adding water to the pond after preparation and before stocking shrimp. The water is typically sourced from rivers, canals, or wells and must be filtered and treated to remove potential pathogens and predators. The water level should be gradually increased to the desired depth, usually between 1-1.5 meters, ensuring that water quality parameters such as salinity, pH, and temperature are within optimal ranges for shrimp growth.

Water Quality Management: Water quality management encompasses the monitoring and regulation of key parameters such as pH, dissolved oxygen, salinity, temperature, ammonia, nitrite, and nitrate levels in shrimp farming. Maintaining optimal water quality is essential to prevent stress, enhance growth rates, and reduce the risk of disease outbreaks. Regular testing and adjustments, such as aeration, water exchange, and chemical treatments, are integral to effective water quality management.

Water Recirculation: Water recirculation systems in shrimp farming reuse water within the farm by treating it to remove waste and maintain quality. Recirculation reduces water usage, minimizes environmental impact, and allows better control of water parameters. Components include filters, biofilters, and pumps to ensure continuous water movement and quality maintenance.

Water Retention: Water retention in shrimp ponds refers to the ability of the pond to maintain water levels without significant loss due to seepage or evaporation. Optimal water retention depends on proper pond construction, including compacted dikes and impermeable pond bottoms. High retention reduces the need for frequent water exchanges, lowering costs and minimizing environmental impact.

Water Source: A reliable and clean water source is essential for shrimp farming, influencing the water quality in the ponds. Water sources may include rivers, lakes, or underground wells, and should be free from contaminants, heavy metals, and pollutants. Salinity levels must be suitable for the specific shrimp species, generally ranging from 10 to 35 ppt for marine species.

CHAPTER 7

FARM ENGINEERING

Aeration Systems: Aeration systems in aquaculture farms are used to increase the dissolved oxygen levels in water, which is crucial for the respiration of aquatic organisms. Common aeration methods include paddle wheel aerators, diffused air systems, and airlifts. Effective aeration improves water circulation, reduces stratification, and enhances overall water quality. For optimal shrimp and fish growth, dissolved oxygen levels should be maintained above 5 mg/L. Aeration also helps in the breakdown of organic matter, contributing to better pond health.

Aeration Techniques: Aeration techniques in aquaculture ensure that dissolved oxygen levels are sufficient to support the metabolic needs of aquatic organisms. Mechanical aerators, such as paddle wheels, diffusers, and venturi systems, are used to increase the oxygenation of water. Diffused aeration is commonly used in tanks, while paddlewheel aerators are preferred in ponds. Aeration is especially critical in high-density farming systems where oxygen demand is high. Optimum dissolved oxygen levels should be maintained above 5 mg/L for most species, with higher levels required during feeding and in warm water.

Aerator Maintenance: Aerator maintenance is critical for maintaining optimal dissolved oxygen levels (5-8 mg/L) in aquaculture ponds. Maintenance includes regular inspection and cleaning of the blades, propellers, or diffusers to prevent the buildup of algae or debris that could reduce efficiency. Electrical components, including wiring and control systems, should be checked for corrosion or wear. Bearings and shafts may require lubrication or replacement to ensure smooth operation. Proper aeration prevents oxygen depletion, which is crucial for the survival and growth of fish/shrimp, especially in densely stocked systems.

Aquaculture: Aquaculture, also known as aquafarming, is the cultivation of aquatic organisms such as fish, crustaceans, mollusks, and aquatic plants under controlled conditions. It involves breeding, rearing, and harvesting these species in freshwater, brackish, or marine environments. Aquaculture systems can be extensive, semi-intensive, or intensive, with varying levels of inputs such as feed, aeration, and disease control. The industry plays a crucial role in food security, providing a significant portion of the world's seafood supply, and includes practices like pond farming, recirculating systems, and offshore farming.

Berm/Bench: A berm or bench in aquaculture is a flat or gently sloping strip of land, usually at the top of a pond dike or along the perimeter, designed to provide stability, reduce erosion, and facilitate access for maintenance. Berms help to buffer wave action, prevent overflow, and support vegetation that can stabilize soil and improve aesthetics. In larger farms, berms may also serve as access paths for vehicles and equipment, enhancing farm logistics and operational efficiency.

Biosecurity Measures: Biosecurity measures in aquaculture are protocols designed to prevent the introduction and spread of pathogens and diseases within a farm. These include controlling access to the farm, disinfecting vehicles and equipment, isolating new stock in quarantine areas, and ensuring proper waste disposal. Water quality management is also crucial, with regular testing for parameters such as pH, ammonia, and pathogens. Biosecurity plans often involve setting up physical barriers like fencing, footbaths, and handwashing stations, as well as maintaining good farm hygiene practices.

Chain Survey: Chain surveying is a simple and fundamental method of land surveying used in aquaculture farm layout design. It involves measuring distances between points on the ground using a chain or tape and then creating a plan of the surveyed area. This method is particularly useful for smaller, flat areas where high accuracy isn't required. By using triangles and other simple geometric figures, a map of the area can be created to assist with pond layout or farm planning. Accuracy is enhanced with proper calibration and alignment during measurements.

Construction Materials: Construction materials for aquaculture farms include a variety of resources used in building ponds, dikes, channels, and water control structures. Common materials include soil, concrete, HDPE liners, PVC pipes, and geomembranes. The choice of materials depends on factors like soil type, water quality, farm scale, and budget. Materials must be durable, resistant to corrosion and erosion, and suitable for creating stable and efficient farm structures that support sustainable aquaculture practices.

Contour Lines: Contour lines are used in aquaculture farm planning to represent the elevation and slope of the land. They are essential for designing ponds, drainage systems, and irrigation channels. Contour mapping ensures proper water flow and helps in calculating earthwork for pond construction. The spacing of contour lines indicates the steepness of the slope, with closer lines representing steeper areas. Proper contouring aids in water management, prevents erosion, and ensures efficient pond layout and drainage.

Contour Mapping: Contour mapping is the representation of the topography of a site by drawing lines that connect points of equal elevation. It is essential in aquaculture for planning pond construction, drainage, and irrigation systems, especially in uneven terrains. Contour maps help in determining the natural flow of water and identifying suitable locations for ponds, dikes, and drainage channels. Ideal pond sites are typically located in areas with gentle slopes (1-2%) to allow for controlled water flow and minimal excavation during construction.

Desilting Basin: A desilting basin is a structure used to trap and remove silt and other fine sediments from water before it enters an aquaculture system. Desilting basins are typically constructed at the inlet of water sources and are designed to slow down the flow of water, allowing sediments to settle out. Regular maintenance, including the removal of accumulated silt, is necessary to prevent clogging and maintain water flow. Desilting basins improve water quality and reduce the risk of sediment build-up in ponds or tanks, which can affect the health of aquatic organisms

Dikes: Dikes are earthen or constructed barriers used in aquaculture farms to contain water in ponds, protect against flooding, and separate different pond units. They are typically built from compacted soil or reinforced materials and are designed to withstand hydraulic pressure and erosion. Dikes must be properly sloped (usually 2:1 or 3:1) and maintained to prevent breaches or seepage, which can lead to water loss and farm failure. Regular inspection and maintenance are crucial to ensure the integrity of dikes and the safety of the aquaculture operation.

Drainage Channel: A drainage channel in aquaculture farms is designed to remove excess water, waste, and effluents from ponds or tanks. These channels are essential for maintaining optimal water levels, preventing flooding, and facilitating water exchange. Proper drainage design includes considerations for channel slope, width, and capacity to handle peak discharge volumes without causing erosion or backflow. Effective drainage systems help maintain water quality and support the overall health of cultured species.

Drainage Systems: Drainage systems in aquaculture are designed to efficiently remove excess water from ponds, tanks, and farm infrastructure, preventing flooding and waterlogging. These systems typically include drainage ditches, pipes, sluices, and culverts that guide water out of the culture area during water exchange or after heavy rainfall. Proper slope design (usually 0.5-1% gradient) ensures gravity-assisted drainage. Regular maintenance of drainage systems is critical to avoid blockages and ensure optimal water flow, contributing to effective pond management and minimizing waterborne diseases.

Dykes: Dykes in aquaculture are embankments constructed around ponds to contain water, prevent flooding, and separate different pond units. Made from compacted soil, rock, or other materials, dykes must be stable, impermeable, and resistant to erosion. They are typically sloped at 2:1 or 3:1 and may include features like berms or vegetation to enhance stability. Properly designed and maintained dykes are essential for protecting farm infrastructure, managing water levels, and ensuring the safety of aquaculture operations.

Earthwork Calculations: Earthwork calculations in aquaculture involve determining the volume of soil that needs to be excavated or filled during pond construction or land grading. These calculations are essential for ensuring proper pond dimensions, embankment stability, and drainage efficiency. The volume of earth moved is typically calculated in cubic meters and involves contour mapping and levelling data. Accurate earthwork calculations optimize construction costs and ensure that ponds are built to the correct specifications, with sufficient depth and slope for water retention and drainage.

Effluent Treatment: Effluent treatment in aquaculture involves the management and treatment of wastewater discharged from farming operations to minimize environmental impact. Treatment systems may include sedimentation ponds, biological treatment units, and constructed wetlands, designed to remove solids, nutrients, and pathogens. Effective effluent treatment is essential for compliance with environmental regulations, protecting surrounding water bodies, and ensuring sustainable farm operations. Parameters such as nitrogen, phosphorus, and suspended solids should be reduced to acceptable levels before discharge.

Electrical Basics: Understanding electrical basics in aquaculture is important for the installation and maintenance of electrical systems that power equipment such as pumps, aerators, and lighting. Knowledge of circuit design, grounding, and load distribution is essential to avoid electrical hazards. Voltage levels (usually 220V or 380V, depending on the region) and amperage need to be matched to equipment specifications to prevent overloading. Proper insulation, earthing, and circuit breakers are critical for safety, and routine checks help prevent equipment failure due to electrical faults.

Electrical Repairs: Electrical repairs in aquaculture involve maintaining and troubleshooting electrical systems that power critical farm equipment such as pumps, aerators, and lighting. This includes inspecting wiring for wear or damage, replacing faulty switches or circuit breakers, and ensuring that grounding and insulation are intact to prevent short circuits. Voltage checks (typically 220-380V) are performed to ensure that equipment is receiving the correct power supply. Proper electrical repairs reduce the risk of equipment failure and accidents, such as electrical fires or shocks.

Electrical Wiring: Electrical wiring in aquaculture involves the proper installation of electrical systems to power equipment such as pumps, aerators, and lighting. Wiring must comply with safety standards, including the use of waterproof insulation, grounding, and circuit breakers to prevent short circuits or electrocution hazards. Wiring should be rated for outdoor use and protected from moisture, UV rays, and mechanical damage. Voltage (220-380V) and amperage should be correctly matched to equipment specifications, ensuring reliable power distribution across the farm.

Emergency Protocols: Emergency protocols in aquaculture refer to predefined procedures and actions to be taken during emergencies such as disease outbreaks, equipment failure, power loss, or extreme weather events. These protocols include backup power systems for critical equipment, isolation of affected stock, and rapid water quality testing. In the event of a power outage, immediate activation of generators is crucial to maintain aerators and pumps, ensuring dissolved oxygen levels do not drop below 5 mg/L. Clear communication channels and trained personnel are essential for timely response and risk mitigation.

Environmental Impact: Environmental impact in aquaculture refers to the effect that farming activities have on the surrounding ecosystems, including water pollution, habitat disruption, and biodiversity loss. Wastewater discharge, nutrient runoff, and the use of chemicals can contribute to eutrophication and harm local wildlife. Sustainable practices, such as integrated multi-trophic aquaculture (IMTA), proper waste management, and the use of eco-friendly feed, help minimize negative environmental impacts. Monitoring and mitigating these impacts is essential for long-term farm viability and regulatory compliance.

Erosion Control: Erosion control in aquaculture focuses on preventing the loss of soil from pond dikes, roads, and drainage channels due to water flow and weather conditions. Techniques include planting vegetation like grasses on pond dikes, using riprap or rocks to stabilize embankments, and constructing properly sloped dikes (with a recommended slope of 1:1.5 to 1:3) to reduce erosion from water runoff. Geotextiles and erosion control mats are also commonly used on steeper slopes to reinforce soil and prevent washout during heavy rains or high water levels.

Farm Design: Farm design in aquaculture involves planning the layout and structure of ponds, channels, dikes, and other components to optimize water management, production efficiency, and environmental sustainability. Key considerations include pond size and shape, water flow patterns, aeration systems, and access for feeding and harvesting. Proper design enhances the farm's ability to maintain water quality, reduce disease risks, and maximize yield. Detailed design plans also address biosecurity, effluent management, and resource utilization.

Farm Fencing: Farm fencing is used in aquaculture to secure the farm perimeter, prevent unauthorized access, and protect ponds from predators such as birds or animals. Fences are commonly made from materials like wire mesh, chain link, or electric fencing, depending on the threat and budget. The height of the fence varies based on the species of predator being excluded, but typically ranges from 1.5 to 2 meters for effective protection. In addition to perimeter fencing, screens are often installed over pond inlets and outlets to prevent escape of stock.

Farm Layout: The layout of an aquaculture farm is designed to optimize the use of space, resources, and water flow while minimizing biosecurity risks. A well-planned farm includes a strategic arrangement of ponds, drainage systems, water supply channels, access roads, and farm buildings. Key considerations include the topography of the site, soil type, and water source proximity. Ponds are usually laid out in parallel rows with drainage systems to ensure easy water exchange and harvesting. Access roads should allow for easy movement of feed, equipment, and harvested stock.

Farm Operation: Farm operation in aquaculture encompasses the day-to-day management of water, feed, stock health, and infrastructure. Successful farm operation requires regular water quality monitoring, feed management to optimize growth while minimizing waste, and maintaining biosecurity to prevent disease outbreaks. In addition, operators must manage energy use, maintain equipment, and monitor growth performance. For shrimp and fish farms, careful record-keeping of parameters such as feeding rates, stocking densities, and water quality ensures optimal productivity and profitability.

Farm Roads: Farm roads in aquaculture farms provide access to ponds, buildings, and other infrastructure for the movement of feed, equipment, and harvest. They must be well-designed to withstand heavy equipment and regular traffic, especially in wet conditions. Farm roads are typically constructed from gravel or compacted soil, with proper drainage systems (culverts, ditches) to prevent erosion and waterlogging. Roads should be wide enough (usually 3-5 meters) to allow for the passage of vehicles and equipment, and placed in locations that minimize disruption to water flow and farm operations.

Farm Structures: Farm structures in aquaculture include the ponds, dikes, drainage systems, buildings, and support infrastructure like feed storage and maintenance workshops. The design and construction of these structures should take into account the local climate, soil type, and water availability. Strong, durable materials that resist corrosion, erosion, and biofouling should be used. For example, pond dikes should be constructed with a 1:1.5 to 1:3 slope to prevent erosion, and buildings should be well-ventilated to store feed and equipment safely.

Filtration Systems: Filtration systems in aquaculture are designed to remove solids, debris, and contaminants from water to maintain a healthy environment for aquatic organisms. These systems can include mechanical filters (such as drum filters), biological filters (like biofilters with media for nitrification), and chemical filters (using activated carbon). Proper filtration is essential to control water quality parameters such as ammonia, nitrite, and suspended solids, ensuring optimal conditions for growth and reducing the risk of disease.

Fish Farm: A fish farm is an artificial aquatic system designed for raising fish in controlled environments for commercial production, conservation, or research purposes. Fish farms can vary in scale from small, family-run operations to large industrial complexes. They typically include a series of ponds, tanks, or cages, equipped with necessary infrastructure such as water supply systems, aeration devices, and feeding equipment. The design of fish farms aims to optimize conditions for growth, health, and yield, ensuring the sustainability and economic viability of the operation.

Fish Grading Systems: Fish grading systems are used to sort fish by size during different stages of growth to ensure uniformity in feeding and reduce competition. Grading is done using mesh screens, nets, or automated mechanical graders, with size ranges typically based on the species being cultured. Regular grading prevents larger fish from dominating smaller ones, which improves overall growth rates and reduces feed waste. Grading is particularly important in high-density systems, where size differences can lead to increased stress and aggression.

Fish Holding Systems: Fish holding systems are temporary enclosures used to store fish before stocking, transport, or sale. These systems can include tanks, raceways, or cages and are designed to maintain water quality and provide adequate aeration. Fish holding systems should maintain stable temperatures (usually 20°-30°C, depending on species), oxygen levels above 5 mg/L, and low ammonia concentrations (<0.02 mg/L). Water recirculation or flow-through systems are often used to keep water clean and oxygenated. Stocking densities are carefully managed to reduce stress and prevent disease outbreaks.

Flood Management: Flood management in aquaculture involves designing systems and structures to protect farms from flooding due to heavy rainfall, river overflow, or tidal surges. Key strategies include constructing levees or dikes, installing drainage systems, and maintaining overflow structures. Ponds should be elevated or located in areas less prone to flooding, and drainage systems must be designed to handle peak water volumes. In regions prone to storm surges, coastal farms may use tidal gates and elevated platforms to protect critical infrastructure from water damage.

Generator Maintenance: In aquaculture, backup generators are essential for ensuring uninterrupted power supply to critical systems like pumps and aerators. Generator maintenance involves regular checks of fuel levels, oil quality, battery charge, and electrical connections. Filters should be replaced periodically, and cooling systems need to be inspected for leaks or blockages. Testing the generator under load conditions at least once a month ensures reliability during power outages. Generator capacity should be matched to the energy demands of the farm, typically calculated in kilowatts (kW) depending on the number of pumps, aerators, and other equipment.

Gravity Flow Systems: Gravity flow systems use the natural force of gravity to move water through an aquaculture system, reducing the need for mechanical pumping. These systems are designed with carefully calculated slopes and water head differences to ensure adequate flow rates (often 0.02-0.1 m/s) for ponds or tanks. Gravity flow is energy-efficient and commonly used in open pond systems or for water exchange between settling ponds and grow-out areas. Proper design is essential to avoid water stagnation or insufficient flow, which could negatively impact water quality.

Hatcheries: Hatcheries are specialized facilities for breeding, hatching, and early-stage rearing of aquatic organisms, such as fish and shrimp. They are designed to provide optimal environmental conditions, including temperature (generally between 26°-30°C for tropical species), salinity (for marine species), and water quality (ammonia <0.1 mg/L, dissolved oxygen >5 mg/L). Hatcheries play a critical role in ensuring the availability of healthy fry and larvae for stocking in grow-out systems. Controlled breeding, often using artificial propagation techniques, ensures high survival rates and uniform growth of the species being cultured.

Hatchery Construction: Constructing a hatchery requires careful planning and the use of appropriate materials to ensure durability and biosecurity. Foundations must be stable to support the weight of tanks and equipment, and construction materials should be non-corrosive and easy to clean. The design must allow for the installation of water supply and drainage systems, including pumps, filters, and aerators. Proper insulation and ventilation systems are needed to maintain optimal temperature and humidity levels within the hatchery. During construction, care must be taken to avoid contamination from external sources.

Hatchery Design: Hatchery design focuses on creating an efficient system for the production of healthy larvae or fry. The design includes considerations for water flow, aeration, and temperature control, ensuring optimum conditions for the species cultured. Tanks are usually constructed from non-toxic, durable materials such as fiberglass or concrete. Water treatment systems, such as biofilters or UV sterilizers, are integrated to maintain water quality. Hatcheries may also include environmental control systems (for lighting, temperature, and humidity) to support breeding and larval development.

Hatchery Layout: A well-planned hatchery layout optimizes space and resource use while minimizing biosecurity risks. Hatcheries typically include separate areas for broodstock management, spawning, incubation, larval rearing, and live feed production. Flow-through or recirculating water systems are used to maintain water quality. The layout must facilitate easy movement of personnel and equipment, with quarantine areas to prevent the spread of diseases. Hatchery tanks are arranged to allow for easy cleaning and water exchange, often incorporating aeration and filtration systems.

Hatchery Maintenance: Hatchery maintenance ensures the smooth operation of breeding, hatching, and rearing activities. Routine tasks include cleaning tanks and water systems, monitoring and adjusting water quality (temperature, pH, ammonia, dissolved oxygen), and inspecting equipment such as pumps, aerators, and filters. Preventive maintenance includes checking structural integrity, repairing leaks, and controlling biofouling in water lines. Additionally, regular health checks on broodstock and larvae, along with disease monitoring and management, are essential to prevent outbreaks and ensure high survival rates.

Hatchery Site Selection: Selecting an appropriate site for a hatchery is crucial for its success. Key factors include proximity to a reliable and clean water source, soil quality (permeability and fertility), and access to power and transportation. Water quality should meet the requirements of the cultured species, including suitable pH (6.5-8.5), salinity (for marine species), and minimal levels of contaminants or pollutants. The site should also be located in an area free from flooding and away from industrial pollution, with sufficient space for future expansion.

Inlets: Inlets are structures or openings through which water enters aquaculture ponds or tanks from the main water supply. They are designed to regulate the volume and flow rate of incoming water, ensuring even distribution and minimizing turbulence that could stress aquatic organisms. Inlets may include screens or filters to prevent the entry of debris, predators, or unwanted species. Proper inlet design is crucial for maintaining water quality and supporting the growth and health of farmed species.

Invasive Species Control: Invasive species control in aquaculture focuses on preventing the introduction and spread of non-native organisms that could disrupt the ecological balance of the farm or surrounding environment. Methods include physical barriers such as screens and nets, biosecurity measures like quarantine for new stock, and regular monitoring of water sources. Invasive species, such as predatory fish or plants that compete for resources, can impact the health of cultured species and reduce farm productivity. Effective control measures are necessary to protect biodiversity and comply with environmental regulations.

Irrigation Ditches: Irrigation ditches are channels used to supply water to aquaculture ponds or farms, typically through gravity flow. These ditches are lined to minimize seepage and constructed with proper slopes (0.1-1% gradient) to ensure efficient water distribution. Ditches also serve as drainage systems during heavy rainfall, preventing waterlogging or flooding. Regular maintenance, such as clearing debris and managing erosion, is necessary to ensure uninterrupted water flow. In large farms, irrigation ditches are often integrated with other water management infrastructure, such as pumps and gates.

Land Survey: A land survey is a systematic process of determining the precise boundaries, topography, and features of a site for aquaculture development. It involves measuring distances, angles, and elevations using tools like theodolites, GPS, and leveling instruments. The data collected helps in planning the layout of ponds, channels, dikes, and other farm structures. Accurate surveys are essential for ensuring efficient water management, optimal pond design, and the prevention of erosion or flooding issues.

Levelling: Levelling in aquaculture is the process of determining height differences between various points on a farm to create a level surface or a gradient for water flow. This is crucial in pond construction and drainage systems, as it ensures that water can flow evenly throughout the farm without pooling. Instruments like dumpy levels or laser levels are used to measure elevations, and gradients are designed to facilitate proper water management. Typical slopes for pond bottoms range from 0.02-0.05% to promote good drainage and ease of harvest.

Main Feeder Channel: The main feeder channel is a primary water conduit in an aquaculture farm, responsible for delivering water from the source (e.g., river, reservoir) to the ponds or tanks. It must be designed to handle the required flow rates, minimize erosion, and maintain water quality. Main feeder channels are typically lined or constructed with materials that prevent seepage and contamination. Their dimensions and slope must be carefully calculated to ensure efficient water distribution throughout the farm.

Mechanical Repairs: Mechanical repairs on an aquaculture farm involve fixing or maintaining equipment such as pumps, aerators, and feed dispensers. Regular inspections are needed to identify wear and tear, and basic tools and knowledge of mechanics are essential for performing repairs on-site. Preventive maintenance helps in minimizing downtime and improving the efficiency of farm operations. For example, a malfunctioning aerator can lead to oxygen depletion in ponds, affecting the health of the stock. Routine tasks include lubricating moving parts, replacing worn-out bearings, and tightening loose bolts.

Nutrient Management: Nutrient management in aquaculture involves the control of inputs like feed, fertilizers, and water quality to maintain a balanced nutrient load within the culture system. Overfeeding or excess fertilizers can lead to nutrient buildup, resulting in water quality issues such as algal blooms and ammonia toxicity. Nutrient levels, especially nitrogen and phosphorus, should be monitored regularly (ideal nitrogen <0.02 mg/L, phosphorus <0.1 mg/L). Managing nutrient inputs through proper feed formulation, feeding schedules, and water exchange helps optimize growth rates while minimizing waste production.

Outlets: Outlets are structures used in aquaculture ponds or tanks to discharge water, facilitating drainage, water exchange, and effluent management.

They are typically equipped with screens, valves, or gates to control the flow and prevent the escape of cultured species. Outlets must be designed to handle the required discharge volumes while maintaining water quality and preventing erosion or backflow. Effective outlet design is essential for water level control and the removal of waste and excess nutrients.

Overflow Structures: Overflow structures in aquaculture are designed to prevent ponds or tanks from flooding by allowing excess water to escape safely during heavy rainfall or water exchange operations. Common types include weirs, spillways, and sluice gates. Overflow structures should be sized to handle peak water flow rates and are usually constructed with non-corrosive materials like concrete or PVC. Properly designed overflow structures ensure that excess water is directed away from ponds without eroding embankments or disrupting the surrounding environment.

Pipe Sizing: Pipe sizing in aquaculture is the process of determining the appropriate diameter of pipes used for water supply, drainage, and aeration systems. The size of pipes affects water flow rate, pressure, and system efficiency. Larger pipes reduce friction losses and are necessary for high-flow systems, while smaller pipes are suitable for low-flow applications. Sizing is based on factors such as flow rate (L/min) and head loss calculations. For most aquaculture applications, pipe diameters range from 50 mm to 200 mm, depending on system needs.

Plane Table Survey: Plane table surveying is a method of topographic surveying in which field observations are plotted directly onto a map. A plane table mounted on a tripod is used as a drawing surface, and the operator measures angles and distances from a known point. This technique is useful in aquaculture for creating detailed maps of farm sites, including pond locations, elevations, and structures. It provides a visual representation of the terrain, helping to guide decisions on pond placement, drainage systems, and other infrastructure.

Plumbing Basics: Plumbing in aquaculture involves the installation and maintenance of water delivery systems, including pipes, valves, and fittings. The layout must ensure efficient water flow with minimal friction losses, typically requiring the use of PVC or HDPE pipes due to their resistance to corrosion. Sizing the pipes according to flow rate requirements (in liters per second or cubic meters per hour) is critical to ensure proper water exchange in ponds or tanks. Regular inspections should be carried out to check for leaks, blockages, or biofouling that could disrupt water circulation and affect water quality.

Pond Depth: Pond depth is an important design consideration in aquaculture, affecting water temperature, dissolved oxygen levels, and species-specific habitat requirements. For fish farming, ponds are typically 1.5 to 3 meters deep, depending on the species and environmental conditions. Shallower ponds are prone to temperature fluctuations and require more frequent aeration, while deeper ponds help buffer against temperature extremes and maintain more stable water conditions. Optimal depth also helps with water management during harvesting and allows for better control of pond stratification.

Pond Excavation: Pond excavation is the process of digging and shaping a pond to the desired size, depth, and slope for aquaculture. It involves removing soil, rocks, and other materials, followed by grading to ensure proper water retention and drainage. Optimal pond depth typically ranges from 1.2 to 2.5 meters, depending on the species being farmed, with deeper ponds used for fish and shallower ponds for shrimp. Proper excavation also involves constructing sloped embankments (1:1.5 to 1:3) and drainage channels to ensure good water circulation and harvesting efficiency.

Pond Lining: Pond lining involves installing a layer of material on the pond bed to prevent water seepage and improve water quality control. Common materials include clay, HDPE (high-density polyethylene) liners, and geotextiles. The choice of material depends on the soil type and cost considerations. HDPE liners are commonly used because they are durable, UV-resistant, and can last for over a decade. Proper pond lining helps maintain water levels, reduces contamination from the underlying soil, and simplifies pond maintenance.

Pond Maintenance: Pond maintenance involves a set of practices aimed at maintaining the water quality, structural integrity, and ecological balance of aquaculture ponds. These activities include monitoring water quality parameters such as pH (6.5-9.0), dissolved oxygen (above 5 mg/L), ammonia, and temperature (optimum range varies by species), along with periodic removal of sludge and detritus from the pond bottom. Regular checks on embankments for erosion, maintaining inlet/outlet structures for proper water flow, and controlling aquatic weeds and algal blooms are essential to avoid water stagnation and oxygen depletion, which could negatively impact fish/shrimp growth.

Pond Reclamation: Pond reclamation is the process of restoring or rehabilitating abandoned or degraded aquaculture ponds for reuse. This may involve removing accumulated sediment, repairing dikes or drainage systems, and treating the soil to improve water retention and quality. Reclamation can also include measures to control pests, remove invasive species, and restore water flow. Proper pond reclamation allows for the sustainable reuse of resources, reduces the need for new pond construction, and can enhance the productivity of previously unused or underperforming areas.

Pond Shape: Pond shape in aquaculture can significantly influence water circulation, aeration efficiency, and ease of management. Common shapes include rectangular, square, and circular ponds, each with specific advantages. Rectangular ponds are easy to manage and construct, while circular ponds promote better water flow and oxygen distribution. The choice of shape depends on site topography, intended stocking density, and the species being cultured. Proper pond shape design helps minimize dead zones and supports uniform growth of aquatic organisms.

Pond Size: Pond size in aquaculture refers to the surface area and volume of the pond, which affect stocking density, water management, and operational costs. Typical pond sizes range from small (0.1-0.5 hectares) for intensive systems to larger ponds (1-5 hectares) for extensive farming. The size must be matched to the capacity of water supply, aeration, and feeding systems to maintain optimal conditions. Larger ponds offer advantages in terms of reduced crowding stress and natural food availability but may require more complex management.

Power Backup: Power backup systems are critical in aquaculture to ensure continuous operation of life-support systems like aerators and pumps during power outages. Backup generators, battery banks, or solar systems provide emergency power. The capacity of backup systems must match the power consumption of essential equipment, typically calculated in kilowatts (kW). For example, a farm with multiple aerators and pumps may require a generator with a capacity of 10-50 kW, depending on equipment size. Regular maintenance and testing ensure backup systems are operational when needed.

Pump Maintenance: Pump maintenance in aquaculture ensures the reliable transfer of water between ponds and other systems. Key maintenance tasks include checking for wear and tear on mechanical components such as impellers and seals, inspecting for leaks, and ensuring the electrical connections are intact. Pumps should be routinely cleaned to avoid blockages from debris and biofouling. Pump efficiency directly affects water circulation, aeration, and filtration, and the operational performance is typically monitored by flow rate (L/min), pressure (typically 2-3 bar), and motor performance.

Pump Selection: Pump selection in aquaculture involves choosing pumps that meet the specific water flow and pressure requirements of the system. Key factors include the total dynamic head (TDH), flow rate (liters per second or cubic meters per hour), and energy efficiency. Centrifugal pumps are commonly used for water circulation, while airlift pumps are preferred for oxygenation. Pump sizing must account for the elevation differences and friction losses in pipes. Oversized pumps can waste energy, while undersized pumps may lead to insufficient water exchange or aeration.

Pump-Fed Ponds: Pump-fed ponds use mechanical pumps to deliver water from a source, such as a river or well, to the aquaculture ponds. This system provides greater control over water levels and quality, allowing for precise management of parameters like salinity, temperature, and flow rates. Pump-fed systems are suitable for areas without reliable tidal influence and are essential for maintaining stable conditions in intensive aquaculture operations. However, they involve higher operational costs due to energy consumption and pump maintenance.

Pumping Systems: Pumping systems are essential components of aquaculture farms, used to move water into and out of ponds, tanks, or other farming units. They include centrifugal pumps, submersible pumps, and axial flow pumps, selected based on factors like flow rate, head, and energy efficiency. Proper pump selection and maintenance are crucial for ensuring consistent water circulation, oxygenation, and waste removal. Pumps should be capable of handling the required flow rates (e.g., 5-10 liters per second per hectare for pond systems) without excessive energy consumption.

Rainwater Harvesting: Rainwater harvesting in aquaculture involves collecting and storing rainwater for use in ponds or tanks, reducing dependence on external water sources. Harvested rainwater can be used to supplement freshwater supplies, especially in regions with limited water availability. Rainwater is typically collected from roofs, gutters, and specially designed catchment systems, and stored in tanks or reservoirs. Proper filtration and treatment may be necessary to ensure water quality. Rainwater harvesting helps in reducing costs, conserving water resources, and maintaining sustainable farm operations.

Safety Measures: Safety measures in aquaculture ensure the protection of workers, equipment, and stock from accidents or health hazards. This includes the use of personal protective equipment (PPE), proper handling of chemicals and machinery, and the installation of safeguards like circuit breakers and emergency shutoff valves. Electrical systems must be properly grounded, and biosecurity protocols followed to prevent disease transmission. Regular training in first aid, fire safety, and emergency response procedures ensures that workers can handle risks effectively and maintain a safe working environment.

Sediment Control: Sediment control in aquaculture is the management of soil and organic particles in ponds, tanks, or water inflows to prevent the buildup of silt and debris, which can reduce water quality and affect fish health. Techniques include sedimentation basins, settling ponds, and filter systems. Regular dredging of ponds and cleaning of inflow channels are essential for effective sediment management. Uncontrolled sedimentation can lead to increased turbidity, reduced light penetration, and clogging of drainage systems, all of which negatively impact farm productivity.

Sedimentation Tanks: Sedimentation tanks are used in aquaculture to remove suspended solids from water before it enters rearing systems or ponds. The water is held in the tank long enough for particles to settle to the bottom, which are then removed periodically. These tanks are critical in recirculating aquaculture systems (RAS) to maintain water clarity and reduce the load on biofilters. The design of sedimentation tanks includes features like sloped bottoms for easier sludge removal and baffles to slow water flow, maximizing particle settling.

Seepage Control: Seepage control in aquaculture.e involves preventing water loss from ponds or tanks through the soil or substrate. Techniques include the use of clay liners, synthetic liners (e.g., HDPE), or concrete sealing to reduce water infiltration. Seepage can lead to significant water loss, increase operational costs, and reduce water quality by allowing the entry of contaminants from surrounding soil. Proper pond construction and the use of effective seepage control methods ensure water retention and reduce the need for frequent water top-ups.

Settling Ponds: Settling ponds serve a similar purpose to sedimentation tanks but are typically larger and used in outdoor systems to remove suspended solids from effluent water before discharge or reuse. These ponds allow water to flow slowly, enabling suspended particles to settle to the bottom. Regular dredging is required to remove the accumulated sludge. Settling ponds are an important component of integrated water management in aquaculture, helping to reduce the environmental impact of waste discharge and improve water quality for reuse.

Site Evaluation: Site evaluation in aquaculture involves assessing the suitability of a location for farming activities based on factors like soil quality, water availability, topography, and environmental conditions. Key considerations include proximity to water sources, risk of flooding, accessibility, and the potential environmental impact of the farm. Soil pH (usually between 6.5 and 8.5) and permeability are important for pond construction, while water quality parameters such as salinity, dissolved oxygen, and temperature should match the species being cultured. A thorough site evaluation ensures optimal farm performance and sustainability.

Site Selection: Site selection for aquaculture farms involves evaluating potential locations based on a range of criteria to ensure optimal conditions for the species being cultured. Key factors include water quality, soil type, topography, access to water sources, proximity to markets, and environmental impact. Ideal sites have gentle slopes (1-2% gradient), minimal flooding risk, good road access, and are located away from industrial pollutants. Soil should have a high clay content (20-30%) for pond construction, while water quality parameters such as pH, salinity, and temperature must meet the specific requirements of the species.

Slope: Slope in aquaculture ponds refers to the gradient of pond bottoms and dikes, which is critical for effective drainage, water flow, and maintenance. Ideal bottom slopes range from 1-2% towards the drainage point to facilitate easy removal of sediments and waste. Dike slopes are generally 2:1 or 3:1 (horizontal to vertical) to ensure stability and prevent erosion. Proper slope design enhances water management, reduces labor for pond cleaning, and supports overall farm efficiency.

Sluice Gates: Sluice gates are adjustable barriers used in aquaculture farms to control the flow of water into or out of ponds, channels, or tanks. They consist of a frame and a sliding gate that can be raised or lowered to regulate water levels and flow rates. Sluice gates are commonly used in tide-fed and pump-fed systems, allowing precise management of water exchange, salinity, and aeration. They are critical for maintaining optimal conditions for aquaculture operations and preventing flooding or drainage issues.

Soil Type: Soil type is a critical factor in aquaculture farm design, influencing pond construction, water retention, and overall farm stability. Suitable soils for pond construction should have a high clay content (20-30%) to minimize seepage and maintain water levels. Sandy soils, which have high permeability, are less suitable unless lined with materials such as HDPE (High-Density Polyethylene). Soil testing for parameters like pH (ideally 6.5-8.5), organic matter, and texture helps in assessing suitability for aquaculture.

Storage Tanks: Storage tanks in aquaculture are used to hold water for emergency use, water exchange, or treatment before use in ponds or tanks. These tanks must be made from materials that are non-reactive and durable, such as HDPE, fiberglass, or concrete, and sized according to the farm's water demand. Storage tanks should be located in elevated positions to allow for gravity-fed distribution when needed. Proper maintenance and cleaning of storage tanks are necessary to prevent the buildup of sediments or algae, which can affect water quality.

Surveying Equipment: Surveying equipment in aquaculture includes tools like theodolites, GPS units, and leveling rods used to measure land elevations, distances, and angles during farm construction. This equipment helps in accurately mapping the farm site, determining pond locations, and calculating slopes for drainage systems. GPS units provide precise location data, while theodolites and leveling rods are used to establish elevation differences. Proper use of surveying equipment ensures that the farm is constructed according to design specifications, optimizing water flow and preventing construction errors.

Tidal Gates: Tidal gates are structures used in coastal aquaculture to regulate water flow between the farm and tidal waters. They allow water to enter during high tide and prevent its exit during low tide, maintaining optimal water levels in ponds. Tidal gates are essential in regions with tidal variations, helping control water exchange without the need for pumping. They must be designed to prevent the entry of predators or contaminants and ensure that water quality parameters, such as salinity (typically 15-35 ppt for shrimp farming), remain stable.

Tidal Influence: Tidal influence refers to the impact of tidal movements on water levels and quality in aquaculture ponds located near coastal areas. Tidal cycles can affect salinity, temperature, and nutrient availability, influencing the growth and health of cultured species. Managing tidal influence involves designing appropriate water control structures, such as sluice gates, to regulate the timing and volume of water exchange. Farms must account for local tidal ranges and patterns to optimize pond management and prevent issues like excessive salinity or sediment buildup.

Tide-Fed Ponds: Tide-fed ponds are aquaculture systems that rely on tidal movements to provide water exchange, aeration, and nutrient inputs. These ponds are typically located in coastal areas with sufficient tidal range to facilitate regular flushing. The design includes structures like gates and sluices to control water inflow and outflow according to tidal cycles. Tide-fed ponds are energy-efficient, utilizing natural forces for water management, but require careful planning to avoid issues with salinity fluctuations and sedimentation.

Topography: Topography refers to the physical features of the land surface, including its contours, elevation, and slope. In aquaculture, understanding topography is critical for designing water flow systems, pond layouts, and drainage structures. Ideal topographical conditions include gentle slopes (1-2%) that facilitate natural water flow and prevent erosion. Topographic maps are used to identify suitable sites, guide construction, and ensure efficient water management, minimizing the need for pumping and reducing operational costs.

Valves: Valves are devices used to control the flow of water within an aquaculture system. Common types include gate valves, ball valves, and check valves, each serving different functions, such as opening/closing flow, regulating flow rates, or preventing backflow. Proper valve placement allows for easy control of water exchange, drainage, and aeration. Valves must be corrosion-resistant, especially in saline environments, and sized appropriately for the pipe diameter to ensure a good seal and long-term durability in aquatic systems.

Waste Treatment Plants: Waste treatment plants in aquaculture are facilities designed to process and treat effluent water before it is discharged back into the environment. These plants use physical, chemical, and biological methods to remove solid waste, excess nutrients, and harmful pathogens from wastewater. Treatment stages may include sedimentation, filtration, and biofiltration to reduce pollutants. By treating wastewater, aquaculture farms can minimize their environmental impact, comply with discharge regulations, and recycle water for reuse in the farm.

Wastewater Management: Wastewater management in aquaculture involves the treatment and disposal of water used in farming operations to prevent environmental pollution. This process includes the removal of organic waste, excess nutrients (nitrogen, phosphorus), and chemical residues from water before it is released back into the environment or recirculated. Techniques include sedimentation, biofiltration, and constructed wetlands. Effective wastewater management ensures compliance with environmental regulations, prevents eutrophication of nearby water bodies, and allows for the reuse of water, thereby reducing overall water demand.

Water Accessibility: Water accessibility refers to the ease with which a site can obtain the required volume and quality of water necessary for aquaculture operations. Ideal sites should be located near reliable water sources such as rivers, lakes, or groundwater. The quality of accessible water should meet the needs of the cultured species, with parameters like temperature, salinity, and pH within acceptable ranges. Additionally, the water source should be sustainable, with sufficient flow rates to support the farm's needs without causing environmental degradation.

Water Control Structure: Water control structures in aquaculture are engineered components such as sluice gates, weirs, and culverts that regulate the flow, level, and distribution of water within the farm. These structures are essential for maintaining optimal water conditions, facilitating water exchange, and preventing flooding or drainage issues. Proper design and placement of water control structures help manage water quality, ensure efficient operation, and protect the farm from environmental variations.

Water Demand Calculation: Water demand calculation in aquaculture involves estimating the volume of water needed for daily farm operations, including filling ponds, water exchange, and maintaining optimal water quality. Factors influencing water demand include stocking density, pond size, evaporation rates, and the frequency of water exchange. For instance, in a recirculating aquaculture system (RAS), water usage may be minimized through filtration and recycling, whereas open pond systems require regular water replenishment due to evaporation and seepage.

Water Distribution System: A water distribution system in aquaculture is a network of structures and channels that deliver water from the source to the farming units and facilitate drainage. This system includes main feeder channels, secondary distribution channels, and drainage channels, designed to ensure efficient water flow and minimize dead zones. Proper design is crucial for maintaining water quality, preventing stagnation, and optimizing oxygen levels. The layout should support the specific needs of the aquaculture species and farm type.

Water Flow Control: Water flow control in aquaculture involves regulating the amount and direction of water moving through ponds, tanks, or irrigation systems. Control mechanisms include valves, gates, and pumps that adjust water flow rates to maintain optimal water quality and oxygen levels. Flow rates should be monitored regularly to avoid excessive or insufficient water movement, with ideal rates varying depending on species and system type. Proper water flow control minimizes the risk of stagnation, nutrient buildup, and oxygen depletion in aquaculture systems.

Water Flow Rate: Water flow rate refers to the volume of water moving through a system over time, typically measured in liters per second (L/s) or cubic meters per hour (m /h). In aquaculture, flow rates are important for maintaining water quality, aeration, and waste removal. Optimum flow rates depend on the species and system type; for example, recirculating aquaculture systems (RAS) often require flow rates of 10-20 L/s to ensure adequate filtration and oxygenation. Excessive flow can stress fish, while insufficient flow can lead to poor water quality and reduced growth.

Water Inflow Control: Water inflow control refers to the management of water entering aquaculture ponds or tanks, ensuring a consistent and adequate supply of clean water to maintain optimal conditions for fish or shrimp. Inflow systems include pipes, valves, and gates that regulate water volume, flow rate, and quality. Flow rates must be calibrated according to species requirements, typically measured in liters per minute (L/min) or cubic meters per hour (m /h). Proper filtration and aeration systems may also be integrated to remove sediments and improve water quality before it enters the culture system.

Water Outflow Control: Water outflow control involves regulating the removal of water from ponds or tanks to maintain water quality and prevent flooding or contamination. This is done through a network of drainage systems, valves, and sluices that allow for controlled discharge. Outflow structures are usually designed to prevent sedimentation and the escape of stock. Proper control of outflow is crucial for water exchange and waste removal, ensuring optimal conditions for the cultured species. Regular maintenance of outflow systems is essential to avoid blockages and backflow.

Water Pressure Management: Water pressure management in aquaculture is critical for ensuring that water distribution systems function efficiently without damaging pipes or equipment. Properly regulated water pressure ensures consistent flow rates for aeration, water exchange, and filtration. High water pressure can cause leaks or bursts in pipes, while low pressure may result in insufficient flow. Pressure regulators, gauges, and valves are used to maintain optimal pressure levels, typically 1-3 bars for most aquaculture systems, depending on the scale and requirements of the farm.

Water Quality Monitoring: Water quality monitoring in aquaculture is the systematic measurement of physical, chemical, and biological parameters to ensure optimal conditions for the health and growth of cultured species. Key parameters include dissolved oxygen (above 5 mg/L), pH (6.5-9.0), temperature (specific to species, generally 24°-30°C for most tropical species), ammonia (<0.02 mg/L), nitrite (<0.1 mg/L), and salinity (for marine species). Monitoring is done using sensors and manual testing kits, and it ensures early detection of imbalances that could lead to stress, disease, or mortality in aquatic organisms.

Water Recycling: Water recycling in aquaculture is the process of treating and reusing water to reduce freshwater demand and minimize environmental impact. Recirculating aquaculture systems (RAS) are a common approach, where water is continuously filtered, aerated, and treated for reuse. Water recycling reduces the need for large volumes of water and helps maintain consistent water quality. Common treatment methods include mechanical filtration, biofiltration, and UV sterilization. Proper recycling systems maintain water quality parameters such as ammonia (<0.02 mg/L) and nitrite (<0.1 mg/L), ensuring a healthy environment for stock.

Water Supply for Hatcheries: A reliable water supply is crucial for hatchery operations, as water quality directly affects the health of aquatic organisms. Water should be free of harmful contaminants, with optimal pH, dissolved oxygen, and nutrient levels tailored to the species being cultured. Hatcheries often use boreholes, rivers, or seawater as their water sources, and these must be filtered and treated before use. Flow rates should be designed to allow for continuous or recirculating systems, ensuring sufficient water exchange without stressing larvae or fry.

Water Supply System: A water supply system in aquaculture farms includes the infrastructure and methods used to deliver water to ponds or tanks, ensuring an adequate and continuous supply of high-quality water. It typically consists of intake structures, pumps, pipelines, and filtration units. The design must account for the required flow rate, water quality parameters (e.g., pH 6.5-8.5, dissolved oxygen >5 mg/L), and potential sources of contamination. Reliable water supply systems are crucial for maintaining optimal growth conditions and managing farm operations effectively.

CHAPTER 8

SHRIMP FARMING

Acclimatization: Acclimatization is the process of gradually adjusting shrimp larvae or post-larvae to new environmental conditions, such as changes in salinity, temperature, or pH, before transferring them from hatcheries to grow-out ponds. Typically, gradual changes in salinity by 2-3 ppt per hour are recommended. Proper acclimatization helps reduce stress, improving survival rates and ensuring healthy growth during the grow-out phase.

Aeration Systems: Aeration systems in hatcheries are used to increase dissolved oxygen levels in water, essential for shrimp larvae survival. Aeration is achieved through devices like air diffusers, air stones, and paddle wheels, which promote oxygen transfer and water circulation. Adequate aeration is necessary to maintain DO levels above 5 mg/L and to prevent the formation of anaerobic zones, which can lead to toxic conditions like ammonia buildup.

Aerobic Bacteria: Aerobic bacteria are microorganisms that thrive in environments with oxygen and play a crucial role in the biofiltration process within hatcheries. Bacteria such as *Nitrosomonas* and *Nitrobacter* help convert toxic ammonia into less harmful nitrate in the nitrogen cycle. These bacteria require well-oxygenated water (DO >5 mg/L) for efficient biofiltration, and their presence is essential for maintaining water quality in shrimp hatcheries.

Algal Culture: Algal culture refers to the controlled growth of microalgae species such as *Tetraselmis*, *Chaetoceros*, and *Isochrysis* in aquaculture hatcheries. Algae serve as an essential food source for shrimp larvae, providing vital nutrients like omega-3 fatty acids, amino acids, and pigments. Optimal culture conditions include light intensity of 3,000–5,000 lux, temperature between 20°–25°C, salinity of 28–35 ppt, and pH levels between 7.5–8.5. Algal cultures are maintained in sterilized seawater enriched with nutrients such as nitrate, phosphate, and trace metals.

Ammonia Control: Ammonia control is vital in shrimp hatcheries as ammonia (NH_3) is toxic to shrimp larvae at elevated levels. Ideally, total ammonia-nitrogen (TAN) should be kept below 0.1 mg/L. Ammonia accumulates due to waste, uneaten feed, and organic matter breakdown. Biological filtration systems, regular water changes, and reduced feeding during high ammonia levels are key strategies for controlling ammonia concentrations.

Anaerobic Bacteria: Anaerobic bacteria function in environments devoid of oxygen and are often involved in the breakdown of organic matter in sedimentation tanks or biofilters. Some anaerobic bacteria can produce harmful byproducts like hydrogen sulfide (H_2S) or methane (CH_4), which can be toxic to shrimp if accumulated. In hatcheries, anaerobic zones should be avoided through proper aeration and waste removal to prevent water quality deterioration.

Artemia: Artemia (*Artemia salina*), commonly known as brine shrimp, is a key live feed used in shrimp hatcheries, especially during the mysis and post-larval stages. Artemia cysts are hatched in salinities of 25–35 ppt and temperatures of 25°–28°C. They are highly valued for their rich protein content (50-60%) and essential fatty acids like EPA and DHA. The bioencapsulation of Artemia with vitamins and minerals can further enhance shrimp larval nutrition.

Bacterial Diseases: Bacterial diseases in shrimp hatcheries, such as vibriosis caused by *Vibrio* species, are a significant threat to larval survival. Symptoms include lethargy, body discoloration, and high mortality. Prevention includes maintaining good water quality, biosecurity measures, and the use of probiotics or antibiotics (under strict regulation). Regular monitoring of bacterial counts in water and shrimp is crucial for early detection.

Better Management Practices (BMPs): Better Management Practices (BMPs) are a set of guidelines that aim to improve productivity, environmental sustainability, and disease control in shrimp hatcheries. BMPs cover water quality management, feed management, disease prevention, and biosecurity protocols. Key BMPs include maintaining stable salinity (30–35 ppt), regular water quality monitoring, optimal stocking densities, and strict biosecurity measures to prevent disease outbreaks.

Biofilters: Biofilters are essential components in shrimp hatcheries, aiding in the biological filtration of water. They house beneficial bacteria, primarily nitrifying bacteria (*Nitrosomonas* and *Nitrobacter*), that convert toxic ammonia into nitrite and then nitrate. Biofilters maintain a healthy nitrogen cycle and ensure low ammonia (<0.1 mg/L) and nitrite (<0.5 mg/L) levels, critical for larval health. The efficiency of biofilters depends on maintaining adequate surface area, water flow, and oxygen supply.

Biosecurity Protocols: Biosecurity protocols in shrimp hatcheries are preventive measures aimed at minimizing the risk of disease introduction and spread. These include the use of pathogen-free broodstock, regular disinfection of tanks and equipment, quarantine of new stock, and restricted access to hatchery areas. Effective biosecurity protocols are essential for maintaining disease-free conditions and ensuring high seed survival rates.

Biosecurity: Biosecurity in shrimp hatcheries refers to the measures taken to prevent the introduction and spread of pathogens. This involves strict control over water sources, disinfecting incoming stock, using pathogen-free broodstock, and enforcing access controls to hatchery facilities. Maintaining high biosecurity standards, such as regular water quality monitoring (e.g., salinity at 30-35 ppt, pH 7.5-8.5) and cleaning protocols, is crucial in minimizing disease outbreaks.

Breeding Programs: Breeding programs in shrimp hatcheries involve controlled reproduction strategies to improve desirable traits such as growth rate, disease resistance, and reproductive performance. Selective breeding and crossbreeding are commonly used to enhance genetic diversity and produce high-quality offspring. Modern breeding programs may also employ molecular tools like marker-assisted selection (MAS) to improve genetic traits.

Breeding Season: The breeding season for shrimp varies by species and environmental factors like temperature and photoperiod. In tropical species like *Penaeus monodon*, breeding can occur year-round under controlled conditions, but natural spawning may peak during warmer months. Hatcheries often manipulate environmental factors to induce year-round breeding, ensuring a steady supply of larvae for aquaculture.

Brine Shrimp: Brine shrimp (Artemia) are commonly used as live feed in shrimp hatcheries due to their high nutritional value, particularly for post-larvae and early juvenile stages. Artemia cysts are hatched in saline water (15-35 ppt) and provide a rich source of protein and essential fatty acids for developing shrimp. Brine shrimp are favored for their ease of storage and cultivation

Broodstock Conditioning: Broodstock conditioning is the process of preparing mature shrimp for spawning by providing optimal environmental conditions and a nutrient-rich diet. Conditioning involves maintaining proper water quality (temperature 28°-30°C, salinity 30-35 ppt), feeding high-protein diets (40-50% protein), and sometimes using hormonal treatments to stimulate gonadal maturation. Proper conditioning leads to high fecundity and quality seed production.

Broodstock Handling: Broodstock handling involves careful management of shrimp broodstock to minimize stress and ensure successful reproduction. This includes proper acclimatization to tank conditions, gentle handling during transfer, and maintaining optimal water quality (salinity 30–35 ppt, temperature 28°–30°C). Overhandling or exposure to poor water conditions can result in reduced fecundity, stress, and even mortality, affecting the overall seed production process.

Broodstock Management: Broodstock management involves maintaining mature shrimp used for breeding in optimal conditions to ensure high reproductive success. It includes selecting disease-free, genetically superior shrimp, feeding them with nutrient-rich diets, and managing water quality (salinity 28-35 ppt, temperature 28°-30°C, pH 7.5-8.5). Monitoring gonadal development and employing techniques like eyestalk ablation to induce maturation are common practices in hatcheries.

Broodstock Nutrition: Broodstock nutrition plays a critical role in enhancing reproductive performance and seed quality in shrimp hatcheries. A well-balanced diet rich in protein (40-50%), lipids, essential fatty acids (e.g., EPA and DHA), vitamins, and minerals is provided to ensure proper gonadal development and high fecundity. Common broodstock feed ingredients include squid meal, fishmeal, and marine oils. Nutrient deficiencies can lead to poor egg quality and reduced hatch rates.

Capital Investment: Capital investment in shrimp hatcheries refers to the initial expenditure required to establish infrastructure, such as tanks, aeration systems, filtration units, and biosecurity equipment. Costs typically include land acquisition, construction of hatchery buildings, and purchase of hatchery technology (e.g., water treatment systems). Depending on the scale, capital investments for a small to medium-sized hatchery can range from INR 10 lakhs to several crores, with a payback period of 3–5 years under good operational management.

Captive Rearing: Captive rearing refers to the practice of breeding and growing shrimp within a controlled hatchery environment. This involves managing water quality, feeding regimes, and disease prevention to ensure the successful development of shrimp from broodstock to post-larvae. Captive rearing helps reduce reliance on wild seed and ensures a consistent supply of shrimp for aquaculture.

Chemical Treatments: Chemical treatments in hatcheries are employed to disinfect water, equipment, and broodstock, or to control disease outbreaks. Common chemicals include formalin for ectoparasite removal, chlorine for water disinfection (concentration 10-15 ppm), and potassium permanganate for treating fungal infections. It is crucial to follow strict dosage and application guidelines to avoid harming shrimp larvae or compromising water quality.

Cold Storage: Cold storage is used to preserve shrimp feed, broodstock diets, and other hatchery materials by slowing down microbial growth and degradation. For shrimp feed, storage temperatures typically range from -20°C to -10°C, depending on the feed type. Cold storage helps maintain the nutritional quality of feed, especially those rich in fatty acids, vitamins, and live feed (e.g., Artemia cysts), extending their shelf life and efficacy.

Culture Systems: Culture systems in shrimp hatcheries refer to the various methods used to rear shrimp from larvae to post-larvae stages. Common systems include flow-through systems, recirculating aquaculture systems (RAS), and semi-closed systems. Each system aims to maintain optimal water quality by controlling parameters like salinity (30–35 ppt), temperature (28°–30°C), and dissolved oxygen (>5 mg/L) while minimizing waste and disease transmission.

Disease Management: Disease management in shrimp hatcheries involves the integrated use of biosecurity, early detection, water quality control, and the administration of treatments to mitigate the impact of infections. Regular monitoring of water parameters (ammonia <0.1 mg/L, nitrite <0.5 mg/L, dissolved oxygen >5 mg/L) and the use of probiotics or antimicrobial agents help in reducing pathogen load and preventing disease spread.

Disease Monitoring: Disease monitoring involves routine observation and testing of shrimp larvae and broodstock for early signs of disease. This includes visual inspections for abnormal behavior or physical symptoms, as well as laboratory tests such as PCR for detecting viral pathogens or bacterial cultures for identifying bacterial infections. Regular monitoring helps in early detection and timely intervention to prevent disease outbreaks.

Disinfection: Disinfection is a critical process in shrimp hatcheries to eliminate harmful pathogens from water, equipment, and broodstock. Common disinfectants include chlorine, iodine, and formaldehyde, used in specific concentrations (e.g., chlorine 5-10 ppm) to sterilize incoming water and hatchery surfaces. Disinfection helps maintain a pathogen-free environment, thereby reducing the risk of disease outbreaks.

Effluent Treatment: Effluent treatment is the process of removing pollutants from hatchery wastewater before it is discharged into the environment. Treatment methods include physical filtration, sedimentation, biological treatment (biofilters), and chemical disinfection (e.g., chlorine or UV). Effluent standards require that ammonia levels are below 0.5 mg/L, and suspended solids are minimized to prevent water pollution and ecosystem damage.

Eyestalk Ablation: Eyestalk ablation is a common practice in shrimp hatcheries used to induce maturation and spawning in female shrimp. The removal of one eyestalk disrupts the release of inhibitory hormones, promoting the production of gonad-stimulating hormones. This accelerates ovarian development, leading to spawning. While effective, it requires careful broodstock management to minimize stress and ensure recovery.

Fecundity: Fecundity refers to the reproductive capacity of shrimp, typically measured by the number of eggs produced by a female in one spawning event. Fecundity in shrimp is influenced by species, broodstock health, diet, and environmental conditions. For example, *Penaeus monodon* females can produce 500,000 to 1 million eggs per spawn. Fecundity is a key indicator of reproductive success in hatcheries.

Feed Additives: Feed additives in shrimp hatcheries are compounds added to larval and broodstock feeds to enhance growth, immunity, and feed efficiency. Common additives include probiotics, prebiotics, enzymes, vitamins (e.g., vitamin C and E), and minerals such as selenium and zinc. Immunostimulants like beta-glucans are also used to enhance disease resistance. Additives are typically included at low concentrations (0.5–2%) in formulated diets.

Feed Management: Feed management in shrimp hatcheries involves providing broodstock, larvae, and post-larvae with nutritionally balanced diets that support growth and reproductive health. Live feeds like algae, rotifers, and Artemia are essential during early larval stages, while formulated feeds are used for post-larvae and broodstock. Proper feed management ensures optimal growth, high survival rates, and successful maturation in broodstock.

Fertilization: Fertilization in shrimp involves the union of sperm and eggs, typically occurring during mating or spawning in the water. In hatcheries, fertilization can be either natural or induced by manipulating environmental factors. After fertilization, eggs hatch into nauplii within 12-24 hours. Fertilization success depends on the quality of broodstock and proper spawning conditions.

Fungal Diseases: Fungal diseases, caused by pathogens like *Fusarium* or *Lagenidium*, primarily affect shrimp eggs and larvae. Symptoms include fuzzy growth on the body and reduced survival rates. Fungal infections are often linked to poor water quality or overcrowding. Preventive measures include maintaining optimal water conditions (pH 7.5–8.5), using antifungal treatments like formalin or potassium permanganate, and ensuring good hygiene in hatchery tanks.

Genetic Improvement: Genetic improvement refers to the long-term enhancement of desirable traits in shrimp populations through selective breeding, hybridization, or advanced molecular techniques like CRISPR or gene editing. In hatchery settings, genetic improvement programs focus on traits such as disease resistance, growth rates, and feed conversion efficiency. These programs contribute to sustainable shrimp production by increasing survival rates and reducing the need for chemical interventions.

Gonadal Maturation: Gonadal maturation refers to the development of reproductive organs in shrimp, leading to the production of viable eggs and sperm. This process is influenced by environmental factors such as temperature, light, and diet. In hatcheries, gonadal maturation is often enhanced by manipulating light cycles and diet, as well as by using techniques like eyestalk ablation, which accelerates maturation by altering hormone levels.

Gonadosomatic Index (GSI): The Gonadosomatic Index (GSI) is a ratio that expresses the relative size of the gonads to the total body weight of a shrimp. It is used as an indicator of reproductive maturity and fecundity. GSI is calculated by dividing the gonad weight by the total body weight and multiplying by 100. A higher GSI indicates a greater reproductive investment and readiness for spawning.

Grow-out Stages: Grow-out stages in shrimp farming refer to the period when shrimp post-larvae are transferred to ponds or tanks for growth to market size. During this phase, shrimp are fed commercial feed, and water quality parameters like dissolved oxygen, ammonia, and pH are closely monitored. It typically takes 3-6 months to grow shrimp to harvestable size, depending on species, stocking density, and environmental conditions.

Hatchery Design: Hatchery design refers to the structural layout and organization of shrimp hatchery facilities. A well-designed hatchery should include separate areas for broodstock, larval rearing, water treatment, and live feed production. Adequate space for tanks, biosecurity measures, and proper water flow systems are important to ensure high survival and growth rates. Hatchery systems must be designed to maintain water quality and minimize contamination risks.

Hatchery Economics: Hatchery economics involves the financial planning and management of shrimp seed production facilities, including capital investments, operational costs, labor, and maintenance. A detailed cost-benefit analysis is necessary to ensure profitability. Factors like seed yield, feed costs, survival rates, and market demand for shrimp post-larvae influence economic viability. Optimizing production efficiency through better management practices (BMPs) and reducing waste can significantly improve profitability.

Hatchery Hygiene: Hatchery hygiene is essential to prevent the spread of diseases and ensure high survival rates in shrimp larvae. Routine disinfection of tanks, equipment, and water lines using chemicals like chlorine, iodine, or ozone is crucial. Workers must adhere to strict sanitation protocols, including wearing protective clothing and regularly washing hands. Good hygiene minimizes the risk of pathogen introduction and contamination.

Hatchery Technology: Hatchery technology refers to the set of systems and tools used in the controlled breeding and rearing of shrimp. Key components include water filtration and recirculation systems, larval rearing tanks, live feed culture units, and automated feeding systems. Advances in hatchery technology help improve seed production efficiency, reduce disease outbreaks, and enhance water quality management.

Hatching: Hatching refers to the process where shrimp eggs develop into nauplii, the first larval stage. In hatcheries, hatching typically occurs within 12-24 hours of fertilization under optimal conditions (temperature 28°-30°C, salinity 30-35 ppt). Proper management of water quality and biosecurity is crucial to ensure high hatching rates and larval survival.

Immunostimulants: Immunostimulants are compounds added to shrimp diets or water to enhance the innate immune response of shrimp, making them more resistant to diseases. Common immunostimulants include beta-glucans, chitosan, and marine-derived polysaccharides. These substances activate immune cells like hemocytes, improving their ability to respond to pathogens. In shrimp hatcheries, immunostimulants are often included in broodstock and larval diets at low concentrations (typically 0.5-1%) to reduce mortality and increase survival rates.

Induced Spawning: Induced spawning in shrimp hatcheries involves using physical or chemical methods to trigger broodstock to spawn. The most common method is eyestalk ablation, where one eyestalk is removed to increase reproductive hormone levels, leading to maturation and spawning. Induced spawning is essential in hatcheries to ensure a steady supply of fertilized eggs for seed production. Proper care of broodstock is required to maintain their health and fertility post-ablation.

Larval Diseases: Larval diseases in shrimp hatcheries are caused by bacterial, viral, fungal, or parasitic infections that affect shrimp during their vulnerable early life stages. Common diseases include White Spot Syndrome Virus (WSSV), Vibriosis, and fungal infections like Fusariosis. Optimal biosecurity, water quality management (salinity 30-35 ppt, temperature 28°-30°C), and the use of pathogen-free broodstock are essential in controlling disease outbreaks.

Larval Feed Formulation: Larval feed formulation is a specialized process that ensures the nutritional needs of shrimp larvae are met for optimal growth and survival. Larval feeds typically contain microalgae (e.g., *Chaetoceros* or *Isochrysis*), rotifers, Artemia, and formulated microparticulate diets. Feed formulations are rich in essential nutrients such as proteins (30-40%), lipids (10-15%), and essential fatty acids to support early development stages. The particle size of feed ranges from 50–200 µm for larvae to ensure proper ingestion.

Life Cycle: The life cycle of shrimp includes several stages: eggs, nauplii, zoea, mysis, post-larvae, juvenile, and adult. After fertilization, eggs hatch into nauplii, which progress to zoea and then mysis stages before metamorphosing into post-larvae. Post-larvae grow into juveniles and finally mature adults capable of reproduction. The entire life cycle can take 3-6 months depending on species and environmental conditions, with hatchery-reared shrimp undergoing controlled growth phases.

Light Regulation: Light regulation in shrimp hatcheries involves controlling the photoperiod (light/dark cycle) to optimize larval development. Shrimp larvae typically require 12-14 hours of light per day, with light intensities around 1000-2000 lux for optimal feeding and growth. Excessive light or sudden changes in lighting conditions can stress larvae and affect survival rates.

Live Feed: Live feed, such as phytoplankton, rotifers, and *Artemia* (brine shrimp), are essential in the early life stages of shrimp larvae. These feeds provide the necessary nutrients for growth and development, especially in the nauplii, zoea, and mysis stages. Hatcheries often culture live feed on-site to ensure a consistent supply and maintain high nutritional value for the larvae.

Maintenance Costs: Maintenance costs refer to the expenses incurred for the upkeep and repair of hatchery infrastructure and equipment, such as aerators, pumps, tanks, and filtration systems. Regular maintenance is essential for preventing system breakdowns and ensuring smooth hatchery operations. Maintenance costs typically range from 5–10% of total operating costs and are necessary for prolonging the lifespan of hatchery assets.

Maturation Stages: The maturation stages in shrimp refer to the sequential development of gonads, from immature to fully mature, ready for spawning. These stages are monitored through visual and microscopic examination of the gonads. Shrimp broodstock typically reach sexual maturity when they are 8-12 months old, with gonads filling and developing color. Proper broodstock conditioning, water quality management, and nutrient-rich diets are critical for successful maturation.

Microalgae Culture: Microalgae culture is crucial for providing live feed to shrimp larvae in the early stages of development. Species like *Chaetoceros, Tetraselmis*, and *Isochrysis* are commonly cultured. Optimal culture conditions include a light intensity of 1000–2000 lux, temperature of 22°–25°C, and a pH of 7.5–8.5. Microalgae provide essential nutrients, including omega-3 fatty acids, pigments, and proteins, necessary for larval growth and survival.

Natural Seed Resources: Natural seed resources refer to shrimp larvae or post-larvae collected from the wild. These resources are abundant in coastal estuaries and mangrove areas. Shrimp larvae are typically gathered using nets or other seed collection methods during the breeding season. However, over-reliance on natural seed resources can lead to ecosystem disturbances, which is why hatcheries are increasingly favored for sustainable seed production.

Nitrogen Cycle: The nitrogen cycle is a biological process in shrimp hatcheries where nitrogenous wastes, primarily ammonia, are converted into nitrite (NO_{2-}) and then into less harmful nitrate (NO_{3-}) by nitrifying bacteria. This cycle is crucial for maintaining a healthy aquatic environment. Biofilters are essential for this process, and effective nitrogen cycling ensures ammonia and nitrite levels remain below toxic thresholds (ammonia <0.1 mg/L, nitrite <0.5 mg/L).

Operation Costs: Operation costs in shrimp hatcheries include all recurring expenses associated with running the facility. These include the cost of broodstock, feed, chemicals (e.g., disinfectants, probiotics), energy (for water pumps, aerators, and lighting), labor, and water. Monitoring and controlling these costs are critical for maintaining profitability. On average, operational costs account for 60–80% of total production expenses in shrimp hatcheries.

Oxygenation: Oxygenation refers to the process of maintaining adequate dissolved oxygen (DO) levels in shrimp hatchery tanks, critical for the respiration and survival of larvae. Ideal DO levels range between 5-7 mg/L. Poor oxygenation can lead to hypoxia, causing stress and increased mortality. Oxygen levels are usually maintained through aeration systems such as air diffusers or paddlewheels, ensuring uniform distribution throughout the tanks.

Packing Methods: Packing methods for shrimp larvae or post-larvae involve careful handling to minimize stress during transport. Shrimp are typically packed in oxygenated plastic bags with clean, aerated water at an optimal temperature (24°–26°C) and salinity (30–35 ppt). Packing density ranges from 1000 to 3000 post-larvae per liter, depending on the duration of transport. Oxygen levels must be maintained above 5 mg/L to ensure high survival rates during transit.

Pathogen Control: Pathogen control involves strategies to prevent the introduction and spread of diseases within shrimp hatcheries. Methods include strict biosecurity protocols, regular disinfection of tanks and equipment, use of pathogen-free broodstock, and quarantine of new stock. PCR-based diagnostic tests are often used to detect pathogens early, and probiotics or immunostimulants can be added to feed to boost shrimp immunity.

Pathogen-Free Broodstock: Pathogen-free broodstock refers to shrimp that are certified free from specific diseases such as White Spot Syndrome Virus (WSSV) and Infectious Myonecrosis Virus (IMNV). These broodstock are critical for producing disease-resistant offspring. Broodstock are usually raised in biosecure environments, and their pathogen-free status is verified through PCR tests and regular health assessments to prevent the introduction of diseases into hatcheries

pH Regulation: pH regulation in shrimp hatcheries is crucial for maintaining optimal physiological function in larvae and preventing stress. The ideal pH range for shrimp larvae is between 7.5 and 8.5. Fluctuations outside this range can negatively affect the health of shrimp by impairing their ability to maintain osmotic balance. pH can be managed by regular monitoring and adjusting through the addition of buffers like sodium bicarbonate if needed.

Phytoplankton Culture: Phytoplankton culture in shrimp hatcheries involves growing microalgae like *Tetraselmis* and *Isochrysis*, which are used as live feed for shrimp larvae. Phytoplankton provides essential fatty acids and nutrients, supporting the development of shrimp in their early stages. The culture is maintained in tanks under controlled conditions, including light, nutrients, and water quality.

Post-Larvae: Post-larvae are the juvenile stage of shrimp that follows the larval stages. This stage is critical as post-larvae are transferred to grow-out ponds for further development. Post-larvae feed on a variety of live and formulated feeds and require optimal water conditions, including salinity (28-35 ppt), temperature (28°-30°C), and dissolved oxygen (>5 mg/L), to ensure survival and growth.

Prophylactic Measures: Prophylactic measures refer to the preventative strategies used to protect shrimp from disease outbreaks. These include maintaining optimal water quality, using pathogen-free broodstock, administering immunostimulants or probiotics, and implementing quarantine protocols. Ensuring stable salinity (30-35 ppt) and temperature (28°-30°C) is key to reducing stress in shrimp, thereby preventing opportunistic infections.

Quarantine Procedures: Quarantine procedures in shrimp hatcheries are essential for isolating newly introduced broodstock or shrimp larvae to prevent the spread of diseases. Quarantine typically lasts 14–21 days, during which the shrimp are monitored for any signs of disease or stress. Key parameters, such as water temperature (28°–30°C), salinity (30–35 ppt), and dissolved oxygen (>5 mg/L), are carefully maintained. Disinfectants like formalin or iodine may be used for prophylactic treatments, and regular health checks are conducted before introducing shrimp into the main culture system.

Rearing Conditions: Rearing conditions in shrimp hatcheries refer to the controlled environment provided for the development of larvae and post-larvae. This includes maintaining optimal water parameters like temperature (28°-30°C), salinity (28-35 ppt), pH (7.5-8.5), and dissolved oxygen (>5 mg/L). Proper feed, light regulation, and water exchange are critical for ensuring healthy larval growth and reducing the risk of disease outbreaks.

Reproductive Hormones: Reproductive hormones like gonadotropin-releasing hormone (GnRH) play a critical role in regulating shrimp reproduction. In hatcheries, the manipulation of these hormones through methods like eyestalk ablation or hormone injections stimulates gonadal maturation and spawning. Maintaining proper water quality and nutrition helps enhance the natural production of reproductive hormones, ensuring high-quality egg and sperm production.

Rotifers: Rotifers (*Brachionus plicatilis*) are microscopic zooplankton widely used as live feed in shrimp hatcheries due to their small size, high digestibility, and rich nutritional profile. They are particularly used in the zoea stage of shrimp larvae. Rotifer cultures require optimal water conditions: temperature 25°–28°C, salinity 15–25 ppt, and oxygen levels above 4 mg/L. Nutritional enrichment of rotifers with essential fatty acids like DHA ensures the proper growth and survival of shrimp larvae.

Salinity Management: Salinity management in shrimp hatcheries is essential for maintaining optimal osmotic balance in shrimp larvae. Shrimp hatcheries typically operate at salinities of 30-35 ppt, though larvae can tolerate slightly lower salinities (down to 15 ppt) during acclimation. Sudden fluctuations in salinity can cause stress, affecting growth and survival rates. Regular monitoring and gradual adjustments help maintain stable conditions.

Sanitary and Phytosanitary (SPS): Sanitary and phytosanitary (SPS) measures in aquaculture include protocols and regulations that ensure that aquatic animals and their products are safe for human consumption and free from pests and diseases. These standards, often set by international bodies like the World Trade Organization (WTO), regulate water quality, disease control measures, and feed safety. Compliance with SPS measures is essential for international trade and sustainable hatchery practices.

Sedimentation Tanks: Sedimentation tanks are used in shrimp hatcheries to remove suspended solids and particulates from water through gravitational settling. Water flows slowly through these tanks, allowing heavier particles to settle at the bottom while cleaner water is drawn off for further filtration or reuse. These tanks help reduce the organic load in hatchery systems, minimizing ammonia production and improving water quality.

Seed Collection Methods: Seed collection methods involve harvesting shrimp larvae or post-larvae from natural waters or hatcheries. Traditional methods include the use of push nets, drag nets, or sieve nets. In hatcheries, larvae are harvested using fine mesh screens or siphoning systems. The collected seed must be carefully handled to avoid physical stress and ensure their survival during transport and stocking.

Seed Grading: Seed grading refers to the process of sorting shrimp post-larvae based on size and health to ensure uniform growth rates in grow-out ponds. Grading minimizes competition for food and space, leading to better survival and growth. Shrimp seeds are typically graded using sieves or mechanical graders, with size uniformity ensuring even development in grow-out ponds.

Seed Packing: Seed packing is the process of preparing shrimp post-larvae for transport to grow-out facilities. Seeds are packed in oxygenated plastic bags filled with water at a salinity of around 30 ppt and oxygen levels above 5 mg/L. Packing density typically ranges from 2000 to 5000 post-larvae per liter, depending on the duration of transport. Proper packing ensures minimal stress and high survival rates during transit.

Seed Production: Seed production in shrimp farming refers to the process of breeding shrimp in hatcheries and nurturing them through various larval stages to post-larvae, ready for stocking in grow-out systems. The process involves the use of pathogen-free broodstock, controlled breeding conditions, and optimal feed management. Critical stages include hatching, larval rearing, and metamorphosis to post-larvae, all of which require precise water quality control (temperature 28°-30°C, salinity 30-35 ppt).

Seed Quality Evaluation: Seed quality evaluation in shrimp hatcheries involves assessing the health, growth potential, and disease resistance of shrimp post-larvae before stocking. High-quality seed should have active movement, uniform size, a well-formed gut, and no signs of deformities or infections. Seed evaluation also includes stress tests, such as salinity shock tests, to ensure robustness for grow-out stages.

Seed Transport: Seed transport involves the careful movement of shrimp post-larvae from hatcheries to grow-out ponds. Transport conditions must maintain optimal water quality, including temperature (24°-28°C), salinity (30-35 ppt), and dissolved oxygen (>5 mg/L). To prevent stress, acclimatization to pond water is essential upon arrival. Long-distance transport often requires continuous aeration and temperature control to ensure seed survival.

Selective Breeding: Selective breeding is a process used to enhance specific traits in shrimp, such as faster growth, higher fecundity, or disease resistance. By choosing the best-performing broodstock, hatcheries can gradually improve the genetic quality of their stock. This method requires careful tracking of broodstock performance and the maintenance of detailed genetic records. In shrimp hatcheries, selective breeding can significantly improve overall yield and seed quality.

Sexual Maturity: Sexual maturity in shrimp is the stage when broodstock can reproduce successfully. In most shrimp species, sexual maturity is reached when the animals are 8-12 months old, with fully developed reproductive organs. Factors influencing sexual maturity include genetics, nutrition, and environmental conditions like temperature and salinity. In hatcheries, this process is often accelerated using techniques like eyestalk ablation to enhance reproductive output.

Shrimp Hatchery Labor: Shrimp hatchery labor includes skilled and unskilled workers responsible for various tasks such as broodstock management, larval feeding, water quality monitoring, and equipment maintenance. Hatcheries may require biologists, technicians, and support staff. Efficient labor management is essential to maintain productivity. Labor costs typically constitute 20-30% of the operational budget, depending on the size of the hatchery and labor availability.

Shrimp Hatchery Standards: Shrimp hatchery standards are a set of guidelines and regulations that ensure the production of high-quality, disease-free shrimp seed. These standards encompass water quality management, broodstock selection, biosecurity protocols, and health assessments. International bodies such as the Global Aquaculture Alliance (GAA) and local regulatory authorities set these standards, which are crucial for sustainable and profitable hatchery operations.

Shrimp Hatchery: A shrimp hatchery is a controlled environment where shrimp seeds (larvae and post-larvae) are produced through artificial breeding techniques. The hatchery consists of broodstock tanks, larval rearing tanks, water filtration systems, and live feed culture units. Optimal water parameters in hatcheries include temperature 28°-30°C, salinity 28-35 ppt, pH 7.5-8.5, and dissolved oxygen >5 mg/L. Proper biosecurity, water quality, and feeding protocols are essential for healthy seed production.

Shrimp Larvae: Shrimp larvae refer to the early developmental stages (nauplii, zoea, mysis) after hatching from eggs. These stages are characterized by rapid morphological changes as larvae grow and prepare to transition to the post-larvae stage. In hatcheries, shrimp larvae are fed live feed like microalgae, rotifers, and Artemia to ensure proper nutrition. Maintaining optimal water quality (temperature 28°-30°C, salinity 30-35 ppt) is essential for their development.

Shrimp Seed Certification: Shrimp seed certification ensures that post-larvae produced in hatcheries meet specific quality standards, including freedom from pathogens, genetic robustness, and overall health. Certification agencies perform audits and tests to verify the disease-free status and genetic quality of shrimp seed. Certified shrimp seeds have higher market value and better survival rates in grow-out systems, contributing to sustainable aquaculture practices.

Shrimp Survival Rates: Shrimp survival rates refer to the percentage of shrimp that survive from larval stages to post-larvae. High survival rates (above 80%) are an indicator of successful hatchery management, including optimal water quality, feed provision, and disease prevention. Factors affecting survival rates include water quality, feeding practices, and biosecurity measures. Hatcheries aim to improve survival through better management practices (BMPs) and genetic improvement programs.

Site Selection: Site selection for shrimp hatcheries is critical for successful seed production. Factors like proximity to seawater, availability of clean and uncontaminated water, stable land conditions, and accessibility to infrastructure are essential. The site should allow for optimal water quality management (salinity 28-35 ppt, pH 7.5-8.5) and biosecurity to prevent the introduction of pathogens.

Spawning: Spawning is the process by which mature shrimp release eggs and sperm into the water for fertilization. In shrimp hatcheries, spawning is often induced through environmental manipulations or eyestalk ablation. Successful spawning requires mature broodstock, proper nutrition, and optimal water conditions (temperature 28°-30°C, salinity 30-35 ppt). Fertilized eggs hatch into nauplii within 12-24 hours.

Stress Management: Stress management in shrimp hatcheries is crucial for minimizing mortality and maintaining growth rates. Stress factors include poor water quality, overcrowding, and handling during transport. Techniques such as gradual acclimatization to new environmental conditions, reducing handling, and maintaining stable water quality parameters (temperature, salinity, pH) help minimize stress and improve larval health.

Sustainable Hatchery Practices: Sustainable hatchery practices aim to reduce environmental impacts while maintaining high productivity. This includes using energy-efficient systems, recycling water through recirculating aquaculture systems (RAS), reducing chemical usage, and maintaining genetic diversity in breeding programs. Sustainable practices not only enhance environmental conservation but also ensure long-term profitability by reducing operational costs and improving product quality.

Tank Systems: Tank systems in shrimp hatcheries are designed for the different stages of shrimp development, from broodstock holding tanks to larval and post-larval rearing tanks. Common tank sizes range from 1 to 10 tons, and they are often made from fiberglass or concrete. Tanks must have proper aeration, filtration, and water circulation systems to maintain stable environmental conditions, such as optimal temperature (28°–30°C) and salinity (30–35 ppt).

Temperature Control: Temperature control in shrimp hatcheries is critical to ensure proper metabolic rates, molting cycles, and overall growth of shrimp larvae. The optimal temperature range is 28°–30°C. Temperatures below 24°C slow larval development, while temperatures above 32°C can induce stress or lead to disease outbreaks. Automated temperature control systems help maintain stable conditions throughout the hatchery.

Transport of Broodstock: Transport of broodstock involves moving mature shrimp from one location (wild or hatchery) to another, often a hatchery facility, for breeding. Special care is needed during transport to minimize stress, maintain water quality, and ensure broodstock arrive in optimal condition for spawning. Conditions such as temperature (22°-26°C) and oxygenation are closely monitored, and sedation may be used to reduce stress during transit.

Viral Diseases: Viral diseases, such as White Spot Syndrome Virus (WSSV) and Infectious Hypodermal and Hematopoietic Necrosis Virus (IHHNV), are major causes of mortality in shrimp hatcheries. These diseases can lead to mass mortality within days if not controlled. Strict biosecurity, regular disease monitoring, and the use of pathogen-free broodstock are key measures to prevent viral outbreaks. PCR diagnostics are commonly used for early detection.

Wastewater Management: Wastewater management in shrimp hatcheries involves the treatment and disposal of used water to minimize environmental impact. Wastewater typically contains organic matter, nutrients, and pathogens. Treatment methods include sedimentation, biofiltration, and chemical disinfection. Recycling treated water in recirculating systems is an effective way to reduce water usage and maintain optimal water quality.

Water Filtration: Water filtration in shrimp hatcheries involves mechanical, biological, and chemical processes to remove particulates, harmful bacteria, and waste from the water. Mechanical filtration typically removes particles larger than 50 microns, while biological filtration facilitates the breakdown of nitrogenous wastes (ammonia and nitrite). Optimal filtration ensures that water quality remains within ideal parameters (e.g., ammonia <0.1 mg/L, nitrite <0.5 mg/L).

Water Quality Management: Water quality management in shrimp hatcheries involves maintaining optimal levels of temperature, salinity, pH, dissolved oxygen, ammonia, and nitrites. Poor water quality can lead to stress, disease outbreaks, and high mortality. Regular water exchange, biofiltration, and chemical treatments are employed to maintain water quality within acceptable ranges (e.g., temperature 28°-30°C, salinity 28-35 ppt, pH 7.5-8.5).

Water Quality Parameters: Water quality parameters are critical in shrimp hatcheries to ensure optimal health and growth of larvae. Key parameters include temperature (28°–30°C), salinity (30–35 ppt), pH (7.5–8.5), dissolved oxygen (>5 mg/L), and ammonia (<0.1 mg/L). Regular monitoring and adjustment of these parameters are crucial for maintaining a healthy rearing environment and preventing diseases or stress-related mortality.

Water Recirculation: Water recirculation in hatcheries involves the reuse of water after it has been filtered and treated. This system reduces water consumption, limits the introduction of pathogens, and maintains stable environmental conditions. Recirculating aquaculture systems (RAS) use mechanical filters, biofilters, and UV sterilization to clean the water. Recirculation systems are efficient for managing water quality parameters such as salinity (30-35 ppt) and temperature (28°-30°C).

Water Treatment Systems: Water treatment systems in shrimp hatcheries ensure the removal of impurities, pathogens, and toxic compounds from water. These systems typically include mechanical filters, biological filters, UV sterilization, and chemical treatments such as chlorination. Proper water treatment ensures that water quality parameters such as ammonia (<0.1 mg/L), nitrite (<0.5 mg/L), and pH (7.5-8.5) are maintained within optimal ranges for shrimp health.

Zooplankton Culture: Zooplankton culture refers to the controlled growth of small aquatic organisms like rotifers and copepods, which serve as live feed for shrimp larvae. Zooplankton are rich in proteins and fatty acids, making them ideal for feeding zoea and mysis stages. Proper culture conditions include maintaining optimal salinity, temperature, and feed for the zooplankton to thrive.

CHAPTER 9

CULTURE OF FISH FOOD ORGANISMS

Aeration Systems: Aeration systems are vital components of aquaculture that ensure adequate oxygen levels in water to support the respiration of cultured organisms. Aeration is typically achieved through mechanical means, such as air diffusers, paddle wheels or blowers, which introduce oxygen by agitating the water or releasing air bubbles. Optimal dissolved oxygen levels for most aquaculture species range between 5-7 mg/L. Proper aeration prevents hypoxic conditions, enhances water circulation, and aids in maintaining water quality by facilitating the breakdown of organic matter and the release of gases like carbon dioxide and ammonia.

Algal Bioreactors: Algal bioreactors are specialized systems used to grow algae in controlled environments for various applications, including aquaculture. These bioreactors provide optimal conditions for algal growth, such as controlled light, temperature, nutrient supply, and aeration. Closed photobioreactors are often used to culture microalgae for use as live feed, biofuels, or as a source of valuable compounds like omega-3 fatty acids. Algal bioreactors can be designed as tubular, flat-panel, or column systems, depending on the scale of production and the specific requirements of the algal species being cultured.

Algal Culture: Algal culture is the process of growing algae under controlled conditions to produce biomass for various applications, including as live feed in aquaculture. Algal culture requires specific environmental conditions such as light intensity (60-200 μmol photons m^{-2} s^{-1}), temperature (20°-30°C), and nutrient-rich media. Different culture systems, including open ponds, photobioreactors, and flask cultures, are used depending on the scale and species of algae being cultured. Algal culture plays a crucial role in larval rearing by providing a nutritious and natural feed source.

Algal Media: Algal media are nutrient solutions used to culture microalgae in aquaculture. These media provide essential nutrients like nitrogen, phosphorus, trace elements, and vitamins to support algal growth. Commonly used algal media include F/2 medium, Walne's medium, and Conway's medium, with specific formulations depending on the microalgal strain being cultured. Proper formulation of algal media is critical for achieving high algal productivity and maintaining the nutritional quality of the culture.

Algal Strains: Algal strains refer to the specific varieties of microalgae used in aquaculture for feeding live food organisms like Artemia and copepods. Common strains include *Chlorella vulgaris*, *Isochrysis galbana*, and *Nannochloropsis oculata*. Each strain has distinct nutritional profiles, with optimal growth conditions varying but typically requiring temperatures between 20°-25°C, light intensities of 100-200 μmol photons/m^2/s, and nutrient-rich media (e.g., F/2 medium).

Aquaculture Ponds: Aquaculture ponds are water bodies specifically designed and managed for the culture of aquatic organisms. Ponds can be either natural or artificial and vary in size and depth depending on the species being cultured and the management practices used. They are typically classified as extensive, semi-intensive, or intensive, based on the level of inputs, such as feed and fertilizers. Water quality management including aeration, temperature control, and nutrient management, is critical in aquaculture ponds to ensure optimal conditions for the growth and health of cultured species.

Aquaculture Systems: Aquaculture systems encompass various methods and technologies used to cultivate aquatic organisms, including ponds, tanks, cages, and recirculating systems. These systems can be classified as extensive (low input, relying on natural productivity) or intensive (high input, controlled environments with supplemental feeding). Each system has specific management practices for water quality, stocking density, feeding, and disease control. For example, pond systems rely on natural plankton productivity, while recirculating systems require precise control of water chemistry. The choice of aquaculture system depends on the species being cultured, environmental conditions, and economic factors.

Artemia Cysts: Artemia cysts are the dormant, encysted embryos of brine shrimp (*Artemia spp.*) that can remain viable for extended periods in dry storage. These cysts are commonly used in aquaculture as a convenient source of live feed for fish and shrimp larvae. When rehydrated and incubated under optimal conditions (25°-28°C, salinity 25-35 ppt, pH 8.0-8.5), the cysts hatch into nauplii within 24-48 hours. Artemia cysts are favored in aquaculture due to their long shelf life, ease of storage, and high nutritional content.

Artemia Hatching: Artemia hatching involves rehydrating dormant cysts in saltwater under specific conditions to induce the emergence of nauplii. Optimal hatching conditions include a salinity of 28-35 ppt, temperatures of 25°-28°C, and continuous aeration to ensure oxygenation and suspension of the cysts. Light exposure (e.g., 2000 lux) is also necessary to trigger hatching. Successful hatching results in a high yield of live nauplii, which are immediately used as feed in larval fish or shrimp culture.

Artemia in India: In India, *Artemia* is primarily used in shrimp hatcheries along the coastal regions, particularly in Gujarat, Andhra Pradesh, and Tamil Nadu. India imports large quantities of *Artemia* cysts, but local production has gained momentum through initiatives to culture and harvest *Artemia* in salt pans and salt farms. With rising demand for aquaculture feed, efforts to enhance domestic *Artemia* production are focused on optimizing conditions such as salinity and nutrient supply for higher cyst yield.

Artemia: Artemia, commonly known as brine shrimp, are small crustaceans that are extensively used in aquaculture as live feed for fish and shrimp larvae. Artemia cysts, which can be stored and hatched on demand, are a major advantage for their use. Artemia thrive in hypersaline environments, with optimal hatching conditions being a salinity of 25-35 ppt, temperatures of 25°-28°C, and pH levels of 8.0-8.5. Decapsulated cysts are often used to eliminate hatching debris, and the nauplii (newly hatched Artemia) are a rich source of protein, lipids, and essential fatty acids.

Artificial Feeds: Artificial feeds are formulated diets designed to provide the essential nutrients required for the growth and health of cultured aquatic organisms. These feeds are typically made from a combination of protein sources (e.g., fishmeal, soybean meal), lipids, vitamins, and minerals, and are available in various forms such as pellets, flakes, or crumbles. The formulation of artificial feeds is tailored to the nutritional needs of specific species and life stages. Feed conversion ratios (FCR) are used to evaluate the efficiency of artificial feeds, with lower FCR values indicating better feed utilization.

Benthic Organisms: Benthic organisms, also known as benthos, are the organisms that live on or in the bottom substrates of aquatic environments, such as mud, sand, or rocks. These include a wide range of species, such as worms, mollusks, crustaceans, and certain fish. In aquaculture, benthic organisms can be both cultured species (e.g., shrimp, clams) and natural food sources for other species. The health of benthic communities is a key indicator of water and sediment quality, and benthic monitoring is often part of aquaculture management practices to assess the impacts of farming activities on the environment.

Biofloc Technology: Biofloc technology (BFT) is an aquaculture method that enhances water quality by promoting the growth of beneficial microbial communities, which convert waste products such as ammonia and uneaten feed, into microbial protein. This microbial floc becomes an additional food source for cultured organisms, improving feed efficiency and reducing the need for water exchange. Biofloc systems typically require high stocking densities, continuous aeration, and controlled carbon-to-nitrogen (C) ratios to maintain water quality and optimize microbial growth. BFT is widely used in shrimp and tilapia farming to enhance sustainability and reduce environmental impacts.

Biosecurity: Biosecurity in aquaculture refers to the measures taken to prevent the introduction and spread of pathogens, parasites, and other harmful organisms within and between aquaculture facilities. Biosecurity practices include quarantine procedures, disinfection protocols, water treatment, and the use of pathogen-free stock. Strict biosecurity measures are critical to minimizing the risk of disease outbreaks, which can lead to significant economic losses and impact the sustainability of aquaculture operations. For example, footbaths, restricted access to facilities, and regular health monitoring are common biosecurity practices in fish hatcheries.

Bioturbation: Bioturbation refers to the process by which benthic organisms, such as worms, burrowing shrimp, and crabs, disturb and mix sediments in aquatic environments. This activity is important in aquaculture because it helps to oxygenate the sediments, prevent the accumulation of toxic gases, and promote nutrient cycling. In ponds or tanks, bioturbation by natural or introduced organisms can improve sediment health and support the growth of cultured species. However, excessive bioturbation by certain species can also lead to sediment instability and water quality issues, requiring careful management in aquaculture systems.

Chironomid Larvae: Chironomid larvae, also known as bloodworms, are the aquatic larval stage of non-biting midges and are a common live feed in aquaculture for fish and shrimp. They are typically red due to hemoglobin-like proteins, which allow them to thrive in low-oxygen environments. Chironomid larvae are cultured in organic-rich sediments with water temperatures of 20°-25°C. These larvae are rich in protein and lipids and are commonly used in the aquarium trade and as a supplementary feed in aquaculture.

Cladocerans: Cladocerans, commonly known as water fleas, are a group of small crustaceans used as live feed in aquaculture, especially for larval and juvenile fish. Daphnia and Moina are popular cladoceran species used in aquaculture. They thrive in freshwater at temperatures between 18°-24°C and pH levels of 6.5-8.5. These organisms feed on phytoplankton and detritus, making them a nutritious option for fish larvae due to their high protein and lipid content. Cladocerans are often cultured in outdoor ponds or controlled indoor tanks.

Collection Methods: Collection methods in the context of fish food organisms refer to the various techniques used to gather live feeds such as plankton, worms, and other small aquatic organisms. Common methods include the use of plankton nets, sieves, and pumps to capture organisms from natural water bodies or cultured environments. Effective collection requires attention to factors like mesh size, water flow, and minimizing

Commercial Live Food: Commercial live food includes various live organisms, such as *Artemia*, rotifers, copepods, and microalgae, that are mass-produced and sold to aquaculture facilities for feeding larval fish and invertebrates. The quality of commercial live food is determined by factors like nutritional content, survival rates, and ease of cultivation. These live feeds are critical for supporting early life stages of aquaculture species, where conventional formulated feeds are unsuitable.

Copepod Classification: Copepods belong to the class Maxillopoda and subclass Copepoda, comprising over 12,000 species classified into three major orders: Calanoida, Cyclopoida, and Harpacticoida. Calanoids are primarily pelagic and are widely used in marine aquaculture due to their higher nutritional content, while cyclopoids and harpacticoids are often benthic or associated with detritus and are more suitable for freshwater aquaculture systems.

Copepod Culture: Copepods are a vital live feed in aquaculture, especially for the larviculture of marine fish species due to their high nutritional content, including essential fatty acids and carotenoids. Copepod culture requires controlled conditions such as a temperature range of 20°-26°C, salinity levels of 25-35 ppt, and an adequate supply of microalgae (e.g., *Isochrysis galbana*). Copepods are often cultured in batch or continuous systems, with careful management of water quality to prevent crashes in the population.

Copepods: Copepods are small crustaceans that serve as a primary food source for many marine fish larvae due to their high nutritional content, including essential fatty acids like DHA and EPA. Copepods are generally cultured at temperatures of 20°-25°C, with salinities ranging from 20-35 ppt for marine species. These organisms are a vital component of marine aquaculture systems, particularly in the early stages of larval development, where they contribute to higher survival and growth rates compared to other live feeds.

Culture Media Sterilization: Culture media sterilization is the process of eliminating all forms of microbial life, including bacteria, viruses, and fungi, from the media used to culture live food organisms in aquaculture. Sterilization methods include autoclaving, filtration, and chemical treatments. Sterilization is essential to prevent contamination and ensure the success of live food cultures, such as microalgae, Artemia, and rotifers. Sterile culture media provides a controlled environment that supports the growth of the desired species while minimizing competition from unwanted organisms.

Culture Techniques: Culture techniques in aquaculture refer to the methods used to grow and maintain live food organisms under controlled conditions. These techniques vary depending on the species and include batch culture, continuous culture, and semi-continuous culture systems. For example, microalgae can be cultured using a photobioreactor system that provides continuous light and controlled nutrient inputs. The choice of culture technique affects the growth rate, biomass yield, and overall quality of the cultured organisms, making it a critical aspect of aquaculture operations.

Culture Units: Culture units in aquaculture refer to the systems or structures used to grow live feed organisms, fish, or other aquatic species. These can range from small laboratory-scale tanks for microalgae to large outdoor ponds for *Artemia*. The design and operation of culture units depend on the species being cultured, with considerations for water quality management, aeration, and space requirements.

Decapsulation Technique: Decapsulation is a technique used to remove the hard outer shell (chorion) of Artemia cysts without affecting the viability of the embryos inside. This process involves treating the cysts with a concentrated solution of sodium hypochlorite (bleach) under controlled conditions. Decapsulated Artemia cysts are preferred in larviculture because they eliminate the risk of shell ingestion by fish larvae and improve hatchability. The process also reduces the risk of bacterial contamination.

Decapsulation: Decapsulation is the process of removing the hard outer shell (chorion) of Artemia cysts to produce clean, debris-free nauplii for use as live feed. This process involves treating the cysts with a solution of sodium hypochlorite (bleach) to dissolve the shell without harming the embryo inside. Decapsulated Artemia cysts can be directly fed to fish larvae or hatched to release nauplii. Decapsulation reduces the risk of contamination and improves the nutritional quality of the feed by eliminating the indigestible cyst shell.

Detritivores: Detritivores are organisms that feed on decomposing organic matter, such as dead plants, animals, and fecal material. In aquaculture, detritivores like certain species of shrimp, worms, and snails are important for breaking down waste and recycling nutrients within culture systems. They help maintain water quality by reducing the accumulation of organic matter on the pond or tank bottom. Detritivores are often incorporated into polyculture systems to improve the overall efficiency of nutrient utilization and enhance the sustainability of aquaculture operations.

Earthworms: Earthworms, particularly species like *Eisenia fetida*, are widely used in aquaculture as a protein-rich live feed for fish. Earthworms are known for their high protein content (50-60% dry weight) and essential amino acids. They can be cultured in vermiculture setups using organic waste as feed, under optimal conditions of 20°-25°C and a moist environment with a pH of 6.5-7.5. Earthworms are particularly favored for their ease of cultivation, rapid reproduction, and ability to convert organic waste into high-quality protein.

Epifauna: Epifauna refers to benthic organisms that live on the surface of the substrate, rather than burrowing into it. Examples include oysters, barnacles, and certain species of crabs and starfish. In aquaculture, epifauna may be cultured directly, as in the case of oyster farming, or they may be part of the natural ecosystem that supports other cultured species. Epifauna play important roles in the cycling of nutrients and organic matter in aquatic ecosystems, and their presence can influence the overall health and productivity of aquaculture systems.

Eutrophication: Eutrophication is the process by which water bodies become enriched with excess nutrients, particularly nitrogen and phosphorus, leading to the overgrowth of algae and other aquatic plants. In aquaculture, eutrophication can result from the accumulation of uneaten feed, fish waste, and fertilizers, leading to harmful algal blooms, oxygen depletion, and the deterioration of water quality. Managing nutrient inputs and implementing practices such as biofiltration, water exchange, and integrated multitrophic aquaculture can help mitigate the risk of eutrophication in aquaculture systems.

Feed Conversion Ratio: Feed conversion ratio (FCR) is a key performance indicator in aquaculture that measures the efficiency of feed utilization, expressed as the ratio of feed input (in kilograms) to biomass gain (in kilograms). For example, an FCR of 1.5 means that 1.5 kg of feed is required to produce 1 kg of fish. Lower FCR values indicate more efficient feed utilization and reduced waste. Achieving optimal FCRs requires careful management of feeding practices, feed quality, and environmental conditions. Species like tilapia have FCRs around 1.2-1.8, while carnivorous species like salmon may have FCRs of 1.0-1.2.

Filter Feeders: Filter feeders are aquatic organisms that obtain their food by filtering suspended particles, such as plankton, detritus, and organic matter, from the water. Examples of filter feeders include bivalves (e.g., oysters and mussels), certain fish species (e.g., silver carp), and zooplankton. In aquaculture, filter feeders play a crucial role in water quality management by removing excess nutrients and organic matter, which helps prevent eutrophication. They are often used in integrated multitrophic aquaculture (IMTA) systems to enhance sustainability by recycling waste and contributing to overall ecosystem balance.

Fish Larvae Feed: Fish larvae feed consists of small, nutritionally dense live feeds like Artemia, rotifers, and copepods, which are essential for the early developmental stages of fish. These feeds provide high levels of protein, essential fatty acids, and other nutrients necessary for the growth and survival of fish larvae. The size and nutritional profile of the feed must match the requirements of the larvae, as they often have limited feeding capabilities and specific dietary needs during their initial growth phases.

Food Webs: Food webs represent the complex network of feeding relationships in an ecosystem, illustrating how energy and nutrients flow between different trophic levels. In aquaculture, understanding food webs is essential for optimizing the management of natural food resources and ensuring a balanced ecosystem within culture systems. For example, in a pond system, primary producers like phytoplankton are consumed by zooplankton, which in turn are eaten by fish. Disruptions in the food web, such as overstocking or poor water quality, can lead to imbalances that affect the health and productivity of cultured organisms.

Freshwater Species: Freshwater species in aquaculture refer to fish, invertebrates, or plants that are cultured in freshwater systems. Common freshwater aquaculture species include tilapia, catfish, and ornamental fish such as guppies. These species require specific water quality parameters, such as a pH range of 6.5-8.0 and dissolved oxygen levels above 5 mg/L. The culture of freshwater species can occur in ponds, tanks, or raceways, depending on the species and production scale.

Growth Pattern: Growth patterns in aquaculture refer to the rate and nature of growth observed in cultured organisms over time. Growth can be influenced by factors such as food availability, water quality, stocking density, and temperature. For example, copepods exhibit exponential growth under optimal conditions, while fish species like tilapia show sigmoidal growth patterns. Understanding these patterns is key to optimizing feed strategies and harvest timings.

Harvesting Techniques: Harvesting techniques in the culture of live food organisms involve methods used to collect mature organisms from culture systems. Common techniques include filtration, centrifugation, and sedimentation for microalgae, while mesh sieves and siphoning are used for zooplankton and worms. Effective harvesting ensures the maximum yield of high-quality organisms while minimizing stress or mortality. For example, microalgae are often harvested by flocculation, where a chemical agent is added to clump cells together for easier collection.

Identification: The identification of plankton involves distinguishing species based on morphological characteristics using microscopes and taxonomic keys. In aquaculture, accurate identification of phytoplankton and zooplankton species is crucial for assessing the nutritional suitability of live feed and monitoring potential harmful species (e.g., toxic algal blooms). Techniques such as DNA barcoding are also increasingly used for precise species identification.

Infauna: Infauna are benthic organisms that live within the substrate of aquatic environments, such as worms, clams, and burrowing shrimp. These organisms play a crucial role in bioturbation, the process of mixing and oxygenating sediments, which helps to maintain sediment health and water quality. In aquaculture, infauna can be both beneficial, as part of the natural food web, and problematic, if they compete with cultured species or disturb sediments. Managing the balance of infauna in culture systems is important for ensuring optimal conditions for aquaculture production.

Infusoria: Infusoria is a collective term for small, ciliated protozoans that are often used as a first feed for very small fish larvae. These microorganisms can be cultured by allowing organic matter, such as hay or vegetable matter, to decompose in water, encouraging the growth of bacteria and subsequently infusoria. Optimal culture conditions include room temperature (20°-25°C) and an oxygen-rich environment. Infusoria are rich in protein and are ideal for feeding newly hatched fish that require small, motile prey.

Isolation Methods: Isolation methods are techniques used to separate and purify specific species or strains of live food organisms from mixed cultures or natural samples. These methods include serial dilution, streak plating, centrifugation, and filtration. Isolation is particularly important in algal culture to prevent contamination and maintain monocultures. For example, microalgae can be isolated by streaking samples on agar plates and selecting single colonies for further culture. Proper isolation ensures the purity and quality of live feed cultures in aquaculture.

Isolation of Live Feed: The isolation of live feed involves selecting specific organisms from mixed cultures or wild collections to create pure cultures for aquaculture. This can be done using methods like micropipetting, filtration, or centrifugation. Isolating live feed species like Artemia or copepods ensures that the cultured organisms are free from contaminants or predators, which is essential for reliable and consistent production.

Larviculture: Larviculture refers to the culture of fish or invertebrate larvae in controlled aquaculture systems. This critical phase involves providing optimal environmental conditions and nutrition to ensure high survival rates and rapid growth of larvae. Parameters such as water temperature, salinity, pH, and dissolved oxygen are closely monitored, along with the provision of suitable live feed (e.g., Artemia, rotifers) to meet the nutritional demands of the developing larvae.

Life Cycle: The life cycle in aquaculture describes the various developmental stages an organism undergoes from conception to maturity. For *Artemia*, the life cycle involves hatching from cysts, developing into nauplii, progressing through several molts to become adults, and then reproducing either ovoviviparously or by producing encysted embryos. Environmental factors such as salinity (optimal range of 30-35 ppt for Artemia) and temperature (20°-30°C) play a critical role in the duration and success of each life stage.

Light Intensity: Light intensity, measured in μmol photons m^{-2} s^{-1}, plays a critical role in aquaculture, particularly in the culture of photosynthetic organisms such as microalgae and plants. Optimal light intensities vary among species; for example, microalgae like Chlorella may require light intensities of 60-200 μmol photons m^{-2} s^{-1} for efficient photosynthesis and growth. Inadequate or excessive light can limit growth and productivity. Artificial lighting, such as LED systems, is often used in indoor aquaculture systems to control light intensity and ensure consistent growth conditions.

Live Food Enrichment: Live food enrichment is the process of enhancing the nutritional quality of live food organisms, such as rotifers, Artemia, and copepods, to improve their nutritional value for larval fish and other cultured species. Enrichment typically involves feeding the live food organisms with diets high in essential fatty acids (e.g., DHA, EPA), vitamins, and minerals, which are then transferred to the cultured species upon consumption. Live food enrichment is particularly important in larval rearing, where the nutritional requirements of the larvae are critical for survival and growth.

Live Food Organisms: Live food organisms are small aquatic organisms cultured to feed larval and juvenile stages of fish and shellfish in aquaculture. These organisms, including phytoplankton, zooplankton, and worms, are preferred due to their natural movement, size, and nutritional profile, which promote better growth and survival rates in fish larvae. Cultivation of live food requires precise control of environmental factors such as light, temperature, and water quality. For example, microalgae thrive at temperatures between 22°-28°C and light intensities of 60-100 μmol photons m^{-2} s^{-1}.

Marine Species: Marine species in aquaculture are those that are cultured in seawater environments. This includes species like shrimp, sea bass, and bivalves (e.g., oysters). Optimal salinity for marine species ranges between 30-35 ppt, and water temperatures typically range from 20°-28°C, depending on the species. Marine aquaculture often involves more complex systems, such as recirculating aquaculture systems (RAS) or sea cages, to maintain the required water quality and environmental conditions.

Mass Culture: Mass culture refers to the large-scale production of live food organisms, such as microalgae, rotifers, or Artemia, to meet the demands of aquaculture operations. These cultures are typically grown in large outdoor ponds, raceways, or photobioreactors, with controlled inputs of light, nutrients, aeration, and temperature. For example, microalgae mass cultures may be maintained at temperatures of 24°-30°C with continuous light exposure and nutrient supplementation. The goal of mass culture is to produce large quantities of live feed in a cost-effective and sustainable manner.

Media Preparation: Media preparation in aquaculture involves creating nutrient-rich solutions or substrates that support the growth of live food organisms, particularly microalgae and bacteria. These media typically contain sources of nitrogen, phosphorus, trace minerals, and vitamins. For microalgae culture, commonly used media include F/2, Walne's, and Conway's, which vary in nutrient concentration and composition. Media preparation requires careful sterilization to prevent contamination, and the composition of the media must be tailored to the specific requirements of the cultured species.

Microalgae Estimation: Microalgae estimation is a process used in aquaculture to determine the concentration of microalgae in culture systems, often measured in cells per milliliter. Techniques such as hemocytometer counting, spectrophotometry, or flow cytometry are used to estimate cell density, which is essential for monitoring algal culture health and ensuring adequate feed availability for live food organisms.

Microalgae Growth: Microalgae growth in aquaculture refers to the proliferation of algae under controlled conditions to produce biomass for feeding live food organisms like Artemia or copepods. Factors influencing microalgae growth include light intensity (optimally 100-200 µmol photons/m^2/s), nutrient availability, pH (around 7.5-8.5), and temperature (20°-25°C). Monitoring and optimizing these parameters are essential to ensure high yields of microalgae with desirable nutritional qualities.

Microalgae: Microalgae are single-celled photosynthetic organisms that play a fundamental role in aquatic ecosystems and aquaculture, primarily as a food source for zooplankton and filter-feeding fish. Commonly cultured microalgae species include Chlorella, Spirulina, and Tetraselmis. Their growth is optimized under conditions of 24°-28°C, light intensities of 60-200 µmol photons m^{-2} s^{-1}, and pH levels between 7.5-8.5. Nutrient-rich media, particularly with adequate nitrogen and phosphorus, are necessary for their growth, making them an essential component of aquaculture systems.

Moina micrura: *Moina micrura* is a species of water flea commonly used in freshwater aquaculture as a live feed for fish larvae due to its high protein content and digestibility. It thrives in water with temperatures ranging from 20°-30°C, a pH of 6.5-8.5, and moderate light. *Moina* can reproduce rapidly under favorable conditions, making it a reliable source of live feed in both hatchery and grow-out systems.

Monoculture Systems: Monoculture systems in aquaculture refer to the practice of cultivating a single species in a culture system. Monoculture allows for specialized management practices tailored to the specific needs of the species, such as feed formulation, water quality control, and disease prevention. While monoculture can lead to higher production efficiency, it also increases the risk of disease outbreaks and environmental impacts due to the concentration of waste products. Common examples of monoculture systems include shrimp farms, tilapia ponds, and salmon net pens.

Morphology: Morphology refers to the physical structure and form of an organism, including aspects such as shape, size, and anatomy. In aquaculture, understanding the morphology of live food organisms helps in their identification, culture, and utilization. For example, the morphology of rotifers, with their characteristic ciliated corona, is important for their feeding behavior and suitability as live feed. Morphological studies also help in distinguishing between different species and strains, which can have different nutritional profiles and growth characteristics.

Natural Food Organisms: Natural food organisms, including plankton, worms, and benthic invertebrates, are the naturally occurring feed available in aquatic environments. In extensive and semi-intensive aquaculture systems, natural food plays a crucial role in providing essential nutrients for cultured species, particularly in the early life stages. For example, phytoplankton and zooplankton are primary food sources for larval fish and shrimp. Enhancing natural food availability through pond fertilization or water management can reduce the need for artificial feeds and improve growth rates and survival in aquaculture operations.

Nematodes: Nematodes are microscopic roundworms that are used in aquaculture as live feed, especially for small fish larvae and shrimp post-larvae. They thrive in moist environments with temperatures ranging from 20°-30°C and pH levels of 6.0-8.0. Free-living nematodes, such as *Panagrellus redivivus*, are commonly cultured on agar plates or in liquid media enriched with bacteria or yeast. Nematodes are highly nutritious, containing proteins, lipids, and essential fatty acids, making them a suitable first feed for many aquaculture species.

Nitrogen Cycle: The nitrogen cycle in aquaculture refers to the transformation of nitrogen compounds in water, which is critical for maintaining water quality. Nitrogen enters the system through feed and waste products, primarily as ammonia, which is toxic to aquatic organisms at high concentrations. The nitrogen cycle involves the conversion of ammonia to nitrite and then to nitrate by nitrifying bacteria, a process known as nitrification. Denitrification, carried out by anaerobic bacteria, converts nitrates into nitrogen gas, removing nitrogen from the system. Managing the nitrogen cycle is essential in aquaculture to prevent the accumulation of toxic nitrogen compounds and ensure a healthy environment for cultured species.

Nutrient Composition: Nutrient composition refers to the proportion of proteins, lipids, carbohydrates, vitamins, and minerals in live feed or cultured species. For instance, Artemia nauplii are rich in essential fatty acids like EPA and DHA, making them a valuable feed for larval fish. Nutrient composition can vary depending on the diet provided to the live feed organisms and their culture conditions. Ensuring optimal nutrient content is critical for supporting the health and growth of aquaculture species.

Nutrient Cycling: Nutrient cycling in aquaculture refers to the movement and transformation of nutrients like nitrogen and phosphorus within the system. Effective nutrient cycling is crucial for maintaining water quality and supporting the growth of primary producers like phytoplankton. In integrated aquaculture systems, waste products from fish can be used as nutrients for plants (e.g., in aquaponics), thereby closing the nutrient loop and reducing the need for external inputs while minimizing environmental impact.

Nutrient Enrichment: Nutrient enrichment in aquaculture refers to the addition of essential nutrients, such as nitrogen, phosphorus, and trace minerals, to support the growth of cultured organisms, including live food organisms like microalgae. Nutrient enrichment is a common practice in pond fertilization, where organic or inorganic fertilizers are added to stimulate the growth of phytoplankton, which serves as food for zooplankton and fish. However, excessive nutrient enrichment can lead to water quality problems, such as eutrophication, making it important to balance nutrient inputs with the needs of the culture system.

Nutritional Value: The nutritional value of live food organisms refers to their content of essential nutrients such as proteins, lipids, carbohydrates, vitamins, and minerals. In aquaculture, organisms like Artemia, rotifers, and microalgae are highly valued for their rich nutritional profiles, which are critical for the growth and survival of fish larvae. For instance, Artemia nauplii contain about 50-60% protein and 10-20% lipids, making them a highly nutritious feed for marine fish larvae. Nutritional value can be enhanced through enrichment techniques to improve the content of essential fatty acids like DHA and EPA.

pH Levels: pH levels refer to the measure of acidity or alkalinity of water, with a range from 0 (acidic) to 14 (alkaline), and a neutral value at 7. In aquaculture, maintaining the optimal pH level is crucial for the health and growth of cultured species. Most freshwater fish thrive in a pH range of 6.5-8.0, while marine species prefer slightly higher pH levels of 7.8-8.5. pH fluctuations can stress organisms, affect metabolic processes, and alter the solubility of nutrients and toxic compounds like ammonia, necessitating careful monitoring and control in aquaculture systems.

Phosphorus Cycle: The phosphorus cycle in aquaculture involves the movement of phosphorus through water, sediments, and living organisms. Phosphorus is a key nutrient for the growth of aquatic plants and algae, but excessive phosphorus can lead to eutrophication and water quality issues. In aquaculture systems, phosphorus enters the cycle through feed and fertilizers and is taken up by primary producers like phytoplankton. Managing phosphorus levels is crucial to prevent the buildup of phosphorus in sediments and reduce the risk of algal blooms, which can deplete oxygen and harm cultured species.

Photoperiod: Photoperiod refers to the duration of light and darkness that an organism is exposed to in a 24-hour cycle. In aquaculture, the photoperiod is carefully controlled to regulate the growth, reproduction, and feeding behavior of cultured species, including live food organisms like microalgae and zooplankton. For example, a photoperiod of 16 hours of light and 8 hours of darkness is commonly used in microalgae culture to optimize photosynthesis and biomass production. Photoperiod manipulation can also influence spawning in species like fish and shrimp.

Phytoplankton Bloom: A phytoplankton bloom is a rapid increase in the population of phytoplankton, microscopic photosynthetic organisms, in a water body. Blooms occur when conditions are favorable, such as the availability of nutrients (nitrogen and phosphorus), optimal light, and suitable temperature. In aquaculture, phytoplankton blooms can provide a rich source of natural food for filter-feeding organisms like bivalves and zooplankton. However, excessive blooms, particularly of harmful algal species, can lead to oxygen depletion, the release of toxins, and other water quality problems that affect the health of cultured species.

Phytoplankton: Phytoplankton are microscopic photosynthetic organisms that serve as the primary producers in aquatic ecosystems. They form the base of the food chain and are crucial in aquaculture as a food source for many fish and invertebrate larvae. Optimal growth of phytoplankton is influenced by water parameters such as light intensity (usually between 50-200 μmol photons m^{-2}

s^{-1}), temperature (20°-30°C), pH (7.0-8.5), and nutrient availability, particularly nitrogen and phosphorus. Commonly cultured phytoplankton species include Chlorella, Isochrysis, and Nannochloropsis.

Plankton Blooms: Plankton blooms are rapid increases in the population of plankton, particularly phytoplankton, in water bodies. In aquaculture, controlled phytoplankton blooms are encouraged in ponds to enhance natural food availability for fish and invertebrates. However, excessive or harmful blooms (e.g., caused by toxic algae like *Karenia brevis*) can deplete oxygen levels, produce toxins, and lead to fish kills. Monitoring nutrient levels (e.g., nitrogen and phosphorus) and water quality is essential to managing blooms.

Plankton Collection: Plankton collection in aquaculture involves gathering microscopic organisms like phytoplankton and zooplankton that serve as a primary food source in larviculture. Collection techniques include using plankton nets with mesh sizes ranging from 10-200 µm, depending on the target organism. The success of collection depends on factors like water movement, plankton density, and the time of day, as many species exhibit diel vertical migration.

Plankton Density: Plankton density refers to the concentration of plankton, including both phytoplankton and zooplankton, in a given volume of water, usually expressed as cells or organisms per milliliter or liter. In aquaculture, managing plankton density is important for optimizing the availability of natural food for filter feeders and larval fish. High plankton densities can support rapid growth and survival of cultured species, but excessive densities may lead to oxygen depletion and water quality problems. Plankton density is often monitored through regular sampling and microscopic analysis to ensure balanced ecosystem dynamics in culture systems

Polyculture Systems: Polyculture systems involve the simultaneous cultivation of multiple species in the same culture system, often at different trophic levels. This approach can enhance resource utilization, improve water quality, and reduce the risk of disease outbreaks. For example, fish and shrimp may be cultured together with filter feeders like bivalves or seaweed, which absorb excess nutrients and help maintain water quality. Polyculture systems mimic natural ecosystems and are often more sustainable and resilient than monoculture systems, particularly in extensive or semi-intensive farming operations.

Primary Productivity: Primary productivity in aquaculture refers to the rate at which photosynthetic organisms, primarily phytoplankton, convert solar energy into biomass. This process is fundamental in pond-based aquaculture systems, where phytoplankton serves as the base of the food chain. Factors such as light availability, nutrient concentrations, and temperature influence primary productivity. High primary productivity can enhance the growth of zooplankton and, subsequently, fish and shrimp, but must be balanced to prevent over-enrichment of the system.

Prospectus of Artemia: Artemia species, commonly known as brine shrimp, are highly valued in aquaculture due to their high nutritional content and ease of cultivation. Their cysts are widely used for larval fish and shrimp feed. The prospects of Artemia in aquaculture include their role as a live food source rich in essential fatty acids, proteins, and micronutrients. Given their adaptability to extreme saline conditions, Artemia can be cultured in a range of environments, making them an essential component in hatchery and nursery operations.

Proximate Composition: Proximate composition refers to the analysis of the basic nutritional components of an organism or feed, including moisture, protein, lipid, ash (minerals), and carbohydrate content. This analysis is crucial in aquaculture to ensure that live feeds or formulated diets meet the nutritional needs of cultured species. For example, high-quality live feeds like Artemia typically contain 40-60% protein, 10-20% lipids, and 20-40% ash. Understanding proximate composition helps optimize the growth, health, and survival of aquaculture species.

Purification Techniques: Purification techniques in aquaculture involve isolating and maintaining pure cultures of microalgae or live feed organisms to prevent contamination from other species or pathogens. Common methods include serial dilution, streak plating, and the use of antibiotics or antifungal agents. Maintaining sterile conditions during culture preparation is essential for the success of these techniques.

Recirculating Aquaculture: Recirculating aquaculture systems (RAS) are highly controlled, closed-loop systems that reuse water by filtering and treating it to remove waste products and maintain water quality. RAS technology allows for high-density fish farming in land-based facilities, minimizing water usage and environmental discharge. Components of RAS include mechanical filters, biofilters, UV sterilizers, and oxygenation systems. Maintaining optimal water quality parameters, such as ammonia levels below 0.5 mg/L and dissolved oxygen above 6 mg/L, is essential for the success of these systems. RAS is particularly valuable in areas with limited water resources or stringent environmental regulations.

Reproduction: In aquaculture, reproduction refers to the process by which organisms produce offspring, either sexually or asexually, ensuring the continuity of species and enabling the mass production of fish or other aquatic species. For species like Artemia or *Moina micrura*, successful reproduction is crucial for maintaining a stable supply of live feed. Optimal conditions for reproduction include water temperatures between 25°-30°C, pH levels ranging from 7.0-8.5, and adequate food supply (e.g., microalgae). Light intensity and photoperiod can also significantly influence reproductive success, especially in live food organisms.

Rotifers: Rotifers are small, wheel-bearing planktonic organisms that are widely used in aquaculture as live feed for fish and shrimp larvae, particularly in marine hatcheries. Their ability to reproduce quickly and their small size (50-300 microns) make them ideal for feeding delicate larvae. Rotifers are cultured under conditions of 20°-28°C, with a salinity range of 15-25 ppt for marine species, and pH levels of 7.0-8.0. They are often enriched with essential fatty acids like DHA and EPA before being fed to larvae.

Salinity: Salinity refers to the concentration of dissolved salts in water, typically measured in parts per thousand (ppt). It is a critical parameter in aquaculture, affecting osmoregulation, growth, and survival of cultured species. Marine species thrive in salinities around 30-35 ppt, while freshwater species require lower salinities (0-5 ppt). Euryhaline species, like certain fish and crustaceans, can tolerate a wide range of salinities. Controlling salinity is particularly important in brackish water systems and during acclimation processes, ensuring that organisms can maintain their internal osmotic balance.

Sexual Dimorphism: Sexual dimorphism refers to the differences in appearance between males and females of a species, including differences in size, coloration, and morphology. In aquaculture, recognizing sexual dimorphism is important for breeding programs and the management of live food organisms. For example, in Artemia, males are typically smaller with specialized appendages for clasping females during mating, while females carry eggs in a brood pouch. Understanding sexual dimorphism helps in identifying the reproductive stages of organisms and optimizing culture conditions for breeding.

Spirulina Culture: Spirulina culture involves the cultivation of Spirulina, a blue-green algae (cyanobacteria), in controlled environments for use in aquaculture and human nutrition. Spirulina is typically cultured in alkaline water (pH 8.5-11) with high temperatures (30°-35°C) and abundant sunlight or artificial light. The culture is often carried out in open raceway ponds or closed photobioreactors. Spirulina is harvested through filtration or sedimentation and processed for use as a protein-rich feed supplement, particularly for fish and shrimp larvae.

Spirulina Harvesting: Spirulina, a cyanobacterium, is commonly harvested from culture systems for use as a nutritional supplement in aquaculture. Harvesting methods typically involve filtration, centrifugation, or flotation techniques to separate the biomass from the culture medium. Spirulina harvesting is followed by drying and processing, ensuring the preservation of its high protein and essential fatty acid content. The timing of harvest is crucial, as it directly affects yield and quality.

Spirulina Nutrition: Spirulina is a highly nutritious microalga used in aquaculture due to its high protein content (up to 70% by dry weight), essential amino acids, vitamins (B_{12}, E), and fatty acids like gamma-linolenic acid (GLA). It also contains pigments like phycocyanin and chlorophyll, which have antioxidant properties. Spirulina is often included in fish diets to enhance growth, immunity, and pigmentation, making it a versatile feed supplement in both freshwater and marine aquaculture.

Spirulina Processing: After harvesting, Spirulina undergoes various processing steps to convert it into a usable product for aquaculture. This typically includes washing, drying (usually by spray drying), and grinding into a powder form. Spirulina processing aims to retain its high protein content, antioxidants (e.g., phycocyanin), and essential vitamins, making it a valuable supplement for fish feed, particularly for species requiring high-quality protein sources.

Spirulina: Spirulina is a genus of blue-green algae (cyanobacteria) known for its high protein content (up to 60-70% dry weight), essential fatty acids, vitamins, and minerals. It is cultivated in aquaculture as a nutritional supplement for fish and shrimp larvae. Spirulina thrives in alkaline water (pH 8.5-11) with temperatures ranging between 30°-35°C. It can be cultured in open raceway ponds or closed photobioreactors with adequate light exposure (100-200 µmol photons m^{-2} s^{-1}). The harvested biomass is dried and processed for use in aquafeeds.

Stock Culture Inoculation: Stock culture inoculation in aquaculture involves introducing a small, controlled population of live food organisms, such as microalgae or zooplankton, into a larger culture system to initiate growth. This process ensures that the culture system is populated with the desired species, while minimizing the risk of contamination by unwanted organisms. Stock cultures are typically maintained under optimal conditions in the laboratory and are periodically inoculated into larger tanks or ponds for mass culture. Proper inoculation techniques are crucial for establishing healthy and productive cultures in aquaculture operations.

Stock Culture Maintenance: Stock culture maintenance involves the long-term cultivation of live feed organisms or microalgae under controlled conditions to ensure a continuous supply for aquaculture operations. This typically involves periodic subculturing, maintaining optimal environmental conditions (e.g., temperature, pH, and light), and monitoring for contamination. Proper stock culture maintenance is crucial for preventing crashes in production and ensuring the health of larvae and juvenile fish.

Stock Culture: A stock culture is a small, controlled culture of live food organisms that is maintained to serve as a continuous source of inoculum for mass cultures. Stock cultures are kept under optimal and stable conditions to preserve the health and genetic integrity of the organisms. For microalgae, stock cultures are typically maintained at lower light intensities (20-50 µmol photons m^{-2} s^{-1}) and temperatures (18°-24°C) in small volumes to minimize contamination and overgrowth. Stock cultures are regularly subcultured to ensure long-term viability.

Stocking Density: Stocking density in aquaculture refers to the number of organisms stocked per unit volume of water. Optimal stocking density is species-specific and influences growth rates, feed efficiency, and water quality. For example, in shrimp aquaculture, stocking densities typically range from

10-30 post-larvae per square meter in extensive systems to over 100 per square meter in intensive systems. High stocking densities can lead to stress, increased disease risk, and poor water quality, requiring more intensive management practices such as aeration and biofiltration.

Taxonomy: Taxonomy is the scientific discipline of classifying and naming organisms based on their evolutionary relationships and characteristics. In the context of aquaculture, taxonomy is important for identifying and categorizing live food organisms such as microalgae, zooplankton, and worms. Accurate taxonomy helps in the selection of appropriate species for culture, understanding their ecological roles, and optimizing their culture conditions. For example, understanding the taxonomy of rotifers allows for the selection of species with desirable traits for larval fish feeding.

Temperature Regulation: Temperature regulation in aquaculture is crucial for maintaining optimal growth rates, metabolic functions, and reproductive cycles of cultured species. Different species have specific thermal preferences; for instance, tilapia grow best at temperatures between 25°-30°C, while cold-water species like trout prefer temperatures around 10°-18°C. Temperature also affects dissolved oxygen levels, with warmer water holding less oxygen. Temperature regulation methods include the use of heaters, chillers, or geothermal systems, as well as shading and water flow adjustments to stabilize conditions in both outdoor and indoor aquaculture systems.

Trophic Levels: Trophic levels refer to the hierarchical levels in a food web, representing the position of organisms based on their source of energy. In aquaculture, understanding trophic levels helps in selecting appropriate species for culture and designing efficient feeding strategies. Primary producers (e.g., microalgae) occupy the first trophic level, followed by primary consumers (e.g., herbivorous fish or zooplankton) and secondary consumers (e.g., carnivorous fish). Cultivating species at lower trophic levels, such as filter feeders or herbivores, can be more sustainable and cost-effective due to lower feed requirements.

Tubifex: Tubifex worms, commonly known as sludge worms, are segmented aquatic annelids that are often used as live feed in aquaculture, particularly for carnivorous fish species. They thrive in nutrient-rich, organic-matter-laden waters, often in oxygen-depleted environments. Tubifex can be cultured in shallow trays with a substrate of decomposing organic material and maintained at a temperature of 15°-25°C. These worms are highly nutritious, containing high levels of protein and essential amino acids, making them a valuable food source for aquarium and farmed fish.

Water Quality: Water quality in aquaculture is critical for the health and growth of cultured species. Key water quality parameters include temperature, dissolved oxygen, pH, ammonia, nitrite, nitrate levels, and salinity (in marine systems). Maintaining optimal water quality ensures the prevention of stress, disease, and mortality in aquaculture species. Regular monitoring and management, such as water exchange, aeration, and biofiltration, are essential for sustaining water quality in both freshwater and marine aquaculture systems.

Zooplankton Estimation: Zooplankton estimation involves quantifying the abundance of zooplankton in culture systems, usually measured in individuals per liter. Sampling methods include using plankton nets, followed by microscopic counting and identification. This data is crucial for managing feeding regimes and ensuring optimal growth conditions for larval fish and invertebrates.

Zooplankton Grazing: Zooplankton grazing refers to the consumption of phytoplankton and other small organisms by zooplankton, which are a key link in aquatic food webs. In aquaculture, zooplankton grazing is important for controlling phytoplankton populations and providing a natural food source for larval fish and other cultured species. Grazing rates can be influenced by factors such as temperature, light, and zooplankton density. Managing zooplankton grazing dynamics is essential in extensive and semi-intensive aquaculture systems to balance primary productivity and support the growth of higher trophic levels.

Zooplankton: Zooplankton are heterotrophic planktonic organisms that serve as a crucial intermediate in aquatic food webs, feeding on phytoplankton and providing nourishment for higher trophic levels, including fish larvae. They include species such as rotifers, copepods, and cladocerans. Optimal culture conditions for zooplankton vary with species but generally include temperatures between 20°-25°C, pH levels around 7.5-8.5, and adequate aeration to maintain oxygen levels. They are commonly used in aquaculture as live feed due to their high nutritional value, particularly in terms of proteins and essential fatty acids.

CHAPTER 10

SHRIMP DISEASES AND HEALTH MANAGEMENT

Acute Hepatopancreatic Necrosis Disease (AHPND): AHPND, also referred to as EMS, is a severe bacterial disease caused by a toxin-producing strain of *Vibrio parahaemolyticus*. It primarily affects the hepatopancreas, leading to rapid mortality within the first 20-30 days of stocking. AHPND is associated with environmental stress, including poor water quality and high stocking densities. Effective control measures involve strict biosecurity, optimal water quality management, and the use of probiotics to reduce pathogenic bacterial loads.

Acute Hepatopancreatitis (AHP): Acute Hepatopancreatitis is a bacterial disease affecting shrimp hepatopancreas, leading to tissue necrosis, poor feed conversion, and mortality. It is associated with poor water quality, temperature stress (above 32°C), and high ammonia levels (above 0.1 mg/L). Proper pond management and biosecurity measures are essential to control the disease.

Aeromonas hydrophila: *Aeromonas hydrophila* is a gram-negative bacterium that causes hemorrhagic septicemia and ulcerative diseases in shrimp and other aquatic species. The pathogen proliferates in poor water quality and high organic matter. Proper pond management with regular water exchanges and aeration is crucial to preventing outbreaks.

Agglutination Test: An agglutination test is a diagnostic method used to detect the presence of specific antigens or antibodies in shrimp hemolymph. The test relies on the clumping (agglutination) of particles when a positive reaction occurs. It is a rapid and simple test used in disease surveillance and health monitoring in shrimp farming.

Amoebic Gill Disease: Amoebic Gill Disease (AGD) is caused by the amoeba *Neoparamoeba perurans* and primarily affects finfish like salmon but can impact shrimp through environmental stressors. The disease leads to hyperplasia of gill tissues, impairing respiration. In shrimp, the disease is less common but can be associated with compromised water quality or overstocking. Keeping dissolved oxygen levels above 5 mg/L and water temperatures between 28°-32°C can prevent AGD outbreaks.

Amoebic Gill Disease: Amoebic Gill Disease, caused by *Neoparamoeba* spp., leads to severe gill inflammation and tissue hyperplasia, impairing respiration. In shrimp, it can occur under stress from poor water quality. Ensuring dissolved oxygen levels remain above 5 mg/L and avoiding overcrowding are important preventive measures.

Annelid Parasitism: Annelid parasitism in shrimp occurs when parasitic worms, such as leeches or polychaetes, infest the body, gills, or exoskeleton, causing damage and increasing susceptibility to secondary infections. Reducing organic matter and ensuring good water quality help prevent these infestations.

Antibiotic Sensitivity Assay: An antibiotic sensitivity assay is a test to determine the effectiveness of antibiotics against bacterial pathogens in shrimp. Bacterial isolates from infected shrimp are cultured, and antibiotics are applied to assess which agents inhibit bacterial growth. This test helps guide appropriate treatment and minimizes the risk of antibiotic resistance development.

Antimicrobials: Antimicrobials are agents used to kill or inhibit the growth of bacteria, fungi, or parasites in shrimp farming. They are applied to control bacterial infections like *Vibrio* spp., but their use must be regulated to prevent the development of antimicrobial resistance and environmental contamination.

Aphanomyces spp.: *Aphanomyces* spp. are water molds (oomycetes) responsible for infections such as ulcerative mycosis in shrimp and fish. These pathogens invade through open wounds, causing ulceration and tissue necrosis. Outbreaks are associated with poor water quality and injury from handling or crowding. Optimal pond management with water parameters such as ammonia levels below 0.1 mg/L and dissolved oxygen above 5 mg/L are important in controlling this disease.

Argulus spp.: *Argulus* spp., commonly known as fish lice, are ectoparasitic copepods that infest the skin and gills of aquatic species, including shrimp. They cause irritation, wounds, and secondary infections. The parasites thrive in poor water conditions, and their proliferation can be controlled by maintaining good water quality and limiting stocking densities to reduce host availability.

Bacterial Septicemia: Bacterial Septicemia in shrimp occurs when bacterial pathogens invade the hemolymph, leading to systemic infections, lethargy, and high mortality rates. It is often caused by opportunistic bacteria like *Vibrio* spp. and occurs in poor water conditions. Maintaining dissolved oxygen levels above 5 mg/L and regular water exchange are crucial for preventing septicemia.

Biochemical Assays: Biochemical assays are laboratory tests used to detect the activity of enzymes, toxins, or other biochemical markers that indicate disease or physiological stress in shrimp. These assays can help in diagnosing metabolic disorders or infections that alter normal biochemical functions.

Biological Assays: Biological assays are tests that use live organisms or cells to assess the presence of pathogens or toxins. In shrimp farming, these assays are employed to detect disease-causing agents or to test the efficacy of treatments, such as antimicrobial agents or immunostimulants.

Black Gill Disease: Black Gill Disease is a condition in shrimp and other aquatic species where the gill tissues become darkened due to environmental stress, pollution, or parasitic infections such as protozoans or fungi. It impairs respiration, leading to decreased oxygen absorption. Optimal water quality parameters for shrimp include dissolved oxygen levels above 5 mg/L, ammonia less than 0.1 mg/L, and nitrite below 0.5 mg/L to reduce stress and prevent disease. Stressors like high levels of organic matter and poor water quality exacerbate the condition.

Black Spot Disease: Black Spot Disease, also known as melanized lesions or shell disease, is characterized by blackened spots on the shrimp's exoskeleton due to bacterial or fungal infections. This occurs when opportunistic pathogens invade damaged shell areas. High organic matter (above 40 mg/L), poor water quality, and injury increase susceptibility. Managing water quality with regular exchanges and reducing stressors can prevent outbreaks.

Blue Shrimp Disease: Blue Shrimp Disease, also known as cyanosis, is caused by viral infections or environmental stress, leading to a bluish discoloration of shrimp due to impaired oxygen uptake. Keeping dissolved oxygen levels above 5 mg/L and avoiding overcrowding are important preventive measures.

Ciliate Infection: Ciliate infections in shrimp are caused by protozoan parasites like *Trichodina* and *Epistylis*, which colonize the gills and body surface, leading to irritation and respiratory distress. Keeping water quality parameters within the optimum range, particularly dissolved oxygen levels above 5 mg/L, is important for preventing these infections.

Clinical Symptoms: Clinical symptoms in shrimp are observable signs of disease, such as lethargy, discoloration, abnormal swimming, reduced feed intake, soft shells, or gill necrosis. Recognizing these symptoms early is crucial for diagnosing and managing diseases in shrimp farms.

Cotton Shrimp Disease: Cotton Shrimp Disease, caused by fungal pathogens like *Fusarium* spp., leads to a cotton-like appearance of shrimp muscles as a result of tissue degradation. It significantly affects shrimp health, leading to soft shells, slow growth, and increased mortality. Poor water quality and high levels of organic matter contribute to the disease's spread. Controlling water parameters, especially organic matter levels and ensuring good pond hygiene, are key to managing this condition.

Crayfish Plague: Crayfish Plague is caused by the oomycete *Aphanomyces astaci*, which infects crayfish and other crustaceans. It results in tissue destruction, lethargy, and high mortality rates. The disease thrives in water temperatures of 18°-22°C and can rapidly spread through waterborne zoospores. Managing water quality and preventing the introduction of infected crayfish are essential control measures.

Cryptocaryon irritans: *Cryptocaryon irritans* is a protozoan parasite that causes marine white spot disease in fish. The parasite infects the skin and gills, leading to white spots, respiratory distress, and mortality. The parasite thrives at temperatures between 23°-27°C, and outbreaks can be controlled by raising water temperature above 30°C. Maintaining good water quality is also crucial in reducing the impact of the disease.

Cutaneous Mycosis: *Cutaneous Mycosis* refers to fungal infections affecting the skin and exoskeleton of shrimp. Fungal pathogens like *Fusarium* spp. invade the cuticle, leading to lesions and weakening of the shell. Controlling organic matter (below 40 mg/L), improving water quality, and maintaining proper biosecurity can reduce infection risks.

Disease Diagnosis: Disease diagnosis in shrimp involves identifying the cause of illness through clinical signs, necropsy, and laboratory testing (e.g., PCR, ELISA, or bacterial cultures). Accurate diagnosis is critical for implementing appropriate treatments and managing disease outbreaks, ensuring shrimp health and farm sustainability.

Disease Outbreak: A disease outbreak in shrimp farming occurs when a significant portion of the population shows signs of infection, often due to the spread of pathogens like WSSV or *Vibrio* spp. Disease outbreaks are typically linked to environmental stress or poor biosecurity, leading to high mortality rates and economic losses.

Disease Transmission: Disease transmission in shrimp occurs through direct contact with infected individuals, contaminated water, or through vectors like parasites. Waterborne transmission is particularly common in shrimp farming, emphasizing the need for clean water and biosecurity measures to prevent outbreaks.

Disease Treatment: Disease treatment in shrimp includes the use of antimicrobials, probiotics, immunostimulants, and water management practices to control and eliminate infections. The choice of treatment depends on the disease diagnosis, pathogen sensitivity, and environmental conditions.

Early Mortality Syndrome (EMS): Early Mortality Syndrome (EMS), also known as Acute Hepatopancreatic Necrosis Disease (AHPND), is a bacterial disease caused by a specific strain of *Vibrio parahaemolyticus* that produces toxins, leading to massive shrimp mortality in the early stages of culture. The disease primarily affects the hepatopancreas, causing tissue necrosis and dysfunction. EMS outbreaks are typically seen within the first 20-30 days after stocking and are associated with poor water quality and high stocking densities. Preventive measures include improving biosecurity, using probiotics, and controlling water parameters such as pH (7.5-8.5) and temperature (28°-30°C).

ELISA (Enzyme-Linked Immunosorbent Assay): ELISA is a highly sensitive biochemical assay used to detect specific antigens or antibodies in shrimp, commonly applied to diagnose viral infections like WSSV. It involves the use of enzyme-labeled antibodies and a color change reaction that quantitatively indicates the presence of the pathogen. ELISA is widely used due to its high specificity and ability to process multiple samples simultaneously.

Enterocytozoon hepatopenaei **(EHP):** EHP is a microsporidian parasite that primarily infects the hepatopancreas of shrimp, leading to significant growth retardation. This parasite does not directly cause mortality but significantly affects shrimp productivity, with infected shrimp exhibiting reduced feed conversion rates and stunted growth. EHP thrives in environments with poor water quality, and controlling organic matter levels (less than 40 mg/L) and maintaining optimal stocking densities are key to reducing outbreaks. No specific treatments exist for EHP, so management is focused on biosecurity and pond hygiene.

Environmental Stressors: Environmental stressors such as low dissolved oxygen (below 3 mg/L), high ammonia (above 0.1 mg/L), temperature fluctuations, and poor water quality can compromise shrimp health, making them more susceptible to diseases. Managing water quality parameters and reducing stressors is crucial for optimal shrimp farming conditions.

Epistylis: *Epistylis* is a ciliated protozoan that attaches itself to the exoskeleton and gills of shrimp and other aquatic animals. While the protozoan itself is not highly pathogenic, it can lead to secondary bacterial infections and respiratory issues. High organic loads and poor water quality can increase the risk of epistylis colonization. Maintaining low organic matter in the water (below 40 mg/L) and sufficient water flow are key management practices to reduce infection.

Exophiala spp.: *Exophiala* spp. are black yeast-like fungi that cause systemic and cutaneous infections in shrimp. These fungi typically thrive in environments with high organic content and poor water quality. Infection can lead to tissue necrosis and mortality. Keeping organic matter below 40 mg/L and maintaining optimal dissolved oxygen levels (above 5 mg/L) are critical to control.

Exoskeleton Erosion: Exoskeleton Erosion in shrimp is the degradation of the outer shell caused by bacterial or environmental factors, often seen in poor water conditions with high organic matter. Ensuring proper pond hygiene and regular water exchanges can prevent the condition.

Fungal Gill Rot: Fungal Gill Rot is caused by fungal pathogens like *Fusarium* spp., which infect the gills of shrimp, leading to tissue damage and impaired respiration. Poor water quality, particularly high levels of organic matter, fosters fungal growth. Maintaining low organic matter (below 40 mg/L) and proper pond aeration are critical in preventing gill rot.

Fusarium solani: *Fusarium solani* is a filamentous fungus that causes fungal infections in shrimp, particularly in the exoskeleton and gills. Infected shrimp exhibit dark, necrotic lesions, reduced feeding, and increased mortality. Fungal outbreaks are often associated with poor water quality, especially high organic matter and low dissolved oxygen levels. Control measures include improving water quality, reducing organic loads, and using antifungal treatments where necessary.

Gill Erosion: Gill Erosion is characterized by the breakdown of gill tissue, often caused by bacterial or environmental factors. Poor water quality, high ammonia levels (above 0.1 mg/L), and stress contribute to the condition. Maintaining proper dissolved oxygen (above 5 mg/L) and controlling ammonia are essential to prevent gill erosion.

Gill Necrosis: Gill Necrosis is a severe condition where the gill tissues degrade due to bacterial, fungal, or environmental stress. It reduces the shrimp's ability to respire, often leading to mortality. Poor water quality, especially low dissolved oxygen (below 3 mg/L), and high ammonia levels (above 0.1 mg/L) contribute to the condition. Maintaining high water quality and aeration is essential to prevent gill necrosis.

Gill-associated Virus (GAV): Gill-associated Virus (GAV) is a nidovirus primarily infecting shrimp species like *Penaeus monodon*, causing gill discoloration, reduced feeding, lethargy, and increased mortality in shrimp farms. It belongs to the same group of viruses as yellow head virus and is associated with significant losses in aquaculture. GAV thrives in stressed environments, with outbreaks more likely to occur in suboptimal conditions, including high stocking densities, poor water quality, and elevated temperatures (optimal temperature for shrimp culture being 28°-32°C).

Health Certification: Health certification is a formal document issued by authorized veterinarians or regulatory bodies verifying that shrimp or aquatic products are free from specific diseases. This certification is essential for trade and stocking purposes, and often requires diagnostic tests, such as PCR or ELISA, to confirm the absence of OIE-listed diseases like white spot syndrome virus (WSSV) or Taura syndrome virus (TSV).

Hemocytic Enteritis: Hemocytic Enteritis is an inflammatory condition in shrimp caused by bacterial infections affecting the intestinal lining. It results in poor nutrient absorption and stunted growth. Optimal water quality and maintaining a balanced diet with probiotics can help prevent the disease.

Hepatopancreatic Parvovirus (HPV): Hepatopancreatic Parvovirus (HPV) is a small, non-enveloped DNA virus that infects the hepatopancreas of penaeid shrimp. It causes reduced growth rates, decreased feed efficiency, and in some cases, mortality. HPV is commonly found in juvenile shrimp and can be transmitted horizontally. Infected shrimp exhibit pale hepatopancreas and sluggish behavior. Preventive measures include good management practices, maintaining optimal water quality, and screening for HPV in broodstock and post-larvae.

Host-Pathogen Interaction: Host-pathogen interaction in shrimp refers to the biological processes through which pathogens (viruses, bacteria, parasites) infect and interact with the shrimp immune system. Understanding these interactions is critical for disease control, as it helps develop strategies to enhance shrimp immunity and reduce pathogen virulence.

Hypoxia Stress: Hypoxia Stress occurs when dissolved oxygen levels in the water fall below optimal levels (less than 3 mg/L), leading to respiratory distress and weakened immune systems in shrimp. Aeration and maintaining oxygen levels above 5 mg/L are essential to preventing hypoxia.

Ichthyophthirius multifiliis: *Ichthyophthirius multifiliis*, also known as "Ich" or white spot disease, is a protozoan parasite affecting freshwater fish. It causes small white cysts on the skin and gills, leading to respiratory distress and tissue damage. The disease proliferates at water temperatures between 20°-25°C, so increasing water temperature to 28°-30°C can help control the parasite. Good water quality and management practices are key in preventing outbreaks.

Immunostimulants: Immunostimulants are compounds or additives used in shrimp feed to enhance the shrimp's immune system, improving resistance to diseases. Common immunostimulants include beta-glucans, vitamins, and probiotics. These are particularly useful in stress conditions and when water quality is compromised.

Impact of Diseases on Shrimp Farming: The impact of diseases on shrimp farming includes economic losses due to mortality, reduced growth, and lower feed conversion efficiency. Outbreaks of diseases like WSSV or *Vibrio* infections can lead to massive losses, necessitating investment in health management practices such as biosecurity, vaccination, and regular monitoring.

Infection Control: Infection control in shrimp farming involves biosecurity measures, water quality management, use of disinfectants, and early disease detection to prevent the spread of pathogens. Quarantine, health certification, and the regular use of diagnostics like PCR are essential tools in controlling infections.

Infection Routes: Infection routes in shrimp can occur through direct contact with contaminated water, infected animals, or via vectors such as parasites and biofilms. Pathogens may enter shrimp through gills, digestive tracts, or wounds, spreading rapidly under poor biosecurity conditions. Quarantine and good hygiene are key to controlling infection routes.

Infectious Hypodermal and Hematopoietic Necrosis Virus (IHHNV): IHHNV is a parvovirus that primarily affects penaeid shrimp, particularly the species *Penaeus vannamei* and *Penaeus stylirostris*. The virus causes stunted growth and deformities, such as the Runt-Deformity Syndrome (RDS), though it does not typically lead to mass mortality. IHHNV infections are most severe in juvenile shrimp and can persist as a chronic condition. The virus can be transmitted vertically and horizontally, and prevention involves stringent biosecurity measures and the use of IHHNV-free broodstock.

Infectious Myonecrosis Virus (IMNV): Infectious Myonecrosis Virus (IMNV) is a RNA virus that causes infectious myonecrosis in penaeid shrimp, particularly *Penaeus vannamei*. The disease is characterized by white necrotic lesions in the muscle tissue, leading to high mortality rates, especially during periods of stress. IMNV outbreaks are associated with water temperatures ranging from 27°-30°C. The virus is transmitted horizontally and can persist in pond sediments. Prevention strategies include strict biosecurity measures, stocking healthy post-larvae, and maintaining optimal water quality.

Lagenidium spp.: *Lagenidium* species are water molds (oomycetes) that cause fungal infections in shrimp, particularly in the larvae and post-larvae stages. Infections result in the formation of hyphal structures in the tissues, leading to necrosis and high mortality rates. *Lagenidium* outbreaks are often associated with poor water quality and high organic matter. Preventive measures include maintaining good water quality, reducing organic loads, and using fungicidal treatments when necessary.

Larval Mycosis: Larval Mycosis refers to fungal infections that affect shrimp larvae, often caused by opportunistic pathogens like *Fusarium* spp. and *Lagenidium* spp. These fungi infect weakened or stressed larvae, leading to high mortality rates. Proper hatchery management with clean water, low organic matter (below 40 mg/L), and good hygiene practices are essential to control.

Leucothrix mucor: *Leucothrix mucor* is a filamentous bacterium that affects shrimp by colonizing the gill filaments and body surfaces, leading to tissue degradation, respiratory problems, and decreased growth rates. The bacterium thrives in environments with high organic load and poor water exchange. Water management practices that keep organic matter levels below 40 mg/L and ammonia below 0.1 mg/L are essential to preventing outbreaks.

Loose Shell Syndrome (LSS): Loose Shell Syndrome is a condition in shrimp where the shell is loosely attached to the body, making the shrimp appear lethargic and slow. It is often associated with nutritional deficiencies, poor water quality, or viral infections like white spot syndrome virus (WSSV). Maintaining water temperature between 28°-32°C, optimal salinity (15-25 ppt), and ensuring a balanced diet with adequate minerals like calcium and phosphorous can help prevent LSS.

Melanization: Melanization in shrimp is a defensive response where damaged tissues turn black due to the deposition of melanin. It can result from bacterial, viral, or parasitic infections and is often seen in areas affected by shell disease or wounds. Maintaining good water quality and biosecurity helps reduce the incidence of melanization.

Microsporidian Gill Infection: Microsporidian Gill Infection in shrimp is caused by parasitic microsporidia, leading to gill tissue degeneration and impaired respiration. These parasites spread through spores in the water and infected tissue. Maintaining good biosecurity, reducing organic matter, and ensuring optimal water quality are key in controlling infections. The optimal water temperature for shrimp farming is 28°-32°C, with a dissolved oxygen level of at least 5 mg/L.

Microsporidiosis: Microsporidiosis in shrimp is caused by microsporidian parasites, primarily *Enterocytozoon spp.*, which infect the hepatopancreas and muscle tissue, leading to tissue degradation and stunted growth. The disease is spread through spore-laden fecal matter and contaminated water. Maintaining good biosecurity, water quality (ammonia less than 0.1 mg/L, pH 7.5–8.5), and stocking density are crucial to mitigating the impact of this disease.

Monodon Baculovirus (MBV): Monodon Baculovirus (MBV) is a DNA virus that infects the epithelial cells of the hepatopancreas in shrimp, especially *Penaeus monodon*. It forms large occlusion bodies that can be seen under a microscope. MBV causes stunted growth and increased susceptibility to other pathogens but rarely leads to mass mortality. The virus is most commonly transmitted through contaminated water or feed. Prevention includes using MBV-free broodstock, improving water quality, and practicing good pond management.

Morphological Assays: Morphological assays in shrimp health management involve examining the physical structures of shrimp, such as gills, hepatopancreas, and exoskeleton, to identify abnormalities associated with disease. These tests are often conducted alongside necropsy to diagnose bacterial, viral, or parasitic infections.

Muscle Necrosis: Muscle Necrosis in shrimp is characterized by the breakdown of muscle tissue, which may turn opaque or white. It is caused by environmental stressors such as low dissolved oxygen (below 3 mg/L), temperature fluctuations, or infections. Keeping water temperatures stable (28°-32°C) and ensuring sufficient aeration are crucial to preventing muscle necrosis.

Mycosis: Mycosis refers to fungal infections affecting shrimp, typically caused by water molds like *Aphanomyces* spp. or fungi like *Fusarium*. These pathogens invade shrimp tissues, leading to lesions, reduced growth, and eventual mortality. Poor water quality, especially with low dissolved oxygen (below 3 mg/L) and high organic load, fosters fungal growth. Proper pond management, including regular water exchange and control of organic matter, is crucial to prevent outbreaks.

Necropsy Examination: Necropsy examination is the post-mortem dissection of dead shrimp to examine internal organs and tissues for signs of disease. It allows for visual detection of lesions, tissue damage, or abnormalities that can point to specific pathogens or environmental issues. It is a key diagnostic tool in shrimp health management.

Necrotizing Hepatopancreatitis (NHP): Necrotizing Hepatopancreatitis (NHP) is a bacterial disease caused by *Hepatobacter penaei*, a gram-negative, intracellular bacterium. NHP primarily affects the hepatopancreas, leading to tissue necrosis, reduced feeding, and high mortality. The disease is most common in warm climates, with outbreaks occurring at water temperatures above 29°C. Prevention includes improving water quality, reducing stress, and using antibiotics or probiotics to control bacterial infections.

OIE Listed Diseases: OIE-listed diseases in shrimp farming are diseases recognized by the World Organisation for Animal Health (OIE) as significant threats to aquaculture. These include WSSV, yellow head virus (YHV), and infectious myonecrosis virus (IMNV). Monitoring and controlling these diseases are critical for international trade and shrimp health management.

Parasitic Copepods: Parasitic copepods, such as *Caligus* spp. and *Lernaea* spp., attach to the skin, fins, and gills of fish and shrimp, leading to irritation, tissue damage, and secondary infections. These parasites are more common in crowded environments with poor water quality. Managing stocking densities (50-100 shrimp/m^2) and maintaining optimal water quality, with ammonia levels below 0.1 mg/L and sufficient dissolved oxygen (above 5 mg/L), are crucial for control.

Parasitic Isopods: Parasitic isopods attach to shrimp, feeding on blood and causing tissue damage, anemia, and secondary infections. These parasites are more common in environments with poor water quality or overcrowding. Managing pond hygiene, limiting stocking densities, and improving biosecurity can reduce infestations.

Pathogen Reservoirs: Pathogen reservoirs refer to sources in the environment, such as sediment, water, or infected shrimp, where disease-causing agents can persist and spread. Managing these reservoirs through proper sanitation and biosecurity is key to preventing disease outbreaks.

PCR (Polymerase Chain Reaction): PCR is a molecular diagnostic technique used to amplify specific DNA sequences of pathogens, enabling early detection of shrimp diseases such as white spot syndrome virus (WSSV) or infectious hypodermal and hematopoietic necrosis virus (IHHNV). PCR is highly sensitive, capable of detecting pathogens even at low concentrations, and plays a crucial role in shrimp disease surveillance and certification programs.

Penaeus stylirostris densovirus (PstDNV): PstDNV is a viral pathogen that causes runt deformity syndrome in shrimp, particularly in *Penaeus stylirostris*. Infected shrimp exhibit slow growth, deformities, and poor survival rates. The virus is highly contagious and is often spread in hatchery and grow-out environments. Proper biosecurity measures and water quality management, with optimal salinity levels of 15-25 ppt and temperatures between 28°-32°C, are necessary to control PstDNV outbreaks.

Perkinsus spp.: *Perkinsus* spp. are protozoan parasites that primarily infect mollusks like oysters and clams, causing perkinsosis. The disease leads to tissue necrosis, reduced feeding, and mortality. Optimal water conditions for affected species, such as salinity levels between 20-30 ppt and dissolved oxygen above 5 mg/L, are critical for controlling the spread of *Perkinsus* infections.

Post Mortem Examination: Post mortem examination is the inspection of dead shrimp to identify the cause of death. This can include external and internal assessments of the exoskeleton, gills, hepatopancreas, and other tissues for signs of disease, stress, or environmental damage. It complements necropsy and is essential for managing disease outbreaks.

Preventive Measures: Preventive measures in shrimp farming involve biosecurity protocols, vaccination, use of probiotics, regular water quality monitoring, and maintaining optimal feeding practices to reduce the risk of disease outbreaks. These measures help strengthen shrimp immunity and create an environment that limits pathogen proliferation.

Probiotics: Probiotics in aquaculture refer to beneficial bacteria added to the water or feed that improve shrimp gut health, enhance immunity, and outcompete harmful pathogens. Regular use of probiotics, such as *Bacillus* spp., helps maintain a healthy microbial balance in the pond, reducing disease outbreaks and improving shrimp health

Protozoan Cutaneous Infection: Protozoan Cutaneous Infection is caused by protozoan parasites like *Trichodina* and *Zoothamnium* that colonize the skin of shrimp, causing irritation, tissue damage, and secondary bacterial infections. Maintaining low organic loads (below 40 mg/L) and regular water exchanges are critical to prevent outbreaks.

Protozoan Gill Infection: Protozoan Gill Infections in shrimp are caused by parasitic protozoans such as *Zoothamnium* and *Trichodina*, which colonize the gills, impairing respiration and causing tissue damage. Optimal water conditions include dissolved oxygen levels above 5 mg/L and organic matter below 40 mg/L. Proper water management is critical to reduce protozoan colonization.

Protozoan Infections: Protozoan infections in shrimp and other aquatic species are caused by various species of parasitic protozoans, including *Zoothamnium*, *Epistylis*, and *Trichodina*. These parasites attach to gills, skin, and body surfaces, leading to irritation, fouling, and respiratory distress. Managing water quality and reducing organic loads (organic matter below 40 mg/L) are critical in preventing protozoan outbreaks.

Pseudomonas Infection: *Pseudomonas* spp. are opportunistic bacteria that cause infections in shrimp, particularly in wounds or damaged tissues, leading to ulcerations, septicemia, and death. Maintaining good water quality (ammonia below 0.1 mg/L, dissolved oxygen above 5 mg/L) and reducing stress are key to preventing infections.

Quarantine: Quarantine is the process of isolating new or diseased shrimp from the main population to prevent the spread of pathogens. This is particularly important when introducing shrimp from different sources into a farm. The duration typically ranges from 2 to 4 weeks, during which water parameters like temperature (28°-32°C), salinity (15-25 ppt), and dissolved oxygen (above 5 mg/L) are closely monitored to ensure optimal conditions.

Red Disease: Red Disease in shrimp is characterized by reddening of the body due to bacterial infections, often caused by *Vibrio* spp. The disease is linked to poor water quality, high ammonia levels, and stress. Maintaining water parameters such as ammonia below 0.1 mg/L, with good water exchange and aeration, helps prevent outbreaks.

Sanitation Protocols: Sanitation protocols in shrimp farms include practices like regular pond cleaning, water treatment, and disinfection of equipment to reduce pathogen loads. Effective sanitation helps prevent the spread of diseases and maintains the overall health of the shrimp population.

Shell Disease Syndrome: Shell Disease Syndrome in shrimp and other crustaceans is caused by bacterial infections that degrade the exoskeleton, leading to lesions, soft shells, and increased susceptibility to other pathogens. The disease is often associated with poor water quality, high organic matter, and stress. Maintaining good pond hygiene, regular water exchanges, and controlling bacterial loads (ammonia less than 0.1 mg/L) are key to managing the condition

Shell Rot Disease: Shell Rot Disease is a bacterial infection that degrades the shrimp exoskeleton, often leading to ulcers and secondary infections. High organic matter, poor water quality, and stress increase susceptibility. Proper water management, regular pond cleaning, and reducing organic loads (below 40 mg/L) are key to controlling this disease.

Shrimp Chronic Soft Shell Syndrome (SCSS): SCSS is a condition in which shrimp shells remain soft over a prolonged period due to poor mineral uptake, water quality, or viral infections. Managing calcium and magnesium levels in the water (around 200-400 mg/L) and ensuring a balanced diet with adequate minerals are important in controlling SCSS.

Shrimp Defense Mechanisms: Shrimp defense mechanisms include innate immune responses such as the production of antimicrobial peptides, phagocytosis, and melanization. These mechanisms help shrimp defend against pathogens, although stressors can weaken their effectiveness, leading to increased susceptibility to diseases.

Shrimp Head Softening Disease: Shrimp Head Softening Disease is a condition where the exoskeleton of the shrimp's head becomes soft and deformed, often due to calcium deficiency, bacterial infections, or water quality issues. Maintaining optimal calcium levels in water (around 200-400 mg/L) and a balanced diet with adequate minerals are essential to prevent this condition.

Stress Factors: Stress factors in shrimp farming include environmental conditions like poor water quality, high stocking density, temperature fluctuations, and poor nutrition that can weaken shrimp immunity and increase susceptibility to diseases. Monitoring and controlling these factors is key to preventing disease outbreaks and maintaining shrimp health.

Taura Syndrome Virus (TSV): Taura Syndrome Virus (TSV) is a RNA virus that primarily affects the shrimp species *Penaeus vannamei*. TSV causes Taura Syndrome, which is characterized by reddening of the exoskeleton, soft cuticles, and high mortality during the molting process. TSV outbreaks are often triggered by environmental stressors such as temperature fluctuations and poor water quality. The disease is prevalent in shrimp farming regions with water temperatures ranging from 25°-30°C. Prevention strategies include the use of TSV-resistant shrimp strains and strict biosecurity protocols.

Tetrahymena spp.: *Tetrahymena* spp. are free-living ciliates that can become opportunistic parasites in shrimp, infecting tissues and leading to systemic infections. These ciliates thrive in environments with high organic matter and poor water quality. Reducing organic load and maintaining clean, well-aerated water are important in preventing *Tetrahymena* infections.

Trichodina spp.: *Trichodina* spp. are ciliate protozoans that infect the gills and skin of fish and shrimp, causing irritation, excessive mucus production, and respiratory distress. They proliferate in poor water conditions, with high organic matter and low dissolved oxygen. Keeping organic matter levels below 40 mg/L and ensuring good water aeration are essential in managing *Trichodina* infections.

Vaccination in Shrimp: Vaccination in shrimp is an emerging strategy to enhance immune responses against specific pathogens like WSSV and *Vibrio* spp. Although still in development, vaccines can potentially reduce the need for antibiotics and minimize disease outbreaks. Vaccination effectiveness depends on timing, dose, and the type of pathogen targeted.

Vibrio harveyi: *Vibrio harveyi* is a luminescent, gram-negative bacterium that causes vibriosis in shrimp, leading to diseases such as luminous vibriosis and septicemia. It is commonly associated with poor water quality, high organic loads, and stress conditions. The bacterium proliferates in water with salinity levels of 15-35 ppt and temperatures between 25°-30°C. Control strategies include improving water quality, reducing organic matter, and using probiotics or antimicrobial treatments.

Vibrio parahaemolyticus: *Vibrio parahaemolyticus* is a gram-negative, halophilic bacterium commonly found in marine environments. It is a major pathogen in shrimp farming, responsible for diseases such as AHPND and vibriosis. The bacterium thrives in brackish water environments with salinities ranging from 10-30 ppt and temperatures between 25°-30°C. Control measures include maintaining optimal water quality, reducing stress in shrimp, and using probiotics to outcompete pathogenic strains.

Vibriosis: Vibriosis is a bacterial disease in shrimp caused by various species of *Vibrio*, including *Vibrio parahaemolyticus*, *Vibrio harveyi*, and *Vibrio vulnificus*. Clinical symptoms include lethargy, lesions, and melanized spots on the exoskeleton. Vibriosis is often triggered by poor water quality, high organic loads, and stress conditions. Preventive measures include maintaining water quality, reducing stressors, and using probiotics or antimicrobial agents to control bacterial populations.

Water Quality Monitoring: Water quality monitoring is essential in shrimp farming to maintain optimal conditions (dissolved oxygen above 5 mg/L, ammonia below 0.1 mg/L, pH 7.5-8.5). Regular checks on parameters like temperature, salinity, and pH are critical for preventing disease and promoting healthy growth.

White Feces Syndromel: White Feces Syndrome (WFS) is a disease condition observed in farmed shrimp, characterized by the presence of white or pale fecal strings floating on the water surface. It is often associated with a combination of bacterial infections, parasitic infestations (e.g., *Gregarine spp.*), and poor water quality. WFS leads to reduced feed intake, slower growth, and, in severe cases, mortality. Preventive measures include regular pond monitoring, maintaining water quality, and using probiotics to enhance gut health.

White Muscle Disease: White Muscle Disease in shrimp is characterized by the whitening of muscle tissues due to muscle degeneration or necrosis. It can be caused by viral infections, environmental stress, or nutritional deficiencies. Optimum water parameters include temperatures of 28-32°C and dissolved oxygen levels above 5 mg/L. Providing a balanced diet with adequate protein can prevent this condition.

White Spot Syndrome Virus (WSSV): White Spot Syndrome Virus (WSSV) is a highly pathogenic virus that affects various species of shrimp, causing high mortality rates. WSSV is a double-stranded DNA virus belonging to the family Nimaviridae. The disease is characterized by white spots on the exoskeleton and cuticle, lethargy, and a rapid decline in shrimp populations. It thrives in a wide range of water temperatures (18°-30°C), with outbreaks typically associated with stress factors like sudden temperature changes and poor water quality. Preventive measures include improved biosecurity and the use of pathogen-free broodstock.

Yellow Head Virus (YHV): Yellow Head Virus (YHV) is a highly virulent RNA virus belonging to the family Roniviridae, causing Yellow Head Disease (YHD) in penaeid shrimp, particularly *Penaeus monodon*. The disease is characterized by rapid mortality, yellowing of the cephalothorax, and cessation of feeding. YHV outbreaks typically occur in water temperatures between 26-32°C. The virus spreads rapidly through horizontal transmission, and prevention focuses on biosecurity measures, including the use of specific pathogen-free (SPF) broodstock and quarantine procedures.

Zoothamnium spp.: *Zoothamnium* is a sessile protozoan that forms colonies on shrimp gills and carapace, causing fouling and impairing respiratory function. It typically proliferates in water with high organic matter and low water exchange. Managing water quality by ensuring regular water exchange, keeping organic matter below 40 mg/L, and avoiding overstocking (optimum density around 50-100 shrimp/m^2) can reduce *Zoothamnium* outbreaks.

CHAPTER 11

FRESHWATER AQUACULTURE

Advantages of All Male Culture: All male culture offers several advantages, including faster growth rates, uniformity in size, and prevention of uncontrolled breeding, which can lead to overcrowding and resource depletion. This approach maximizes pond productivity and reduces competition for food. In species like tilapia, males grow 30-40% faster than females, making all-male culture economically advantageous. This method also simplifies management and harvest, as fish reach market size simultaneously.

Agro-industrial Waste: Agro-industrial waste refers to by-products from agricultural and industrial processes, such as rice bran, oilseed cakes, and brewery waste, which can be used as feed or fertilizer in aquaculture. These materials are often rich in nutrients and provide a cost-effective alternative to conventional feeds. However, their use requires careful management to avoid water quality issues like eutrophication. Properly processed and balanced, agro-industrial waste can enhance aquaculture productivity and reduce costs.

Algal Bloom Control: Algal bloom control in aquaculture involves managing excessive growth of algae, which can lead to oxygen depletion and fish kills. Control methods include reducing nutrient inputs, using algicides like copper sulfate (applied at 0.1-0.2 ppm), increasing water circulation, and biological control using filter feeders like tilapia or silver carp. Maintaining balanced nutrient levels and regular monitoring of water quality parameters are key to preventing and managing algal blooms.

Algal Blooms: Algal blooms are the rapid proliferation of algae in aquatic systems, often triggered by nutrient overloads, such as from excess fertilizers or organic matter. While some algae are beneficial as primary producers in aquaculture, blooms of toxic or nuisance species can deplete oxygen and release harmful substances, leading to fish kills. Regular monitoring of water quality and controlled fertilization are essential for preventing algal blooms.

All Male Culture: All male culture is an aquaculture practice that involves cultivating only male fish, especially in species like tilapia, to enhance growth rates and prevent unwanted reproduction. Male fish often grow faster and reach market size quicker, improving production efficiency. Techniques like manual sexing, hybridization, or hormonal treatments are used to achieve all-male populations. Ideal conditions include water temperatures of 25°-30°C, pH 7.0-8.0, and dissolved oxygen levels above 5 mg/L.

Ammonia Control: Ammonia control is crucial in aquaculture as ammonia is toxic to fish and prawns even at low concentrations. Ammonia levels should be kept below 0.05 mg/L to avoid stress and potential mortality. Control measures include regular water exchange, proper aeration, maintaining optimal pH (6.5-7.5 to keep ammonia in the less toxic ionized form), and using biofilters or nitrifying bacteria to convert ammonia into less harmful compounds like nitrites and nitrates.

Aquaculture Marketing: Aquaculture marketing encompasses the activities involved in promoting and selling aquaculture products, including fish, prawns, and other aquatic organisms. This involves market research, product differentiation, value addition (e.g., processing or packaging), and distribution strategies to reach target consumers. Understanding market demand, setting competitive prices, and ensuring consistent product quality are key aspects of successful aquaculture marketing. Additionally, sustainability and traceability are increasingly important factors influencing consumer choices and market access.

Aquaculture Prospects: Aquaculture prospects refer to the future potential and opportunities for growth in the sector. Freshwater aquaculture is expected to expand due to the global need for sustainable protein sources, employment opportunities, and the availability of vast inland water resources. Enhancing productivity through integrated farming systems, technological innovation, and improved feed and disease management are key to maximizing these prospects.

Aquatic Weeds: Aquatic weeds are unwanted plant species that grow in aquaculture ponds, competing with cultured species for nutrients and oxygen, and obstructing water flow and management practices. Common aquatic weeds include water hyacinth, duckweed, and algae. These weeds can adversely affect water quality and fish growth if not controlled. Regular monitoring, manual removal, and biological or chemical control methods are employed to manage weed populations and maintain optimal pond conditions.

Biofertilizer: Biofertilizers in aquaculture are organic substances containing living microorganisms that promote the growth of phytoplankton and zooplankton, which are natural food sources for fish and prawns. Common biofertilizers include composted manure, Azolla, and microbial inoculants like Rhizobium or Azotobacter. They improve nutrient availability in the pond, enhance water quality, and reduce the need for chemical fertilizers. Optimal application rates depend on pond conditions, but typical doses range from 500-1000 kg/ha.

Biomass Estimation: Biomass estimation in aquaculture is the process of calculating the total weight of fish or prawns within a pond or culture system. This is typically done using sampling methods, where a representative sample is weighed and extrapolated to estimate the entire stock. Accurate biomass estimation is crucial for managing stocking densities, feeding rates, and harvest schedules. Regular biomass assessments help in optimizing production and ensuring that the carrying capacity of the culture environment is not exceeded.

Breeding Techniques: Breeding techniques in aquaculture encompass methods used to induce and manage reproduction in fish and prawns, including natural spawning, hormonal induction, and selective breeding. Techniques vary by species but generally involve providing optimal conditions such as temperature, photoperiod, and water quality to stimulate breeding. Hormonal injections (e.g., human chorionic gonadotropin) are commonly used for species with difficult spawning requirements. Successful breeding techniques result in high-quality offspring and improved genetic stock.

Carrying Capacity: Carrying capacity in aquaculture ponds is the maximum biomass of fish or aquatic organisms that the system can support without degrading water quality or causing stress. It depends on pond size, water exchange rates, aeration, and the species cultured. For semi-intensive systems, carrying capacities range from 1,500 to 5,000 kg/ha. Overloading beyond this limit can result in oxygen depletion and water quality deterioration.

Chemical Fertilizers: Chemical fertilizers, like urea (46% nitrogen) and single super phosphate (16-18% P_2O_5), are used in freshwater aquaculture to boost primary productivity by enhancing the growth of phytoplankton. These fertilizers are applied at doses of 30-50 kg/ha/month, depending on pond conditions and water quality. Fertilizer use must be carefully managed to avoid eutrophication and algal blooms.

Commercial Fish Species: Commercial fish species refer to those raised in aquaculture for economic purposes, including species like tilapia, catfish, carp, and trout. These species are chosen based on their growth rates, market demand, feed conversion efficiency, and adaptability to farming conditions. Optimal rearing conditions for these species include specific temperature ranges (e.g., 22-28°C for tilapia), pH levels of 6.5-8.0, and high dissolved oxygen levels (>5 mg/L).

Commercial Prawn Species: Commercial prawn species include freshwater prawns like *Macrobrachium rosenbergii* and marine species such as *Penaeus* *monodon* and *Penaeus vannamei*. These species are farmed for their high market value, rapid growth rates, and adaptability to aquaculture systems. Ideal culture conditions include water temperatures of 28°-32°C, salinity of 10-35 ppt (for marine prawns), and dissolved oxygen levels above 5 mg/L. Proper pond management and disease control are critical for successful prawn aquaculture.

Composite Fish Culture: Composite fish culture is an aquaculture practice where multiple fish species with complementary feeding habits are cultured together in the same pond to maximize resource utilization and production. Typically, combinations of Indian major carps (Catla, Rohu, Mrigal) and minor carps are used, with each species occupying different ecological niches (surface, column, and bottom feeders). Optimum stocking density and feeding are critical to maintaining water quality parameters such as dissolved oxygen (>5 mg/L), pH (7.0-8.5), and temperature (25°-30°C).

Culture Practices: Culture practices in aquaculture refer to the methods and strategies employed to raise aquatic organisms like fish, prawns, or shellfish in controlled environments. This includes selecting species, designing ponds or tanks, managing water quality, feeding, disease prevention, and optimizing conditions for growth and reproduction. Optimal levels of water temperature, dissolved oxygen (5-7 mg/L), pH (6.5-8.5), and ammonia (<0.05 mg/L) are crucial. Different practices such as monoculture, polyculture, and integrated multi-trophic aquaculture (IMTA) are used based on species compatibility, environmental conditions, and market demand.

Depuration Process: The depuration process is a purification method used to cleanse bivalves or other aquaculture products of contaminants, including bacteria, pathogens, or toxins accumulated in their tissues. This process involves holding the organisms in clean, controlled water systems with optimal conditions for filtration (e.g., salinity of 20-35 ppt, temperature of 15°-25°C, and sufficient dissolved oxygen levels above 5 mg/L). Depuration times vary but typically last between 24-48 hours to ensure safety for human consumption.

Detritus Food Chain: The detritus food chain in aquaculture involves the decomposition of organic matter, such as dead plants, animals, and waste, by microorganisms like bacteria and fungi. This decomposition releases nutrients that are consumed by detritivores (organisms that feed on detritus) such as certain fish, prawns, and invertebrates. This food chain plays a crucial role in nutrient cycling and maintaining pond ecosystem health, providing a natural food source that supports the growth of cultured species.

Disease Prevention (Prawn): Disease prevention in prawn farming involves proactive measures to minimize the risk of infections and outbreaks. Key practices include maintaining optimal water quality (temperature 28°-32°C, pH 7.0-8.5, dissolved oxygen > 4 mg/L), regular pond cleaning, and biosecurity protocols to prevent pathogen introduction. The use of probiotics, immunostimulants, and regular health monitoring also plays a significant role in disease prevention. Proper management and early detection of issues are crucial to minimizing losses.

Disease Treatment (Prawn): Disease treatment in prawn farming involves identifying symptoms, diagnosing the causative agents (bacterial, viral, fungal, or parasitic), and applying appropriate treatments such as antibiotics, antifungals, or antiparasitics, often in consultation with a veterinarian. Maintaining optimal water quality (temperature 28°-32°C, pH 7.0-8.5, dissolved oxygen > 4 mg/L) is crucial during treatment to support prawn recovery. Treatment effectiveness is enhanced by removing affected individuals and minimizing stress factors.

Dissolved Oxygen: Dissolved oxygen (DO) is the concentration of oxygen gas present in water, crucial for the respiration of fish, prawns, and other aquatic organisms. Optimal DO levels in aquaculture are typically above 5 mg/L, with levels below 2 mg/L considered stressful or potentially lethal. Maintaining adequate DO involves proper aeration, avoiding overstocking, and controlling organic matter and feed inputs to prevent oxygen depletion caused by microbial activity and decomposition.

Economics of Fish Farming: The economics of fish farming involve the analysis of costs, revenues, and profitability associated with aquaculture operations. Key factors include the cost of inputs (seed, feed, labor, pond construction), productivity (growth rates, survival rates), market prices, and management efficiency. Economic assessments help determine the viability of different species and farming systems, guiding decisions on stocking density, feed strategies, and investment in technology. Profit margins vary widely depending on species, scale, and market conditions.

Exotic Carps: Exotic carps are non-native carp species introduced into a region for aquaculture, including common carp (*Cyprinus carpio*), grass carp (*Ctenopharyngodon idella*), silver carp (*Hypophthalmichthys molitrix*), and bighead carp (*Hypophthalmichthys nobilis*). These species have been integrated into composite culture systems to enhance productivity and utilize different trophic levels, such as grass carp for weed control and silver carp for plankton feeding. Optimal conditions include temperatures of 20°-30°C, pH 6.5-8.5, and dissolved oxygen levels above 5 mg/L.

Feeding (Prawn): Feeding in prawn farming is a critical management aspect that directly influences growth and survival. Prawns are typically fed a high-protein diet consisting of commercial feeds, supplemented with natural foods such as plankton, detritus, and organic matter. Feeding rates and frequencies are adjusted based on prawn size, water temperature (28°-32°C), and overall pond conditions, with typical feeding frequencies of 3-4 times per day. Proper feeding management ensures optimal growth rates and feed conversion efficiency.

Feeding Regimes: Feeding regimes in aquaculture refer to the schedules, quantities, and types of feeds provided to cultured species. A well-planned feeding regime ensures optimal growth, feed conversion, and minimal waste.

Regimes are typically adjusted based on the species, size, water temperature, and biomass, with common practices including multiple daily feedings (2-4 times) and adjusting feed types from starter feeds to grower feeds as fish or prawns mature. Overfeeding should be avoided to prevent water quality deterioration.

Fertilizer Requirement: Fertilizers in aquaculture are used to enhance primary productivity, mainly phytoplankton, which forms the base of the food web. Both organic (e.g., cow manure) and inorganic fertilizers (e.g., urea and superphosphate) are used. Typical application rates are 50-100 kg/ha for nitrogenous fertilizers and 25-50 kg/ha for phosphates. The aim is to maintain a balanced nutrient level that promotes plankton growth without leading to excessive nutrient buildup and algal blooms, keeping water parameters within optimal ranges.

Fish Farm Visits: Fish farm visits are essential components of aquaculture management, involving inspections by experts to assess farm operations, water quality, stock health, feeding practices, and overall management. These visits help identify potential issues, provide recommendations for improvements, and ensure compliance with best practices and regulatory standards. Regular farm visits contribute to better disease control, efficient resource use, and enhanced production outcomes in aquaculture systems.

Fish Farming Systems: Fish farming systems encompass the various methods used in aquaculture, including extensive, semi-intensive, and intensive systems. Extensive systems rely on natural productivity with minimal inputs, while semi-intensive systems use supplemental feeding and fertilization. Intensive systems involve high stocking densities, complete feed, and advanced management practices. Each system has different requirements for water quality, feed, and capital investment, with intensive systems offering higher yields but requiring more careful management.

Fish Farming: Fish farming involves the controlled cultivation of fish species in ponds, tanks, or other aquatic systems for commercial purposes. The practice includes breeding, nursery rearing, grow-out, and harvesting, with a focus on optimizing growth rates, feed conversion, and marketable size. Key species in freshwater fish farming include carps, tilapia, and catfish.

Fish Grading: Fish grading is the process of sorting fish by size, which helps optimize growth rates, prevent cannibalism, and ensure uniformity in stocking densities. Grading is typically done using mesh screens or grading boxes, separating larger and smaller individuals. Regular grading (every few weeks) in grow-out ponds allows for tailored feeding strategies and improves overall production efficiency by reducing competition for food and space among different size classes.

Fish Growth Study: Fish growth studies involve monitoring and analyzing growth rates, feed conversion, and overall health of fish populations under varying conditions. This includes measuring weight and length at regular intervals, assessing environmental factors like temperature, pH, and dissolved oxygen, and evaluating feed quality and quantity. Growth models, such as the von Bertalanffy growth function, are used to predict future growth and optimize feeding and management strategies to achieve the best production outcomes.

Fish Health Monitoring: Fish health monitoring involves regular observation and assessment of fish behavior, appearance, and overall health to detect signs of disease, stress, or suboptimal conditions. This includes checking for abnormal swimming patterns, lesions, discoloration, and appetite changes. Water quality testing, parasite checks, and sampling for pathogens are part of comprehensive health monitoring. Early detection of health issues allows for timely interventions such as treatment, water quality adjustments, or improved management practices.

Fish Storage: Fish storage refers to the methods used to preserve fish post-harvest to maintain quality and extend shelf life. Key storage practices include chilling (0°-4°C), freezing, or using ice in insulated containers. Proper hygiene, temperature control, and minimizing handling are essential to prevent spoilage and maintain the freshness of fish. For longer-term storage, methods like vacuum packing or modified atmosphere packaging can be used to further preserve quality and reduce bacterial growth.

Fish Transport: Fish transport involves moving fish or prawns from hatcheries to farms, between ponds, or to markets. Proper transport techniques minimize stress and mortality, including using aerated or oxygenated containers, maintaining optimal temperature (often 20°-25°C for tropical species), and ensuring clean water conditions. Transport density and duration should be managed carefully, with stress-reducing additives like salt or anesthetics sometimes used to improve fish welfare during transit.

Freshwater Aquaculture: Freshwater aquaculture refers to the farming of aquatic organisms, such as fish, crustaceans, and plants, in inland freshwater systems like ponds, lakes, rivers, and reservoirs. The primary focus is on culturing species that thrive in non-saline environments, such as Indian major carps (Catla, Rohu, Mrigal) and freshwater prawns. Optimal water parameters include pH 6.5-8.5, dissolved oxygen above 5 mg/L, and temperatures between 25°-30°C for most species.

Freshwater Prawn Farming: Freshwater prawn farming involves the cultivation of species such as *Macrobrachium rosenbergii*, known as the giant river prawn, in controlled freshwater environments. This practice requires well-managed pond systems with water temperatures between 28°-32°C, pH levels of 7.0-8.5, and dissolved oxygen levels of 4-6 mg/L. Prawns are omnivorous and require a diet rich in protein, often supplemented with commercial feed. Key practices include water quality management, regular feeding, and disease prevention.

Grow-out Culture: Grow-out culture is the phase in aquaculture where juvenile fish or prawns are raised to market size in ponds or tanks. This stage focuses on maximizing growth through proper feeding, water quality management, and disease prevention. Growth rates vary by species, with common harvest times of 6-12 months for fish like carps and tilapia, and yields of up to 6,000 kg/ha in intensive systems.

Grow-out Pond Management: Grow-out pond management encompasses the practices involved in raising aquaculture species to marketable size in larger ponds. This includes regular water quality monitoring, feeding management, disease prevention, and periodic harvesting. Optimum conditions typically involve maintaining dissolved oxygen levels above 4 mg/L, pH between 6.5-8.5, and temperatures suitable for the species being cultivated. Proper aeration, water exchange, and waste management are essential to sustain high productivity and healthy stock.

Harvesting Techniques: Harvesting techniques in aquaculture involve the methods used to capture fish or prawns from ponds, cages, or tanks for market sale. Common methods include netting, draining ponds, or using traps. Harvesting should be done during cooler parts of the day to reduce stress, and proper handling techniques are crucial to minimize damage and maintain product quality. Gradual harvesting or partial harvests can help manage market supply and extend the production season.

Harvesting: Harvesting in freshwater aquaculture refers to the process of collecting fish or prawns when they reach marketable size. Methods include netting, draining the pond, or partial harvesting, depending on species and farming system. Timing of harvest is crucial to optimize size and quality, with yields of 4-5 tons/ha achievable in well-managed systems.

Hatchery Management: Hatchery management includes the processes and practices involved in breeding, incubating, and rearing fish or prawn larvae until they are ready for stocking in nursery or grow-out ponds. Key components include maintaining optimal water quality, temperature control (e.g., 28°-30°C for many species), appropriate feeding of larvae with live feeds like Artemia or rotifers, and disease prevention. Efficient hatchery management ensures high survival rates, healthy larvae, and consistent production of quality seed stock.

Indian Major Carps: Indian major carps are a group of freshwater fish species, including Catla (*Catla catla*), Rohu (*Labeo rohita*), and Mrigal (*Cirrhinus mrigala*), which are widely cultured in South Asia due to their fast growth rates, high market demand, and suitability for polyculture. They thrive in pond systems with water temperatures ranging from 25°-32°C, pH levels between 7.0-8.5, and dissolved oxygen levels of 5-7 mg/L. These species have varied feeding habits, with Catla as a surface feeder, Rohu as a column feeder, and Mrigal as a bottom feeder.

Inland Saline Soils: Inland saline soils are characterized by high concentrations of soluble salts, typically found in arid or semi-arid regions far from coastal areas. These soils can be challenging for traditional agriculture but are increasingly used for aquaculture, especially with salt-tolerant species like certain tilapia and prawns. Salinity levels in these waters can range from 1-15 ppt, with optimal conditions for aquaculture involving temperature ranges of 25°-30°C and pH levels of 7.0-8.5. Proper management of water quality and salinity is crucial for successful aquaculture in these environments.

Isolation Ponds: Isolation ponds are used to separate diseased or injured fish from the main stock to prevent the spread of disease and allow for individual treatment. These ponds provide a controlled environment with optimal water conditions (temperature 25°-30°C, pH 7.0-8.0, dissolved oxygen above 5 mg/L) for recovery. Isolation ponds are part of biosecurity measures in aquaculture, ensuring that health issues are contained and managed without affecting the entire population.

Lakes: Lakes in aquaculture refer to large natural or artificial bodies of water that can be used for extensive fish farming. Natural lakes provide significant resources for open-water aquaculture, where species like carps or tilapia are stocked and harvested. Lake-based aquaculture involves minimal control over environmental conditions, requiring stocking and harvesting strategies adapted to the lake's natural productivity.

Lime Requirement: Lime is used in aquaculture ponds to adjust pH, improve water quality, and enhance the availability of essential nutrients. The requirement for lime depends on the soil pH and alkalinity of the pond water. Generally, agricultural lime (calcium carbonate) is applied at rates ranging from 200-2,000 kg/ha, depending on the initial pH level (target pH: 6.5-8.5). Lime application helps to neutralize acidity, increase buffering capacity, and promote a favorable environment for fish and plankton growth.

Magur (Clarias batrachus): Magur, also known as the walking catfish, is a hardy air-breathing freshwater fish native to Southeast Asia. It is highly valued for its taste, nutritional value, and adaptability to low-oxygen environments. Magur prefers stagnant or slow-moving waters with abundant vegetation and can tolerate a wide range of water quality parameters, including temperatures between 20°-30°C and pH levels from 6.5 to 8.0. It is omnivorous, feeding on small fish, insects, and organic detritus, making it a resilient species for aquaculture.

Mahua Oil Cake Application: Mahua oil cake is an organic pesticide and fertilizer derived from the seeds of the Mahua tree (*Madhuca indica*), commonly used in aquaculture ponds to control predatory fish and parasites. It contains saponins that are toxic to fish and pests but degrade within 10-15 days, making it safe for the cultured species afterward. The typical application rate is 250-500 kg/ha, depending on water volume and infestation levels, with care taken to apply when the pond is devoid of target species.

Major Carps: Major carps are a group of fast-growing freshwater fish species, including Catla (*Catla catla*), Rohu (*Labeo rohita*), and Mrigal (*Cirrhinus cirrhosus*). These species are highly valued in Indian aquaculture for their large size, high market demand, and suitability for polyculture systems. Major carps are typically raised in ponds at densities of 5,000-10,000 fish/ha, with yields of 3-5 tons/ha.

Manures: Manures, such as cow dung or poultry droppings, are commonly used in freshwater aquaculture to enhance pond productivity by promoting plankton growth, which serves as a natural food source for fish. Application rates range from 500 to 1,500 kg/ha, depending on the type of manure and pond conditions. Over-application can lead to oxygen depletion and water quality issues.

Minor Carps: Minor carps are freshwater fish species belonging to the family Cyprinidae, often found in rivers, lakes, and ponds. Unlike Indian major carps (such as Catla, Rohu, and Mrigal), minor carps are smaller in size and less commercially significant but still important in regional aquaculture due to their adaptability to a wide range of water conditions. Common examples include *Labeo calbasu* (Black Carp), *Labeo bata* (Bata), and *Labeo gonius*. These species thrive in water with a temperature range of 25°-30°C, pH between 7.0-8.5, and dissolved oxygen levels above 5 mg/L.

Monoculture of Prawns: Monoculture of prawns refers to the farming of a single prawn species, such as *Macrobrachium rosenbergii*, in aquaculture systems. This method allows for focused management of feeding, water quality, and disease control. Optimal conditions include water temperatures of 28°-32°C, pH levels of 7.0-8.5, and dissolved oxygen levels above 4 mg/L. Monoculture simplifies the culture environment but requires careful monitoring to prevent overcrowding and water quality deterioration.

Monoculture: Monoculture in aquaculture refers to the farming of a single species in a controlled environment. It allows for specialized management of feeding, water quality, and disease control, leading to potentially higher yields. However, it can also increase vulnerability to diseases and environmental stress. Common species in freshwater monoculture include carps, tilapia, and freshwater prawns.

Murrels (Channa species): Murrels, also known as snakeheads, are predatory freshwater fish native to Asia and Africa, known for their elongated bodies and ability to breathe atmospheric oxygen, allowing them to survive in low-oxygen environments. They are highly valued in aquaculture due to their rapid growth and high protein content. Murrels prefer water temperatures between 26°-30°C, pH levels of 6.5-8.0, and require well-oxygenated water for optimal growth, though they can survive in less favorable conditions.

Nursery Culture: Nursery culture refers to the initial phase of aquaculture where young fish or prawns are reared in a protected environment before being transferred to grow-out ponds. This stage is critical for ensuring high survival rates, with optimal stocking densities and water management required. Water temperatures of 28°-30°C and dissolved oxygen levels above 5 mg/L are ideal for most nursery systems.

Nursery Pond Management: Nursery pond management involves the rearing of fish larvae or juvenile prawns in specially designed ponds until they reach a size suitable for stocking in grow-out ponds. Key practices include maintaining water quality with optimal pH (6.5-7.5), dissolved oxygen levels above 6 mg/L, regular water exchange, and the application of starter feeds or natural food like zooplankton. Stocking densities are kept high initially and reduced as juveniles grow to minimize competition and enhance survival rates.

Nursery Rearing: Nursery rearing is a critical phase in aquaculture where fish larvae or juvenile prawns are raised in controlled conditions before being transferred to grow-out ponds. This stage focuses on providing optimal conditions for growth and survival, including stable water temperatures (26°-30°C), high dissolved oxygen levels (5-7 mg/L), and appropriate pH (7.0-8.5). Intensive feeding with high-quality, size-appropriate feed, and rigorous monitoring of water quality are essential to reduce mortality and ensure healthy development.

Organic Cycling: Organic cycling in aquaculture refers to the natural process of nutrient recycling within the pond ecosystem, involving the breakdown of organic matter (e.g., uneaten feed, fish waste) by microorganisms. This process releases nutrients such as nitrogen and phosphorus, which support the growth of phytoplankton and zooplankton, forming the base of the aquatic food web. Maintaining balanced organic cycling is essential for water quality and the health of cultured species, as excessive organic matter can lead to oxygen depletion and water quality issues.

Pangasius (Pangasianodon hypophthalmus): Pangasius, also known as the iridescent shark or sutchi catfish, is a popular freshwater fish species in aquaculture due to its rapid growth, high yield, and tolerance to high stocking densities. Native to the Mekong Basin in Southeast Asia, Pangasius thrives in warm waters with temperatures between 26°-30°C, pH levels of 6.5-8.0, and moderate to high dissolved oxygen levels. It is an omnivorous species that feeds on a range of plant and animal matter, making it versatile for various feeding regimes.

pH Levels: pH levels indicate the acidity or alkalinity of water and are a vital water quality parameter in aquaculture. The optimal pH range for most freshwater species is between 6.5 and 8.5. Deviations outside this range can cause stress, reduce growth, and impair physiological functions. Regular pH monitoring and management, using lime to raise pH or acids to lower pH, help maintain stable conditions conducive to the health and growth of cultured species.

Plankton Estimation: Plankton estimation is a crucial process in aquaculture to assess the abundance and composition of plankton in the water, which directly affects the food availability for filter-feeding species like fish and prawns. Estimation methods include plankton nets, water sampling, and counting under a microscope. An optimal plankton density is around 30,000-50,000 cells/mL for phytoplankton and 10,000-20,000 organisms/L for zooplankton, which supports healthy growth rates in aquaculture species.

Polyculture with Carps: Polyculture with carps is an aquaculture practice that combines different carp species in the same pond to utilize various ecological niches and optimize resource use. Common species include Catla (surface feeder), Rohu (column feeder), and Mrigal (bottom feeder), often supplemented with exotic carps like grass carp (herbivore) and silver carp (plankton feeder). This system enhances overall productivity and maintains balanced pond ecosystems. Ideal conditions include water temperatures of 25°-32°C, pH 7.0-8.5, and dissolved oxygen levels above 5 mg/L.

Polyculture: Polyculture involves raising multiple species in the same pond, typically combining species that occupy different ecological niches (e.g., surface, mid-water, and bottom dwellers) to maximize the use of available resources. In freshwater systems, Indian major carps are commonly cultured with tilapia or catfish, optimizing production. Polyculture systems tend to be more resilient and productive, with yields of 4,000-6,000 kg/ha in semi-intensive systems.

Pond Culture: Pond culture is a common method in freshwater aquaculture where fish or prawns are raised in managed ponds. It involves managing the water environment, stocking densities, feed inputs, and water quality. Ponds provide a controlled environment that allows for better management of growth conditions, with productivity levels ranging from 1,500 to 5,000 kg/ha/year, depending on intensification.

Pond Preparation (Prawn): Pond preparation for prawn farming involves a series of steps to create an optimal environment for prawn growth. Key steps include drying the pond to eliminate predators, applying lime (100-300 kg/ha) to adjust pH levels to 7.0-8.5, and fertilizing with organic and inorganic fertilizers to promote the growth of natural food sources. Water quality is managed to maintain temperatures of 28°-32°C and dissolved oxygen levels above 4 mg/L. Proper pond preparation is critical for enhancing prawn survival and growth rates.

Pond Preparation: Pond preparation is the process of conditioning the pond environment before stocking fish or prawns. This includes removing weeds, drying and liming the pond, and fertilizing to promote plankton growth. Proper pond preparation ensures a balanced ecosystem, reduces pathogen loads, and enhances the survival and growth of stocked species.

Pond Productivity: Pond productivity refers to the biological carrying capacity and yield potential of a pond in aquaculture. Factors affecting productivity include water quality, feed input, stocking density, and management practices. Well-managed ponds can reach yields of 3,000-6,000 kg/ha annually in semi-intensive systems, with productivity influenced by nutrient cycling, plankton growth, and organic inputs.

Ponds: Ponds are artificial or natural bodies of freshwater used extensively in aquaculture. These range in size from small homestead ponds to large commercial facilities, typically 0.1 to 2 hectares in area. Optimal water depth for fish culture is usually 1-2 meters. Proper pond construction and management, including water inlet and outlet controls, are crucial for maintaining water quality and ensuring high yields.

Prawn Growth Study: Prawn growth studies focus on assessing the growth performance of prawns, typically measured through periodic sampling of weight and size under controlled conditions. These studies involve monitoring water quality parameters, feeding regimes, and stocking densities to optimize growth. Key factors influencing prawn growth include temperature (optimal 28°-32°C), dissolved oxygen (>5 mg/L), and feed quality. Growth studies help in adjusting management practices to maximize yield and improve overall production efficiency.

Prawn Handling: Prawn handling involves the post-harvest management of prawns to maintain quality and minimize stress and damage. Key practices include gentle handling, rapid sorting, and immediate chilling (4°C) to slow metabolic processes and reduce spoilage. Proper hygiene, clean equipment, and quick transfer to cold storage are essential to prevent bacterial growth and ensure the prawns reach the market in optimal condition. Effective handling practices are critical to maintaining prawn quality and market value.

Prawn Harvesting: Prawn harvesting is the final stage in the prawn farming cycle, typically occurring 4-6 months after stocking, depending on the species and growth conditions. Harvesting involves draining the pond or using nets to collect prawns, ensuring minimal stress and damage to the prawns. Optimal water conditions (temperature 28°-32°C, pH 7.0-8.5) are maintained until harvest. Post-harvest handling, including immediate chilling, is crucial to preserve prawn quality and reduce spoilage.

Predatory Fish Identification: Predatory fish identification involves recognizing species that prey on commercially valuable aquaculture stock, such as tilapia, catfish, or prawns. Common predatory fish in freshwater systems include snakehead (Channa spp.), gar (Lepisosteus spp.), and certain cichlids. Identifying these predators is crucial for implementing control measures like pond netting, predator traps, or selective harvesting to prevent stock losses and maintain optimal growth rates.

Predatory Fishes: Predatory fishes, such as snakeheads and catfishes, can be detrimental in aquaculture ponds as they prey on stocked fish or fingerlings. Effective management includes screening pond inlets and outlets to prevent their entry, selective eradication using piscicides like mahua oil cake, or manual removal. The presence of predators can severely reduce survival rates in nursery and grow-out ponds.

Production Trends: Production trends in freshwater aquaculture indicate the growth patterns and yield outputs over time. In recent years, global freshwater aquaculture has seen significant growth due to rising demand for fish protein, technological advancements, and improved management practices. India and China are major contributors, with India producing over 3 million tons annually. Trends show increasing reliance on polyculture and intensification to meet growing demand.

Quarantine Ponds: Quarantine ponds are isolated ponds used to hold newly introduced fish or prawns before they are transferred to main culture ponds. This practice helps prevent the introduction of diseases and parasites into the aquaculture system. Fish are monitored for signs of illness and treated if necessary. Water quality parameters, such as temperature (25°-30°C), pH (7.0-8.5), and dissolved oxygen levels (above 5 mg/L), are carefully managed. Quarantine periods typically last 2-4 weeks, depending on the species and health status.

Rearing Pond Management: Rearing pond management refers to the practices used in intermediate stages of aquaculture where juvenile fish or prawns are grown to fingerling or post-larval size before transfer to grow-out ponds. This includes monitoring water quality, feeding with size-appropriate diets, controlling pests, and maintaining ideal conditions for growth (temperature 26°-30°C, pH 6.5-7.5, and dissolved oxygen above 5 mg/L). Regular grading and stocking density adjustments help prevent overcrowding and ensure uniform growth.

Reservoirs: Reservoirs are large, man-made water bodies created by damming rivers, often used for fish stocking in extensive aquaculture systems. They provide vast freshwater resources for the culture of species like carps and catfish. Reservoir aquaculture benefits from natural productivity but may face challenges related to water flow, depth, and ecosystem management. Stocking density is generally lower due to the extensive nature of the system.

Sex Segregation: Sex segregation in aquaculture involves separating male and female fish to prevent unwanted reproduction, reduce competition, or optimize growth. This practice is common in species like tilapia, where males grow faster than females. Techniques include manual sexing, genetic methods, or hormonal treatments. Maintaining optimal water conditions (temperature 25°-30°C, pH 7.0-8.0, dissolved oxygen above 5 mg/L) and tailored feeding regimes are essential for the success of sex-segregated populations.

Singhi (Heteropneustes fossilis): Singhi, or stinging catfish, is a species of air-breathing freshwater fish known for its resilience and ability to survive in low-oxygen and adverse water conditions. Native to South and Southeast Asia, it inhabits ponds, ditches, and swamps. Singhi thrives in temperatures ranging from 25°-32°C and can tolerate pH levels from 6.5 to 8.5. It is a carnivorous fish, feeding mainly on insects, small fish, and detritus, making it suitable for cultivation in diverse aquaculture systems.

Species Cultured: Species cultured in freshwater aquaculture include fish and invertebrates that are suitable for non-saline environments. Common species include Indian major carps (Catla, Rohu, Mrigal), tilapia, Pangasius, and freshwater prawns (*Macrobrachium rosenbergii*). The choice of species depends on the local environment, market demand, and the aquaculture system in use.

Species Selection: Species selection in aquaculture involves choosing the most suitable species for farming based on factors such as growth rate, market demand, environmental adaptability, disease resistance, and feeding habits. Considerations include water temperature (e.g., warmwater vs. coldwater species), salinity tolerance, and compatibility with existing species in polyculture systems. Optimal species selection enhances production efficiency, economic returns, and sustainability, tailored to specific environmental and market conditions.

Stocking Density: Stocking density refers to the number of fish or prawns introduced into a unit area of water in aquaculture. Optimal stocking density depends on the species, pond management system, and water quality. For Indian major carps, densities range from 5,000 to 10,000 fingerlings per hectare in semi-intensive systems, while more intensive systems may support up to 20,000 fish/ha.

Stocking Practices: Stocking practices in aquaculture involve the introduction of fish or prawns into culture systems at appropriate densities to optimize growth, minimize stress, and prevent disease outbreaks. Stocking densities vary by species and stage of growth; for example, tilapia fingerlings may be stocked at 10,000-15,000 per hectare. Proper acclimatization, gradual adjustment to water temperature, and pH are essential to reduce shock during stocking. Monitoring post-stocking behavior and health is crucial for early detection of any issues.

Stocking Ratio: Stocking ratio refers to the proportion of different species stocked in a polyculture system, ensuring efficient use of resources and minimizing competition. For example, in the composite fish culture system in India, the typical stocking ratio for Indian major carps might be 30% Catla, 30% Rohu, and 40% Mrigal, with adjustments based on pond productivity and market demand.

Stocking: Stocking involves introducing juvenile fish or prawns into ponds, tanks, or other aquaculture systems. Stocking rates are determined by species, farming intensity, and water quality, with typical densities for Indian major carps ranging from 5,000-10,000 fish per hectare. Successful stocking requires careful handling to minimize stress and ensure high survival rates.

Supplementary Feed Effects: Supplementary feeding in aquaculture provides additional nutrition to enhance the growth and health of fish and prawns beyond natural food availability. Common feeds include pellets, grains, and formulated diets containing protein, fats, vitamins, and minerals. The effects of supplementary feeding include improved growth rates, higher feed conversion efficiency (typically 1.2-1.5 FCR for fish), and reduced time to market size. Overfeeding, however, can lead to water quality issues, so careful management of feeding rates and schedules is essential.

Supplementary Feeding: Supplementary feeding involves providing additional food, such as formulated pellets or agricultural byproducts (e.g., rice bran, oil cakes), to enhance fish growth beyond the natural productivity of the pond. Feeding rates are based on body weight, typically ranging from 3-5% of the biomass daily, depending on water temperature, species, and pond management practices.

Tanks: Tanks are controlled, confined water bodies used for the intensive culture of aquatic organisms, often within hatcheries or small-scale aquaculture systems. Tanks are typically made of concrete, plastic, or fiberglass and are used for breeding, nursery rearing, or as grow-out facilities. Water quality parameters can be easily controlled, with dissolved oxygen maintained at 5 mg/L or higher and frequent water exchange to prevent waste buildup.

Tilapia: Tilapia refers to several species of cichlid fish originating from Africa, widely farmed worldwide due to their fast growth, high adaptability to various water conditions, and mild-tasting flesh. They thrive in warm freshwater environments with temperatures between 24°-30°C, pH levels of 6.5-8.5, and dissolved oxygen levels above 5 mg/L. Tilapia are omnivorous and can efficiently convert feed, including algae, plant matter, and commercial feeds, making them a cost-effective option for aquaculture.

Wastewater-fed Aquaculture: Wastewater-fed aquaculture utilizes treated municipal, industrial, or agricultural wastewater as a resource for fish farming. This system capitalizes on the nutrients in wastewater to promote the growth of algae and other natural food sources for fish, reducing the need for external feed and fertilizers. Key challenges include managing water quality to prevent contamination and ensuring the safety of the final product. Species like common carp and tilapia are often used due to their tolerance to variable water conditions.

Water Exchange: Water exchange in aquaculture involves the periodic replacement of pond or tank water to maintain water quality, remove waste products, and reduce harmful substances like ammonia and nitrites. The rate and frequency of water exchange depend on the species, stocking density, and water quality conditions but typically range from 5-10% daily or higher in intensive systems. Effective water exchange helps stabilize environmental conditions, promotes healthy growth, and minimizes disease risks.

Water Quality Management: Water quality management is crucial for maintaining optimal conditions for fish and prawn growth. Key parameters include dissolved oxygen (>5 mg/L), pH (6.5-8.5), ammonia levels (<0.02 mg/L), and temperature (25°-30°C). Regular monitoring and management of these factors help prevent stress, improve feed efficiency, and reduce disease outbreaks in aquaculture systems.

Water Quality Parameters: Water quality parameters in aquaculture are critical factors that influence the health and growth of cultured species. Key parameters include temperature (optimal range depends on species, e.g., 25°-30°C for tilapia), pH (6.5-8.5), dissolved oxygen (>5 mg/L), ammonia (<0.05 mg/L), nitrite (<0.1 mg/L), and salinity (if applicable). Regular monitoring and adjustments are necessary to maintain these parameters within the optimal ranges, ensuring a healthy environment for aquaculture species.

Water Treatment Methods: Water treatment methods in aquaculture are essential for maintaining optimal water quality and include mechanical, chemical, and biological processes. Mechanical methods involve filtration and sedimentation to remove solids, while chemical treatments use disinfectants (e.g., chlorine, ozone) to control pathogens. Biological methods, such as biofilters and aquatic plants, help remove ammonia and other toxins through natural processes. Regular monitoring of parameters like pH (7.0-8.5), dissolved oxygen (>5 mg/L), and ammonia levels (<0.1 mg/L) is crucial for effective water treatment.

Weed Control: Weed control in freshwater aquaculture is essential for maintaining productivity, as aquatic weeds can compete with cultured species for nutrients and oxygen, and may harbor pathogens. Methods include mechanical removal, biological control using herbivorous fish (such as grass carp), and chemical herbicides like copper sulfate, though careful management is needed to avoid negative effects on water quality.

Weedicide Application: Weedicide application in aquaculture involves the use of chemical herbicides to control the growth of unwanted aquatic plants and algae in ponds. Effective management requires selecting appropriate, fish-safe weedicides, such as copper sulfate or diquat, and applying them at recommended doses (e.g., 2-4 ppm for copper sulfate) under calm weather conditions to minimize water movement and ensure effectiveness. It is crucial to follow safety guidelines to prevent toxicity to cultured species and maintain water quality standards.

Wintering Ponds: Wintering ponds are specialized ponds used in aquaculture to overwinter juvenile fish or prawns, protecting them from harsh winter conditions. These ponds are designed to maintain stable temperatures (above 10°C) and oxygen levels (5-7 mg/L) to ensure the survival of young stock. Proper management includes monitoring water quality, providing adequate feed, and ensuring protection from predators. Wintering ponds help in maintaining stock continuity and preparing fish for the next growing season.

Yield: Yield refers to the total biomass of fish or prawns harvested per unit area or volume of water. In semi-intensive freshwater aquaculture systems, yields typically range from 3,000 to 6,000 kg/ha/year, depending on species, stocking density, feeding practices, and water quality management. High-yield systems require careful attention to pond maintenance and resource input.

CHAPTER 12

FRESHWATER FISH SEED PRODUCTION AND HATCHERY MANAGEMENT

Aeration Systems: Aeration systems in hatcheries are used to maintain adequate dissolved oxygen levels in water, essential for the health and growth of broodstock, eggs, and larvae. These systems include air stones, diffusers, paddle wheels, or mechanical aerators that introduce air into the water, enhancing oxygenation and circulation. Proper aeration ensures oxygen levels remain above critical thresholds (>5 mg/L) to prevent hypoxia, reduce stress, and support optimal metabolic functions in aquatic organisms.

Artificial Fertilization: Artificial fertilization is a controlled breeding technique where eggs and sperm are manually stripped from fish and mixed outside the body to achieve fertilization. This method allows precise control over the breeding process, maximizing fertilization rates and ensuring the quality of progeny. Artificial fertilization is often used in conjunction with induced spawning, particularly in species with difficult natural spawning behaviors.

Breeding Techniques: Breeding techniques in aquaculture include both natural and induced methods used to produce fish seed. These techniques vary from species to species and can involve hormone administration (e.g., HCG, GnRH analogues), environmental manipulation (temperature, photoperiod), and manual stripping of eggs and sperm. Breeding techniques are critical for optimizing reproductive output and ensuring the availability of high-quality fish seed for aquaculture operations.

Broodstock Availability: Broodstock availability refers to the accessibility of mature, healthy fish suitable for breeding throughout the year. Consistent availability depends on factors such as seasonal breeding cycles, broodstock management practices, and access to wild or captive broodstock sources. Ensuring a reliable supply of broodstock is critical for uninterrupted hatchery operations and meeting production targets.

Broodstock Conditioning: Broodstock conditioning involves preparing fish for spawning by optimizing their health, nutrition, and environmental conditions. This includes acclimatizing broodstock to breeding conditions, enhancing their diet with nutrient-rich feeds, and managing stress through stable water quality (temperature, dissolved oxygen, pH). Proper conditioning ensures high-quality gametes, increased spawning success, and healthier progeny, forming the foundation of successful hatchery operations.

Broodstock Conditioning: Broodstock conditioning is the process of preparing adult fish for spawning by optimizing their health, reproductive readiness, and environmental conditions. This involves regulating their diet, water quality, and environmental parameters, including temperature (25°-28°C), photoperiod (12-16 hours of light), and water flow. A nutritionally balanced diet rich in proteins (35-40%) and essential fatty acids is provided to enhance gonadal development and increase fecundity. Broodstock conditioning can take several weeks to months, depending on species.

Broodstock Management: Broodstock management encompasses the practices and strategies used to maintain healthy and productive broodfish for breeding purposes. This includes selection of high-quality broodstock, proper nutrition, regular health checks, management of environmental conditions (temperature, dissolved oxygen, pH), and stress reduction. Effective broodstock management ensures high reproductive output, including the production of viable eggs and sperm, thereby supporting the success of hatchery operations.

Broodstock Nutrition: Broodstock nutrition focuses on providing a balanced diet rich in proteins, lipids, vitamins, and minerals to adult fish intended for breeding. Proper nutrition enhances reproductive performance, egg quality, and fry viability. Diets are often supplemented with essential fatty acids (e.g., omega-3s) and Vitamins like E and C to improve gamete quality and immune function. Feeding rates and frequencies are adjusted based on species, size, and reproductive cycles to ensure optimal conditioning.

Broodstock Selection: Broodstock selection involves choosing adult fish for breeding based on desirable traits such as size, health, growth rate, and reproductive performance. Selection criteria may also include genetic diversity and disease resistance to ensure high-quality offspring. Effective broodstock selection is crucial for improving hatchery output and maintaining the genetic integrity of cultured species.

Broodstock Transport: Broodstock transport involves the movement of mature fish from one location to another, such as from ponds to hatcheries, with minimal stress and mortality. Key considerations include maintaining optimal water quality, temperature control (typically 22°-26°C for tropical species), and adequate oxygenation (using aeration or oxygen tablets). Proper transport protocols reduce stress, prevent injuries, and ensure that broodstock arrive in good condition, ready for breeding.

Broodstock: Broodstock refers to the adult fish that are used in aquaculture for breeding purposes. These fish are selected based on desirable traits such as size, growth rate, disease resistance, and reproductive efficiency. Proper management of broodstock is crucial for successful seed production, including maintaining optimal environmental conditions such as water temperature (typically 24°-30°C for tropical species), dissolved oxygen (above 5 mg/L), and pH (7.0-8.5). Broodstock are fed a high-quality diet rich in protein, vitamins, and essential fatty acids to ensure the production of viable eggs and sperm.

Chinese Hatchery: The Chinese hatchery system is a widely used aquaculture method for the mass production of fish seed, particularly carps. It consists of circular tanks for broodstock conditioning, spawning, egg incubation, and larval rearing. The system uses a continuous water flow to keep eggs aerated and free of waste, enhancing survival rates. Chinese hatcheries are designed for large-scale operations, capable of producing millions of fry in a single breeding cycle.

Cryopreservation: Cryopreservation is a technique used to preserve fish gametes (sperm, and in some cases, eggs) at extremely low temperatures (-196°C) using liquid nitrogen. This method allows long-term storage of genetic material, facilitating selective breeding and conservation of genetic diversity. Cryoprotectants, such as dimethyl sulfoxide (DMSO), are used to protect cells from ice damage during freezing. Cryopreservation is especially valuable for maintaining broodstock and managing breeding programs.

Disease Management: Disease management in hatcheries involves the identification, prevention, and treatment of diseases affecting fish or prawn populations. This includes regular monitoring of health status, maintaining optimal water quality, using vaccines or prophylactic treatments, and applying therapeutic measures when outbreaks occur. Effective disease management is crucial for minimizing losses and ensuring the production of healthy seed.

Egg Evaluation: Egg evaluation is the process of assessing the quality and developmental potential of eggs collected from female broodstock. Criteria include size, color, shape, and the presence of visible oil droplets. Fertilized eggs are often evaluated based on developmental staging, with early cleavage and blastula formation being indicators of good quality. Optimal water conditions for egg incubation generally include a temperature range of 24°-28°C and stable pH levels (7.0-8.0).

Egg Handling: Egg handling involves the careful management of fish or prawn eggs during collection, incubation, and transfer to ensure high hatching success. Proper handling techniques include gentle collection, avoiding physical damage, maintaining optimal incubation conditions (temperature, oxygenation), and preventing contamination. Effective egg handling is essential to maximize hatch rates and minimize early-stage mortality.

Egg Hatchability: Egg hatchability refers to the percentage of fertilized eggs that successfully hatch into viable larvae. High hatchability rates are indicative of good broodstock health, optimal environmental conditions, and effective hatchery management. Factors influencing hatchability include water temperature (26°-30°C), dissolved oxygen levels (>6 mg/L), gentle aeration, and proper handling of eggs to prevent physical damage or fungal infections. Monitoring and maintaining these parameters help achieve high hatch success.

Egg Staging: Egg staging refers to the classification of eggs based on their developmental stages, from fertilization through hatching. Staging is used to monitor embryo development, optimize incubation conditions, and predict hatching times. Typical stages include cleavage, blastula, gastrula, and organogenesis. Correctly staged eggs ensure that hatchery operations can be managed efficiently, with temperature and water quality adjustments made to support each developmental phase.

Embryo Development: Embryo development in fish encompasses the stages from fertilization of the egg to the hatching of the larvae. This process involves cleavage, gastrulation, organogenesis, and differentiation of the embryo. Optimal conditions for embryo development include stable temperatures (26°-30°C for many species), high dissolved oxygen levels, and minimal disturbances. Monitoring and managing these conditions are critical to ensuring high hatch rates and the production of healthy, viable larvae.

Environmental Control: Environmental control in hatcheries refers to the regulation of physical and chemical parameters, such as temperature, pH, salinity, dissolved oxygen, and light, to create optimal conditions for breeding and rearing aquatic organisms. Proper environmental control enhances the health, growth, and survival rates of broodstock, eggs, and larvae, reducing stress and improving overall hatchery efficiency.

Exotic Carps: Exotic carps refer to non-native carp species introduced into new regions for aquaculture, including Common Carp (*Cyprinus carpio*), Grass Carp (*Ctenopharyngodon idella*), and Silver Carp (*Hypophthalmichthys molitrix*). These species are valued for their rapid growth and adaptability to diverse aquaculture environments. However, their introduction must be managed carefully due to potential ecological impacts. Exotic carps are typically bred in hatcheries using induced spawning techniques.

Fertilization: Fertilization in aquaculture refers to the union of male and female gametes (sperm and eggs) to form a zygote, initiating the development of a new organism. In hatcheries, fertilization can be achieved naturally or through artificial methods, such as hand-stripping and manual mixing of gametes. Optimal fertilization rates depend on factors such as gamete quality, timing, and environmental conditions, with success rates commonly exceeding 80% in well-managed hatchery operations.

Fertilized Eggs: Fertilized eggs in aquaculture are those that have been successfully fertilized during the spawning process, leading to the development of embryos. These eggs are typically characterized by a clear, translucent appearance with visible cell division when observed under a microscope. Optimal conditions for incubation include maintaining a temperature of 26°-30°C, dissolved oxygen levels above 6 mg/L, and gentle water flow to keep eggs suspended and prevent fungal infections.

Fingerling Rearing: Fingerling rearing refers to the phase where fish fry are grown to fingerlings (usually 5-10 cm in length) before being transferred to grow-out ponds. This stage focuses on providing nutritionally balanced feeds and maintaining water quality with parameters such as temperature (25°-28°C), pH (6.5-7.5), and dissolved oxygen (>5 mg/L). Grading and regular health checks help prevent cannibalism, disease outbreaks, and ensure uniform growth among fingerlings.

Fingerling Transport: Fingerling transport involves the movement of fingerlings (young fish that have developed beyond the fry stage) from nurseries to grow-out facilities. Optimal transport conditions include water temperatures of 20°-25°C, dissolved oxygen above 5 mg/L, and gentle handling to reduce stress. Proper transport techniques help maintain fingerling health and minimize losses, ensuring successful stocking in grow-out systems.

Fish Seed Economics: Fish seed economics involves the analysis of costs, revenues, and profitability associated with the production, management, and distribution of fish seed (fry and fingerlings). Key factors include broodstock costs, hatchery operation expenses, feed and water quality management, labor, and transportation. Profitability is influenced by seed quality, survival rates, market demand, and pricing strategies. Efficient management and economies of scale are critical to maximizing returns in fish seed production.

Fry Grading: Fry grading is the process of sorting fry by size to ensure uniform growth and reduce competition among individuals. Grading helps improve overall survival and growth rates by allowing similarly sized individuals to be reared together, thus minimizing cannibalism and size-related stress. Grading is typically performed manually or using automated systems, and is a critical step in nursery management.

Fry Marketing: Fry marketing includes strategies for promoting and selling fish fry to aquaculture farms and other stakeholders. Effective marketing relies on understanding market demand, ensuring high-quality fry with good survival and growth potential, and building strong relationships with buyers. Marketing also involves logistics coordination, price negotiation, and meeting regulatory standards for biosecurity and transportation. Branding and certification for disease-free status can enhance marketability and premium pricing.

Fry Packaging: Fry packaging refers to the methods used to prepare fish fry for transport, ensuring they remain healthy and stress-free. Common techniques include packing fry in oxygenated water-filled plastic bags or containers with optimal fry-to-water ratios, typically around 100-150 fry per liter depending on species and size. Oxygen levels are maintained with supplemental oxygen, and temperatures are controlled to match transport conditions, often around 20°-25°C to reduce metabolic rates and stress.

Fry Rearing: Fry rearing is the process of growing fish larvae into fry, typically in specially prepared nursery ponds or tanks. This stage involves intensive feeding with appropriate diets (e.g., live feeds such as Artemia or formulated microdiets) and maintaining optimal water quality (temperature 26°-28°C, dissolved oxygen >5 mg/L, and pH 6.5-7.5). Regular monitoring and management of stocking densities are critical to avoid overcrowding, reduce stress, and enhance growth and survival rates.

Fry Supply: Fry supply encompasses the processes involved in producing, handling, and distributing fish fry from hatcheries to aquaculture farms. This includes optimizing hatchery production, ensuring healthy fry through proper feeding and water quality management, and coordinating logistics for safe transport. Reliable fry supply chains are essential for consistent and timely stocking in grow-out operations, directly impacting overall production efficiency and profitability.

Fry Transport: Fry transport refers to the movement of young fish (fry) from hatcheries to nursery or grow-out ponds. Key considerations include maintaining stable water temperature (usually 22°-26°C), adequate oxygenation, and minimizing handling stress. Transport containers are typically oxygenated bags or tanks with controlled water quality parameters to ensure the fry arrive in good health and ready for further rearing.

Gametogenesis: Gametogenesis is the biological process by which gametes (sperm in males and eggs in females) are produced in the gonads. In fish, spermatogenesis (formation of sperm) occurs in the testes, while oogenesis (formation of eggs) occurs in the ovaries. This process is regulated by reproductive hormones such as Follicle-Stimulating Hormone (FSH) and Luteinizing Hormone (LH). The completion of gametogenesis is essential for successful spawning and fertilization, with optimal environmental parameters including stable temperatures and good nutrition.

Gonadal Development: Gonadal development refers to the growth and maturation of reproductive organs (gonads) in fish, including the testes in males and ovaries in females. This process involves multiple stages, from immature (undeveloped) to fully mature gonads capable of producing gametes. The rate of gonadal development is influenced by factors such as age, nutrition, photoperiod, and water temperature (optimal 25°-30°C for many species). Proper gonadal development is critical for successful breeding and is closely monitored in broodstock management.

Grow-out Relationship: The grow-out relationship refers to the connection between hatchery practices and grow-out performance, highlighting how early-stage management impacts final production outcomes. Proper hatchery management, including optimal larval care, feed quality, and disease control, directly affects the growth, survival, and overall health of fish in grow-out ponds. A strong grow-out relationship emphasizes the importance of high-quality seed and robust hatchery protocols to achieve maximum yield and profitability in aquaculture.

Growth Monitoring: Growth monitoring involves the regular measurement of growth parameters (length, weight) of fish or prawn larvae, fry, or fingerlings to assess their development and adjust management practices accordingly. Monitoring growth helps optimize feeding rates, stocking densities, and environmental conditions, ensuring efficient use of resources and achieving target sizes within the desired timeframe.

Hapa: A hapa is a fine mesh enclosure used in aquaculture for the breeding, nursing, or holding of fish and prawns in ponds or tanks. Hap's allow water exchange and are typically used to confine broodstock, fry, or fingerlings, providing a controlled environment for spawning, early-stage rearing, or temporary holding. They are particularly useful for protecting fish from predators, reducing stress during handling, and facilitating easy observation and management of stock.

Hatchery Biosecurity: Hatchery biosecurity encompasses the measures and practices implemented to prevent the introduction and spread of diseases within a hatchery. This includes controlling access to the facility, disinfecting equipment and water supplies, regular health checks, and quarantining new or sick broodstock. Effective biosecurity protocols reduce the risk of outbreaks, safeguard the health of stock, and improve overall hatchery productivity.

Hatchery Design: Hatchery design refers to the planning and construction of facilities for breeding, hatching, and rearing aquatic species, optimized for efficient production. Key elements include well-designed spawning tanks, larval rearing units, incubation systems, and water quality management infrastructure like filtration and aeration systems. The design should facilitate easy movement of stock, water flow management, biosecurity measures, and scalable operations to meet production targets while maintaining optimal environmental conditions.

Hatchery Economics: Hatchery economics encompasses the financial aspects of operating a hatchery, including costs of broodstock, feed, labor, equipment, energy, and other inputs, as well as revenue from the sale of seed. It involves analyzing production efficiency, cost-benefit ratios, and profitability to guide decision-making and improve economic viability. Understanding hatchery economics is crucial for sustainable and profitable aquaculture operations.

Hatchery Equipment: Hatchery equipment includes a wide array of tools and systems necessary for successful breeding, incubation, and larval rearing. Key equipment includes spawning tanks, incubation trays or jars, water pumps, aerators, filtration systems, temperature control devices, oxygen injectors, and egg collection nets. These tools are designed to maintain optimal environmental conditions (e.g., water quality, temperature, oxygen levels) and ensure the safe handling of sensitive life stages such as eggs and larvae.

Hatchery Hygiene: Hatchery hygiene involves maintaining clean and sanitary conditions in hatchery facilities to prevent the spread of diseases and contaminants that could affect broodstock, eggs, and larvae. This includes regular cleaning and disinfection of tanks, equipment, and water systems, as well as controlling biofilm and algal growth. Strict biosecurity measures, such as restricting access, quarantine procedures, and using pathogen-free water sources, are critical to ensuring the health and productivity of hatchery operations.

Hatchery Protocols: Hatchery protocols are standardized procedures that govern the operation of aquaculture hatcheries, covering all aspects from broodstock management to larval rearing. These protocols ensure consistency in water quality management, feeding regimes, disease prevention, and handling practices. Adhering to established protocols is crucial for optimizing hatchery performance, improving seed quality, and ensuring the economic viability of aquaculture operations

Hatchery Site Criteria: Hatchery site criteria involve selecting an appropriate location for establishing a fish or prawn hatchery, focusing on factors such as proximity to clean and abundant water sources, soil quality, ease of access, and protection from environmental hazards like floods. The site should allow for efficient water supply and drainage, with water parameters (pH 7.0-8.5, temperature 25°-30°C) suitable for the species being cultured. Additionally, the site should have good road connectivity for transport and logistics.

Hatchery Staff Training: Hatchery staff training is essential for the efficient operation of fish hatcheries, encompassing all aspects of broodstock management, water quality monitoring, disease management, spawning techniques, and larval rearing. Regular training programs should focus on developing technical skills, biosecurity protocols, and the use of advanced hatchery technologies. Continuous professional development ensures that staff can effectively handle the sensitive life stages of fish, respond to emergencies, and optimize production processes.

Hatchery Systems: Hatchery systems in aquaculture encompass the various methods and technologies used for the breeding, incubation, and rearing of fish and shellfish larvae. These systems include static water, flow-through, and recirculating aquaculture systems (RAS), each offering different levels of water quality control and operational complexity. The choice of system depends on species requirements, production scale, and economic considerations, with the goal of maximizing survival and growth rates.

Hatchery: A hatchery is a facility designed for the breeding, hatching, and early rearing of aquatic species such as fish and prawns. It provides controlled conditions that optimize spawning, egg incubation, and larval development. Key components of a hatchery include broodstock holding tanks, spawning areas, incubators, and larval rearing tanks. Hatcheries manage water quality, temperature (25°-30°C for most freshwater species), and aeration to create optimal conditions for the growth and survival of early life stages, thus playing a critical role in the aquaculture supply chain.

Hormone Administration: Hormone administration in hatcheries is the controlled use of reproductive hormones, such as GnRH analogues or HCG, to induce spawning in broodstock. Proper dosing and timing of hormone injections or implants are critical for stimulating gamete release and achieving successful breeding. Hormone administration requires careful handling and monitoring to avoid stress and ensure effective reproduction.

Hormone Analogues: Hormone analogues are synthetic compounds designed to mimic the action of natural hormones involved in reproduction. In aquaculture, analogues of GnRH, such as Ovaprim and Ovatide, are commonly used to induce spawning in a controlled manner. These analogues are administered through injections or implants and are effective in stimulating gamete release, even in species that do not naturally spawn in captivity.

Incubation: Incubation in aquaculture refers to the process of maintaining fertilized eggs under controlled conditions until they hatch into larvae. Key parameters include maintaining stable temperature (typically 26°-30°C for tropical species), high dissolved oxygen levels (>6 mg/L), and gentle water movement to keep eggs aerated and free from sediment. The incubation period varies by species, ranging from a few days to weeks, with careful monitoring to prevent fungal infections and other issues that can affect hatchability.

Incubator Design: Incubator design in aquaculture involves creating a controlled environment for the development of fish eggs until they hatch into larvae. Incubators must maintain optimal water flow, temperature (typically 24°-28°C), and oxygen levels to support embryo development. Common designs include upwelling and downwelling systems, which ensure constant water circulation around the eggs, reducing the risk of fungal infections and improving hatching success.

Indian Major Carps: Indian Major Carps are a group of economically important freshwater fish species in South Asia, including Catla (*Catla catla*), Rohu (*Labeo rohita*), and Mrigal (*Cirrhinus mrigala*). These species are characterized by their rapid growth rates, high fecundity, and adaptability to various aquaculture systems. They typically spawn during the monsoon season in rivers and reservoirs. Optimal conditions for breeding include water temperatures of 24°-30°C and a pH range of 7.0-8.5.

Induced Breeding: Induced breeding refers to the artificial stimulation of fish to spawn under controlled hatchery conditions. This is typically achieved by administering hormones such as carp pituitary extract (CPE), human chorionic gonadotropin (HCG), or synthetic analogs like Ovaprim, to trigger gonadal maturation and ovulation in mature broodstock. The process is essential for species that do not spawn readily in captivity. In freshwater aquaculture, optimal water temperature (24°-30°C) and dissolved oxygen (DO) levels (5-7 mg/L) are crucial for successful breeding outcomes.

Induced Spawning: Induced spawning is a technique used in aquaculture to stimulate fish to release eggs and sperm through the administration of hormones or hormone analogues. This method is often used for species that do not spawn readily in captivity. Common hormones used include Human Chorionic Gonadotropin (HCG) and synthetic GnRH analogues, administered at doses that vary depending on species and size. Induced spawning allows for precise timing of reproduction, maximizing egg and sperm collection and improving hatchery efficiency.

Jar Hatchery: A jar hatchery is a simple hatchery system that uses glass or plastic jars with a constant flow of water to incubate fish eggs. The jars are typically aerated to keep the eggs suspended and provide a steady supply of oxygen. This system is widely used for small-scale hatchery operations and is ideal for species with non-adhesive eggs. Jar hatcheries are easy to manage, allowing for precise control over incubation conditions and high hatching rates.

Larval Feed: Larval feed consists of specially formulated diets or live feed (such as rotifers, artemia, or microalgae) provided to fish or prawn larvae during their early developmental stages. Larval feed must meet the nutritional needs of the species, with high protein content (40-50%), essential fatty acids, and appropriate particle size. Proper feeding regimes support rapid growth and high survival rates during the critical larval stage.

Larval Nutrition: Larval nutrition is critical during the early stages of fish development, where the diet needs to provide adequate proteins (45-55%), lipids (10-15%), and essential micronutrients. Live feeds such as Rotifers, Artemia nauplii, and copepods are typically used in hatcheries for their high nutritional content and digestibility. As larvae grow, they are gradually transitioned to formulated diets. Proper larval nutrition promotes healthy growth, survival, and resistance to diseases, with optimal feeding rates of 10-15% of body weight per day, depending on species and developmental stage

Larval Stages: Larval stages in fish refer to the early developmental phases immediately following hatching, characterized by high sensitivity to environmental conditions and nutritional requirements. During this period, larvae transition from endogenous feeding (yolk sac absorption) to exogenous feeding (live or formulated feeds). Critical parameters include optimal water temperature (26°-30°C for many tropical species), dissolved oxygen (>5 mg/L), and pH (6.5-7.5). Proper management during these stages is essential to ensure high survival rates and healthy development.

Milt Evaluation: Milt evaluation involves assessing the quality and viability of sperm collected from male broodstock. Key parameters include sperm concentration, motility, and morphology. Evaluating milt quality is essential in artificial fertilization to ensure high fertilization rates. Techniques such as microscopic examination and specialized tests for sperm viability are used, with optimal motility rates being above 80% for most aquaculture species.

Murrels: Murrels, also known as snakeheads, are predatory freshwater fish belonging to the family Channidae, with species such as *Channa striata* and *Channa marulius* being common in aquaculture. They are valued for their high protein content and fast growth rates. Murrels are capable of air-breathing, making them highly adaptable to varying water conditions. They reach sexual maturity within 1-2 years and can spawn multiple times a year under optimal conditions of 26°-30°C water temperature.

Natural Breeding: Natural breeding refers to the process where fish spawn under conditions that mimic their natural environment without human intervention or hormonal induction. This approach relies on creating optimal conditions such as appropriate water temperature, photoperiod, substrate, and shelter to stimulate natural courtship and spawning behaviors. It is used primarily for species that breed easily in captivity, ensuring genetic diversity and reducing stress on broodstock compared to induced methods.

Natural Fertilization: Natural fertilization occurs when fish spawn naturally, releasing eggs and sperm into the water where fertilization occurs externally. This process is typical in species that spawn in response to environmental cues, such as temperature changes or seasonal variations. Natural fertilization requires maintaining optimal spawning conditions, including water temperature (25°-30°C), appropriate substrate, and water flow, to facilitate the meeting of eggs and sperm.

Nursery Pond Design: Nursery pond design involves planning and constructing ponds specifically for rearing fry and fingerlings. Key features include proper pond depth (usually 1-1.5 meters), gentle slopes for ease of harvest, and adequate aeration and water exchange systems to maintain water quality. Ponds are typically lined with clay or plastic to retain water and prevent seepage, and designed to facilitate easy management and monitoring.

Nursery Pond Preparation: Nursery pond preparation involves conditioning the pond environment to create optimal conditions for rearing fry or fingerlings. This includes cleaning and drying the pond, applying lime to adjust pH levels (target pH 7.0-8.0), fertilization with organic and inorganic fertilizers to promote natural food growth, and filling with water free from predators and contaminants. Proper pond preparation ensures high survival rates and promotes rapid growth of fry.

Nursery Tank: Nursery tanks are specialized tanks used to rear fish larvae into fry before they are transferred to grow-out systems. The tanks are usually shallow, with a capacity of 500-1,000 liters, and are equipped with aerators, filtration units, and heating systems. Optimal water conditions, including temperature (28°-30°C), pH (6.8-7.5), and DO levels (5-7 mg/L), are maintained for effective growth and survival of fry. The stocking density typically ranges from 1,000 to 5,000 fry per cubic meter, depending on the species and tank design.

Nutritional Requirements: Nutritional requirements refer to the specific dietary needs of broodstock, which include high levels of protein (35-40%), essential fatty acids, vitamins, and minerals to promote optimal health and reproductive performance. Proper nutrition enhances egg quality, fecundity, and overall broodstock vitality. Diets are often supplemented with live feed, such as earthworms or crustaceans, and formulated feeds designed to meet the high energy and nutrient demands of reproducing fish.

Oxygenation: Adequate oxygenation in hatchery systems ensures proper respiration for broodstock, eggs, and larvae. Oxygen levels should be maintained at 5-8 mg/L to avoid hypoxia, which can lead to stress, developmental issues, and high mortality. Aeration systems, air stones, surface agitators, and liquid oxygen injectors are commonly used to regulate dissolved oxygen levels. Oxygenation becomes even more critical during peak hatchery operations when fish biomass is high, and oxygen demand escalates.

Photoperiod Management: Photoperiod management involves manipulating the light-dark cycle in hatcheries to control reproductive cycles in fish. Extended daylight hours (12-16 hours of light) are typically used to stimulate gonadal development and induce spawning in species sensitive to day length, such as tilapia and catfish. This practice is essential in out-of-season breeding programs and allows hatcheries to produce seed year-round. Artificial lighting systems, with timers and dimmers, are employed to maintain consistent photoperiods.

Pituitary Extract: Pituitary extract is a natural hormone preparation derived from the pituitary glands of fish, commonly used to induce spawning in broodstock. The extract contains gonadotropins that stimulate the reproductive organs, triggering ovulation and spawning. Dosage depends on the species, size, and maturity of the broodstock, but typically ranges from 2-5 mg/kg of body weight for common freshwater species like carp. Proper handling and injection techniques are essential to ensure effective results.

Pituitary Gland: The pituitary gland in fish is a small endocrine gland located at the base of the brain, responsible for secreting hormones that regulate various physiological processes, including growth, metabolism, and reproduction. In aquaculture, pituitary glands from donor fish (often carp) are used as a source of gonadotropins to induce spawning in broodstock via injection, stimulating ovulation and spermiation in species that do not spawn easily under captive conditions.

Prawn Hatchery Protocols: Prawn hatchery protocols refer to the standardized set of procedures and best practices employed to optimize the breeding, hatching, and rearing of prawn larvae. These protocols include maintaining optimal water quality (temperature 28°-31°C, salinity 12-15 ppt, dissolved oxygen above 5 mg/L), feeding schedules, biosecurity measures, and disease management strategies. The goal is to maximize survival and growth rates of prawn larvae by ensuring a stable and conducive environment throughout the hatchery process.

Prawn Hatchery: A prawn hatchery is a specialized facility designed for breeding and rearing prawns, particularly during their larval and juvenile stages. Key components include broodstock tanks, larval rearing tanks, and water quality control systems (maintaining salinity of 15-25 ppt for most species). The hatchery environment is carefully managed to provide optimal conditions for spawning, hatching, and larval development, including stable temperatures (28°-30°C), high dissolved oxygen, and appropriate feeding with microalgae or Artemia.

Prawn Larval Rearing: Prawn larval rearing involves the cultivation of prawn larvae from hatchlings to post-larvae stages under controlled conditions. This process requires maintaining optimal salinity (15-25 ppt), temperature (28°-30°C), and water quality with high dissolved oxygen (>6 mg/L). Feeding typically includes microalgae, Artemia nauplii, and formulated feeds, provided in appropriate quantities to support rapid growth and high survival rates. Regular monitoring and adjustments to rearing conditions are essential for successful larval development

Rearing Pond Design: Rearing pond design focuses on creating optimal environments for the growth of fingerlings or juvenile fish until they are ready for market or further grow-out. Ponds should have a uniform depth (1.5-2.5 meters), be easily drainable, and include aeration systems to maintain oxygen levels. Water quality management, including regular monitoring of temperature, pH, and ammonia levels, is essential to support healthy fish growth.

Reproductive Hormones: Reproductive hormones, such as Gonadotropin-Releasing Hormone (GnRH), Follicle-Stimulating Hormone (FSH), and Luteinizing Hormone (LH), play a critical role in regulating gametogenesis and spawning in fish. These hormones are used in aquaculture to induce spawning, especially in species that do not readily breed in captivity. Synthetic versions, such as GnRH analogues and HCG, are commonly used to synchronize and enhance reproductive processes.

Seed Transportation: Seed transportation involves moving fish fry or fingerlings from hatcheries to rearing or grow-out sites. Key factors include minimizing stress, maintaining optimal water quality, and ensuring adequate oxygenation during transit. Transport containers are usually equipped with aeration systems or oxygen packs, and temperature control is maintained to prevent thermal shock. Proper handling and careful acclimatization upon arrival are crucial to reduce mortality rates.

Sexual Dimorphism: Sexual dimorphism in aquaculture refers to the physical differences between male and female fish of the same species, which are often used for sex identification and selective breeding. These differences can include variations in size, coloration, fin shapes, or the presence of secondary sexual characteristics such as tubercles or breeding nodules. Recognizing sexual dimorphism is crucial for effective broodstock management, allowing accurate pairing and enhanced breeding success.

Sexual Maturity: Sexual maturity is the stage in an organism's life cycle when it becomes capable of reproducing. In fish, this is characterized by the development of mature gonads capable of producing eggs in females and sperm in males. The age and size at which fish reach sexual maturity can vary widely between species; for instance, Indian Major Carps typically mature at 2-3 years of age, depending on environmental conditions and nutrition. Sexual maturity is crucial for broodstock selection and successful breeding in aquaculture.

Spawn Care: Spawn care involves the management of newly spawned eggs and larvae to ensure high survival and hatching rates. Key practices include maintaining optimal water quality, controlling temperature (usually 26°-30°C), providing adequate aeration to prevent egg clumping and fungal infections, and monitoring for any signs of poor health or deformities. Gentle water flow is also maintained to simulate natural conditions and keep the eggs suspended, preventing sedimentation and suffocation.

Spawn Collection: Spawn collection refers to the process of collecting fertilized eggs after the induced spawning event. In hatcheries, eggs are often collected using soft nets, sieves, or trays designed to minimize damage. The collected eggs are then transferred to incubation systems, such as hatchery jars or trays, where they are kept under controlled environmental conditions (temperature 26°-30°C, oxygen 5-8 mg/L) to ensure successful hatching. Proper spawn collection techniques reduce egg mortality and enhance the hatch rate.

Spawn Harvesting: Spawn harvesting is the process of collecting fertilized eggs or newly hatched larvae from breeding tanks or hapas. This operation requires careful handling to minimize stress and damage, with gentle techniques such as using soft nets or siphoning methods. Proper timing is crucial, as delays can lead to fungal infections or predation. Harvested spawn are then transferred to incubation or rearing facilities where they are monitored and managed under optimal conditions for growth and survival.

Spawn Packaging: Spawn packaging is the process of preparing and packing newly hatched larvae (spawn) for transport to nurseries or grow-out facilities. Packaging involves placing spawn in oxygenated bags with water of appropriate quality and temperature, often supplemented with oxygen or transport aids to reduce stress. Correct packaging methods are critical to ensuring high survival rates during transport.

Spawn Survival: Spawn survival refers to the proportion of hatched larvae that survive through the early stages of development, usually measured from hatching to the fry or fingerling stage. Key factors affecting spawn survival include water quality, nutrition, stocking density, and disease management. Optimal conditions, such as maintaining a stable temperature (24°-28°C) and good oxygen levels (>5 mg/L), are critical to minimizing mortality and promoting healthy development.

Spawning Tank Design: Spawning tank design in hatcheries is crucial for creating an environment conducive to successful breeding. Tanks are usually circular or rectangular, with capacities ranging from 500 to 5,000 liters, depending on species and broodstock size. They are equipped with gentle water flow systems, aerators, and egg collection devices. For freshwater fish, the water depth is maintained between 0.5-1.5 meters, and water temperature is regulated to match the spawning requirements of the species, typically 26°-30°C for carps and catfish.

Stocking Density: Stocking density refers to the number of individuals per unit volume of water in a rearing system. Optimal stocking density varies by species and stage of development; for example, carp fry are often stocked at densities of 5-10 fry per liter in nursery ponds. Maintaining appropriate stocking densities is crucial to minimizing stress, reducing competition for resources, and promoting healthy growth.

Stress Management: Stress management in hatcheries involves minimizing stressors such as poor water quality, handling, overcrowding, and temperature fluctuations to improve the health and survival of broodstock, eggs, larvae, and juvenile fish. Techniques include maintaining stable environmental conditions, gentle handling practices, and providing adequate nutrition. Effective stress management enhances immune function and reduces disease susceptibility.

Stripping: Stripping is a manual technique used in aquaculture to collect eggs and sperm from broodstock without killing the fish. Fish are typically anesthetized, and gentle pressure is applied to the abdomen to release gametes. Stripping allows for precise control over fertilization, facilitating artificial breeding programs. This method is commonly used with species that are induced to spawn using hormonal treatments, ensuring high yields of viable eggs and sperm.

Survival Rates: Survival rates in hatchery management refer to the percentage of individuals that survive from one stage of development to the next, such as from egg to fry or fry to fingerling. High survival rates indicate effective hatchery practices, including good broodstock health, proper feeding, optimal environmental conditions, and effective disease management. Typical survival rates vary by species but should exceed 80% under well-managed conditions.

Synthetic Compounds: Synthetic compounds in aquaculture refer to artificial hormones or analogs, such as human chorionic gonadotropin (HCG) and Ovaprim (a combination of GnRH analog and a dopamine antagonist), used to induce spawning in broodstock. These compounds mimic natural reproductive hormones, enhancing ovulation and spawning rates. They are preferred for their consistency, ease of use, and controlled dosing, with application rates tailored to species-specific requirements and broodstock conditions.

Synthetic Hormones: Synthetic hormones are artificially produced compounds that mimic natural reproductive hormones, such as GnRH analogues and HCG. These hormones are used in aquaculture to induce spawning in fish that are otherwise reluctant to breed in captivity. Dosages and administration methods vary depending on species and desired outcomes, with careful management required to avoid overdosing and ensure effective spawning.

Unfertilized Eggs: Unfertilized eggs are eggs that have not been successfully fertilized and therefore do not develop into embryos. These eggs are usually opaque or milky in appearance and may sink or clump together, making them prone to fungal infections. In hatcheries, unfertilized eggs are often removed to prevent contamination and competition for space and oxygen with developing fertilized eggs, ensuring better hatchability rates for viable embryos.

Water Filtration: Water filtration in hatcheries is the process of removing suspended solids, organic matter, and potential pathogens from the water supply to maintain optimal water quality. Common filtration methods include mechanical filters, biological filters, and UV sterilization. Effective filtration helps prevent disease, improves water clarity, and ensures a healthy environment for broodstock, eggs, and larvae.

Water Flow Management: Water flow management in hatcheries is crucial for maintaining optimal environmental conditions for eggs and larvae, involving the control of water movement through tanks, incubators, or hapas. Proper flow helps keep eggs suspended, facilitates waste removal, and ensures uniform distribution of temperature and oxygen. Flow rates are adjusted to species-specific needs, typically gentle enough to prevent damage to eggs and larvae while ensuring effective water exchange and cleanliness.

Water Quality Management: Water quality management in hatcheries involves monitoring and maintaining key water parameters, such as temperature, pH, dissolved oxygen, ammonia, and nitrite levels, within optimal ranges suitable for fish reproduction and larval growth. For instance, ideal conditions often include a temperature of 26°-30°C, pH of 6.5-7.5, and dissolved oxygen above 6 mg/L. Effective management strategies include regular water testing, filtration, aeration, and water exchange to ensure a healthy environment for broodstock, eggs, and larvae.

Water Temperature Control: Water temperature is a critical factor in hatcheries, as it directly influences the metabolic rate, embryonic development, and hatching success of fish eggs. Optimal temperature ranges vary by species, but for most freshwater species like carps and catfish, a temperature range of 26°-30°C is generally maintained. Automated heaters, chillers, and thermostats are used to maintain these precise conditions, especially in closed recirculatory systems where temperature fluctuations can impact egg survival and fry development.

CHAPTER 13

COASTAL AQUACULTURE AND MARICULTURE

Abalone Culture: Abalone culture focuses on species like *Haliotis diversicolor* or *Haliotis midae* grown in land-based tanks or coastal net cages. Abalone prefer cold-water temperatures of 15°-20°C, salinity levels of 30-35 ppt, and high dissolved oxygen levels (>5 mg/L). They are herbivorous, feeding on macroalgae such as kelp, and their culture requires the provision of suitable feed and clean, oxygenated water. Water quality management, including maintaining ammonia levels below 0.1 ppm and pH between 7.5-8.5, is crucial to preventing stress and disease outbreaks.

Aeration Techniques: Aeration is the process of adding oxygen to water to ensure sufficient dissolved oxygen for aquatic organisms. In aquaculture, aeration techniques include mechanical aerators (e.g., paddle wheels, diffusers) and natural methods like water circulation or wave action. Effective aeration promotes gas exchange, prevents stratification, and enhances waste decomposition. It is particularly important in intensive systems where high stocking densities can quickly deplete oxygen levels.

Agar-Yielding Seaweed: Agar-yielding seaweeds, primarily from species like *Gracilaria* and *Gelidium*, are cultivated for the extraction of agar, a gelatinous substance used in food, pharmaceuticals, and microbiological media. These seaweeds thrive in clear coastal waters with optimal salinity levels of 30-35 ppt and water temperatures of 20°-28°C. Seaweed farming typically involves using longlines or net systems. Agar is extracted through boiling and purification processes, and the quality and yield depend on the species and environmental conditions, especially nutrient levels and light availability.

Algin-Yielding Seaweed: Algin-yielding seaweeds, such as *Laminaria* and *Macrocystis*, are cultivated for alginate, a polysaccharide used in the food industry, biotechnology, and textiles. These seaweeds grow in cold-water environments with temperatures ranging from 5°-15°C and salinity levels between 30-35 ppt. Algin extraction is done by alkaline treatment of the seaweed biomass, and the quality of the alginate depends on factors like species, growth conditions, and the maturity of the seaweed. Seaweed farming for algin is sustainable and contributes to nutrient cycling in coastal ecosystems.

Ammonia Control: Ammonia (NH_3) is a toxic nitrogenous waste product generated from the metabolism of aquatic organisms and the decomposition of organic matter. In aquaculture, excessive ammonia can lead to toxic buildup, causing gill damage, reduced growth, and mortality. The optimum ammonia concentration should be below 0.05 mg/L to prevent toxicity. Effective ammonia control involves proper biofiltration, water exchange, and maintaining a healthy microbial community that converts ammonia to less harmful forms like nitrate through nitrification.

Apparatus for Sea Farming: The apparatus used in sea farming includes nets, cages, pens, rafts, and anchors, along with mooring systems to secure the structures in offshore locations. In seaweed farming, ropes, lines, and floats are crucial for suspending the seaweed in the water column. Specific apparatus such as water quality probes, sensors for salinity and temperature, and GPS devices are employed to monitor environmental conditions and ensure the structural integrity of sea farming units.

Aquaculture Biosecurity: Biosecurity in aquaculture involves implementing measures to prevent the introduction, spread, and impact of pathogens and diseases within aquaculture facilities. Effective biosecurity practices include quarantine of new stock, disinfection of equipment, regular health monitoring, and proper waste management. Maintaining optimal water quality, controlling vectors, and minimizing human and animal traffic in production areas are crucial components of biosecurity.

Biomass Estimation: Biomass estimation in aquaculture refers to calculating the total weight of cultured organisms in a system, which is crucial for feed management, growth monitoring, and harvest planning. Biomass is estimated using sample weights and extrapolating to the total population based on stocking densities. For example, in shrimp farming, biomass estimation is done by weighing a random sample of shrimps and multiplying by the total number in the pond. Accurate biomass estimation ensures that feed is provided according to the growth rate and prevents overfeeding or underfeeding.

Brackishwater Finfish Identification: Brackishwater finfish identification is crucial for selecting species like milkfish, mullet, and pearlspot, which are well-adapted to varying salinity levels. Identification involves recognizing species based on morphological characteristics such as body shape, fin configuration, scale patterns, and coloration. For example, mullets have a streamlined body with silver-grey scales, while pearlspots exhibit distinctive dark spots on their body. Proper identification ensures effective culture techniques, as different species have varying tolerance levels to salinity (5-30 ppt) and water conditions.

Cage Culture Unit: A cage culture unit is a structure used to farm fish in marine or brackish waters, where the fish are confined within floating or submerged cages. These cages allow water exchange and provide a controlled environment for species like seabass, cobia, or grouper. Cage dimensions typically range from 5-20 meters in diameter with a depth of 5-10 meters, and they are made from materials like polyethylene or nylon. Cage culture units require regular maintenance to prevent fouling and ensure water quality parameters such as dissolved oxygen (above 5 mg/L) remain within optimal levels for fish growth.

Cage Culture: Cage culture involves the farming of fish in floating or fixed enclosures made of netting or other materials. These cages are placed in water bodies such as lakes, rivers, or coastal areas. Fish are stocked at densities of 10-50 kg/m , depending on species and cage size, with water flowing freely through the cage to provide oxygen and carry away waste. Optimum dissolved oxygen levels are 5-7 mg/L, with temperature controlled at 25°-30°C for most marine species.

Cage Design: Cage design in mariculture is critical for ensuring the proper flow of water, oxygenation, and containment of the cultured species. Typically, cages are made of high-density polyethylene (HDPE) or nylon nets, with buoyancy provided by plastic or metal floats. The size and shape of the cage depend on the species being farmed, ranging from 5-20 meters in diameter, with a depth of 5-10 meters for finfish species. The mesh size varies according to fish size, usually 15-30 mm for grow-out stages. Robust designs also account for tidal currents and wave action.

Carrageenan-Yielding Seaweed: Carrageenan is extracted from red seaweeds, mainly from species like *Kappaphycus* and *Eucheuma*. These seaweeds are farmed in tropical waters with temperatures of 25°-30°C and salinity levels of 30-35 ppt. Carrageenan is widely used as a gelling, thickening, and stabilizing agent in the food, pharmaceutical, and cosmetic industries. Farming techniques include longlines or raft cultures, where the seaweed is suspended in the water column. Environmental conditions such as water quality, nutrient availability, and sunlight are crucial to maximizing carrageenan yield.

Clam Culture: Clam culture, involving species like *Ruditapes philippinarum* (Manila clam), is practiced in coastal intertidal zones, typically using bottom culture techniques. Clams thrive in water temperatures of 10°-25°C, with salinity levels of 20-35 ppt. They are filter feeders and depend on high-quality, plankton-rich waters with dissolved oxygen levels above 5 mg/L. Proper sediment management is crucial as clams burrow into sandy or muddy substrates, and growth can be affected by sediment pollution or low pH levels.

Coastal Aquaculture: Coastal aquaculture involves the cultivation of aquatic organisms such as fish, shellfish, and seaweed in the coastal zones, utilizing the brackish water environment. This type of aquaculture benefits from natural tidal systems and nutrient inflows. Optimal salinity levels range from 10-35 ppt, depending on the species, with water temperatures ideally between 25°-30°C. Coastal aquaculture plays a significant role in producing species like shrimp, seabass, and seaweed, contributing to local economies and food security.

Coastal Aquafarm Visit: A coastal aquafarm visit involves observing the layout, operations, and management practices of aquaculture facilities located along coastlines. These farms typically cultivate species such as shrimp, fish, or mollusks in systems like ponds, cages, or pens. During the visit, key aspects such as site selection, water quality management, feed protocols, stocking density, and disease control strategies are analyzed. Coastal aquafarms play a critical role in enhancing production while maintaining environmental sustainability through practices like effluent treatment and biosecurity measures.

Cobia Culture: Cobia (*Rachycentron canadum*) is a fast-growing marine species cultured in offshore cages in tropical and subtropical waters. Cobia thrive in salinities of 25-35 ppt and temperatures of 25°-30°C. The species is carnivorous, requiring high-protein feeds (40-45%) and stocking densities of 10-20 kg/m . Cobia can reach market size (5-6 kg) within a year, making it an attractive species for commercial mariculture

Cockle Culture: Cockle culture involves the farming of species like *Cerastoderma edule* in intertidal or shallow subtidal zones. Cockles are typically cultured in sandy or muddy substrates with optimal water temperatures between 10°-20°C, salinity levels of 20-35 ppt, and dissolved oxygen levels above 5 mg/L. They are filter feeders, and good water quality with low pollution and abundant phytoplankton is essential. Cockle farming uses bottom culture techniques, and careful sediment management is necessary to avoid siltation and maintain healthy growth.

Construction Specifications: Construction specifications refer to the design and structural requirements for various aquaculture systems, including ponds, cages, and pens. In cage culture, for instance, materials like high-density polyethylene (HDPE) are used for buoyancy, while nets are constructed from nylon or polyethylene to ensure durability in marine conditions. Ponds typically have a depth of 1.5-2.0 meters and are lined with materials that prevent seepage.

Crayfish Culture: Crayfish culture, often involving species like *Procambarus clarkii* (red swamp crayfish), is practiced in freshwater ponds, paddy fields, or tanks. The optimal water temperature for crayfish is 20°-25°C, with pH levels ranging from 6.5-8.5 and dissolved oxygen maintained above 5 mg/L. Crayfish are omnivores and can be fed with plant material, animal-based feed, and agricultural by-products. Regular monitoring of water quality is essential to control ammonia (below 0.5 ppm) and nitrite levels, preventing diseases like crayfish plague (caused by *Aphanomyces astaci*).

Detritus Removal: Detritus consists of organic matter, such as uneaten feed, fecal material, and dead organisms, that accumulates at the bottom of aquaculture systems. If not removed, detritus can decompose and release harmful substances like ammonia, sulfides, and methane. Regular removal of detritus is essential to maintain water quality and prevent the buildup of toxic substances. Detritus can be removed through siphoning, sedimentation tanks, or biological filtration systems.

Disease Management in Mariculture: Disease management in mariculture involves the prevention, early detection, and treatment of diseases that affect cultured marine organisms. Key strategies include maintaining optimal water quality, regular monitoring for signs of disease, vaccination, and the use of probiotics or immunostimulants. Quarantine measures and biosecurity practices help prevent the spread of pathogens. Disease outbreaks are managed through appropriate treatments, such as antibiotics or antiparasitics, and by addressing environmental stressors like poor water quality.

Dissolved Oxygen: Dissolved oxygen (DO) refers to the concentration of oxygen gas in water, essential for respiration in aquatic organisms. In aquaculture, DO levels should be maintained at optimal levels (typically 5–8 mg/L) to ensure healthy metabolism and growth. Low DO levels (<3 mg/L) can lead to hypoxia, causing stress, reduced feeding, and even mass mortality. Proper aeration, water circulation, and regular monitoring are required to maintain adequate DO levels, particularly in dense or closed systems.

Echinoderms: Echinoderms, such as sea cucumbers and sea urchins, are cultured in coastal areas or aquaculture ponds. Optimal water conditions for sea cucumbers include salinity levels of 30-35 ppt, temperatures of 20°-25°C, and dissolved oxygen levels above 4 mg/L. They are detritivores, feeding on organic matter and sediments. Sea cucumbers are particularly valuable in integrated multi-trophic aquaculture systems (IMTA) as they help maintain sediment quality. Sea urchins prefer similar conditions but feed on algae, requiring clean and nutrient-balanced waters for optimal growth.

Eel Culture: Eel culture, particularly of species like *Anguilla japonica*, is widely practiced in Asia. Eels are cultured in ponds, tanks, or recirculating systems with a salinity range of 0-10 ppt and water temperatures of 25°-30°C. They require high-protein feeds (45-55%) and are stocked at densities of 10-20 kg/m^2. Eel farming involves complex breeding and larval rearing techniques, making it a highly specialized form of aquaculture.

Eutrophication: Eutrophication is the excessive enrichment of water bodies with nutrients, leading to an overgrowth of algae and phytoplankton.

In aquaculture, it can occur due to over-fertilization or nutrient runoff from feed and waste, causing oxygen depletion and the production of harmful algal blooms. Eutrophication leads to poor water quality, fish kills, and increased disease risk. Preventing eutrophication involves controlling nutrient inputs, regular water exchange, and maintaining balance in the aquatic ecosystem.

Fertilization in Aquaculture: Fertilization involves adding nutrients (primarily nitrogen and phosphorus) to aquaculture ponds to promote the growth of phytoplankton, which serves as the base of the food web. The proper balance of fertilizers supports zooplankton and higher trophic levels, benefiting filter-feeding species. Fertilization should maintain nutrient levels in an optimum range (e.g., 0.5–1.5 mg/L of nitrate and 0.01–0.05 mg/L of phosphorus) to prevent overgrowth of algae and eutrophication, which can deplete oxygen levels and harm aquatic life.

Fertilizer Dosage Calculation: Fertilizer dosage calculation in aquaculture involves determining the optimal amount of nutrients needed to enhance primary productivity (plankton growth) in ponds or tanks. Fertilizers such as urea, ammonium nitrate, or phosphate are applied based on water volume, target species, and initial nutrient levels. For example, for shrimp ponds, 2-4 kg/ha of nitrogen and 0.5-1.0 kg/ha of phosphorus may be applied. Proper dosage promotes algae growth for filter feeders while preventing over-fertilization, which could lead to harmful algal blooms.

Flatfish Culture: Flatfish culture, commonly involving species such as flounder, halibut, or sole, is conducted in land-based tanks or coastal net pens. These fish thrive in water temperatures ranging from 12°-20°C, and salinity levels of 30-35 ppt. Their diet is primarily composed of high-protein fish meal, and growth rates are monitored closely. Flatfish require clean, sandy substrates, and low stocking densities to prevent stress and reduce disease incidences like viral encephalopathy.

Fouling Organisms: Fouling organisms, such as barnacles, algae, and mussels, are unwanted species that attach to aquaculture structures, cages, and the cultured species themselves. They compete for nutrients and oxygen, obstruct water flow, and can harbor pathogens, negatively affecting the health and growth of cultured organisms. In mariculture, fouling can reduce the efficiency of net pens or longlines, necessitating regular cleaning or the use of antifouling technologies. The degree of fouling depends on water temperature, nutrient levels, and the proximity to natural fouling communities.

Freshwater Shrimp Farming: Freshwater shrimp farming, particularly of *Macrobrachium rosenbergii* (giant freshwater prawn), takes place in ponds with freshwater inputs. Optimal water temperatures range from 28°-31°C, pH levels between 7.0-8.5, and dissolved oxygen levels above 5 mg/L. Salinity should remain below 3 ppt. Freshwater shrimp are omnivores and require balanced diets rich in protein (30-35%), while careful management of pond water quality is crucial for maintaining healthy growth and preventing diseases like bacterial infections.

Grouper Culture: Groupers (*Epinephelus spp.*) are carnivorous fish commonly cultured in marine cages or ponds. They require water temperatures of 26°-30°C and salinities of 25-35 ppt. Groupers are fed a high-protein diet (45-55%) and are typically stocked at densities of 15-20 fish/m . Hatchery-bred fry are used for stocking, and the species is highly valued in both local and export markets due to its firm, white flesh.

Growth Monitoring: Growth monitoring in aquaculture involves regularly measuring the size, weight, and overall health of cultured organisms to assess their development and adjust feeding, stocking density, and water management practices accordingly. For instance, in shrimp culture, growth is monitored bi-weekly by sampling and measuring individual shrimp for weight and length. Growth rates are compared against expected benchmarks for the species under specific environmental conditions, such as water temperature, salinity, and feed quality. Effective growth monitoring ensures optimal production efficiency and timely harvest.

Health Monitoring: In aquaculture, health monitoring involves regularly assessing the physical condition of cultured species to detect early signs of disease or stress. Parameters such as body condition, feeding behavior, and water quality are checked. Health monitoring also includes periodic screenings for pathogens like bacteria, viruses, or parasites, using diagnostic tools. Early detection and response help maintain stock survival rates and optimize production, with an emphasis on water quality parameters like dissolved oxygen (5-8 mg/L), ammonia (below 0.05 mg/L), and temperature stability.

Integrated Aquaculture: Integrated aquaculture involves the farming of multiple species at different trophic levels within the same system, such as fish, shellfish, and seaweed. This approach maximizes resource use and minimizes waste, as one species' by-products serve as nutrients for another. For example, fish excreta provide nutrients for seaweed, and seaweed, in turn, helps purify the water. Salinity levels range from 10-35 ppt, depending on the species involved, with optimal water temperatures varying between 20°-30°C. This method enhances sustainability and promotes ecosystem balance in aquaculture systems.

Intensive Aquaculture: Intensive aquaculture systems aim to maximize production using high stocking densities (15-30 fish/m^2), artificial feeds, and sophisticated water management. These systems require strict control of water quality parameters, such as dissolved oxygen (6-8 mg/L), ammonia (<0.5 mg/L), and temperature (25°-30°C). Intensive systems can achieve high yields (5-10 tons/ha/year), but they require significant investment in aeration, water exchange, and waste management to minimize environmental impact.

Lobster Culture: Lobster culture, particularly of *Panulirus* or *Homarus* species, is primarily conducted in sea cages or land-based systems. Optimal water conditions include temperatures ranging from 20°-25°C and salinity levels between 30-35 ppt. Lobsters are carnivorous, requiring diets high in protein (40-50%), with regular water quality monitoring to keep ammonia below 0.1 ppm and dissolved oxygen levels above 5 mg/L. Disease management and protection from predators are critical for successful culture.

Low Water Exchange System: A low water exchange system in aquaculture refers to practices where water is exchanged minimally (less than 5% per day), relying on filtration, biofilters, or microbial treatments to maintain water quality. These systems are designed to reduce water usage, prevent nutrient loss, and control pollutants like ammonia and nitrites, which must remain under 0.5 ppm. It is commonly used in shrimp and fish farming, with dissolved oxygen levels maintained above 5 mg/L for optimal growth.

Mariculture: Mariculture is the farming of marine organisms in open oceans or marine enclosures such as bays, lagoons, and estuaries. It includes the culture of finfish, shellfish, and seaweeds, using cages, pens, rafts, or racks in coastal waters. Salinity levels are typically 30-35 ppt, with temperature requirements varying between 18°-30°C for different species. Mariculture helps increase fish production and reduce pressure on wild stocks while promoting sustainable marine resource use.

Marine Finfish Identification: Marine finfish identification in aquaculture is essential for selecting species that are suitable for farming based on environmental compatibility and market demand. Identification involves recognizing key species such as seabass, groupers, and yellowtails, which are often identified by features like body shape, fin structure, coloration, and scale pattern. Accurate species identification ensures proper husbandry, feeding, and disease management protocols in marine cages or tanks.

Marine Hatcheries: Marine hatcheries are controlled facilities designed for the breeding, hatching, and early rearing of marine species, particularly fish and shellfish like shrimp, oysters, and clams. These facilities play a critical role in aquaculture by providing larvae or juvenile stock for grow-out operations. Hatcheries maintain optimal conditions such as temperature (24°–30°C), salinity (28–34 ppt), and pH (7.5–8.5) to ensure proper spawning and larval development. Marine hatcheries also control water quality, nutrient levels, and photoperiod to mimic natural conditions that promote growth.

Marine Molluscs: Marine molluscs encompass a wide range of species such as mussels, oysters, clams, and abalones. They are typically cultured in coastal areas using various farming techniques like raft culture, longline, and bottom culture. Molluscs filter-feed on phytoplankton, so water quality must be high, with clean, nutrient-rich waters (10-30 ppt salinity) and low pollution levels. Growth is affected by water temperature (10°-25°C), pH (7.5-8.5), and dissolved oxygen levels (above 5 mg/L), all of which must be maintained within optimal ranges.

Milkfish Culture: Milkfish (*Chanos chanos*) is a popular species in brackish and marine aquaculture, particularly in Southeast Asia. It is typically cultured in ponds or pens with a salinity range of 10-35 ppt and temperatures between 26°-32°C. Milkfish are omnivorous, feeding on algae, detritus, and supplemental feeds, and their culture involves both extensive and semi-intensive systems, with yields ranging from 1.5 to 3 tons/ha/year.

Mud Crab Fattening: Mud crab fattening involves short-term rearing of *Scylla serrata* or related species to enhance their market size and meat content. It is usually conducted in coastal ponds or cages with high salinity levels (20-30 ppt) and water temperatures between 25°-30°C. Crabs are fed a high-protein diet (40-50%), typically including fish or mollusks. The fattening process lasts 15-30 days, and water quality parameters such as dissolved oxygen above 4 mg/L and ammonia levels below 0.1 ppm are essential for success.

Mullet Culture: Mullets, particularly *Mugil cephalus*, are euryhaline species capable of thriving in both freshwater and marine environments, with ideal salinity levels of 5-30 ppt. Mullets are omnivores, feeding on detritus, algae, and formulated feeds in semi-intensive and extensive systems. They are typically grown in ponds, coastal lagoons, and brackish water systems, with moderate stocking densities of 2-4 fish/m^2.

Mussel Farming: Mussel farming primarily involves species like *Mytilus galloprovincialis* or *Perna viridis*. It is practiced in coastal regions using suspended rope culture or longline systems. Mussels thrive in water temperatures of 10°-25°C, salinity levels of 30-35 ppt, and areas with high plankton density. Water quality is crucial, with pH between 7.5-8.5 and dissolved oxygen levels above 5 mg/L. Mussel farming is environmentally sustainable, as mussels filter water and help maintain ecological balance, but they are vulnerable to pollutants and harmful algal blooms.

Nitrite Levels: Nitrite (NO_2^-) is an intermediate product of nitrification, which can accumulate in aquaculture systems and become toxic to aquatic organisms by interfering with oxygen transport in blood (causing "brown blood disease"). Nitrite toxicity varies by species but generally, nitrite levels should be kept below 0.5 mg/L. Proper management includes maintaining biofilters and encouraging bacterial growth to convert nitrite into nitrate, a less harmful compound.

Nursery Rearing: Nursery rearing is the process of raising fish or shellfish juveniles from hatchlings to fingerlings in a controlled environment before transferring them to grow-out systems. Nursery tanks or ponds are usually stocked at densities of 50-100 fry/m^2, and water quality is strictly controlled. Feeds are provided at 8-12% of body weight per day, and optimal water temperature (26°-30°C) is maintained to ensure rapid growth and survival.

Off-Bottom Culture: Off-bottom culture is a method used for farming bivalves like oysters and mussels, where organisms are suspended in the water column using rafts, ropes, or cages rather than being placed directly on the substrate. This technique ensures better water flow, reduces predation, and increases access to food. Water quality parameters must be carefully monitored, with optimal salinity levels between 25-35 ppt, water temperatures of 10°-25°C, and dissolved oxygen levels above 5 mg/L. This method is popular in areas with muddy or sandy substrates unsuitable for bottom culture.

On-Bottom Culture: On-bottom culture refers to the practice of placing molluscs like oysters, clams, or cockles directly on the seabed or in intertidal zones. These organisms rely on natural substrates, and the culture site must have appropriate sediment conditions, such as sandy or muddy bottoms. Water quality must be maintained, with salinity levels of 20-35 ppt, water temperatures of 10°-25°C, and dissolved oxygen above 5 mg/L. On-bottom culture is cost-effective but more susceptible to environmental changes and predation.

Open-Sea Farming: Open-sea farming, or offshore aquaculture, involves growing aquatic species like fish, shellfish, and seaweed in deep waters, away from the coast. This method reduces the environmental impacts seen in coastal aquaculture, such as nutrient buildup and habitat destruction. It requires robust systems to withstand strong currents and waves, with optimal water conditions including temperatures between 10°-30°C, salinity levels of 30-35 ppt, and dissolved oxygen levels above 5 mg/L. Species like salmon, mussels, and seaweeds are commonly cultured in open-sea systems, offering large-scale production potential.

Oyster Farming: Oyster farming involves the cultivation of species like *Crassostrea gigas* (Pacific oyster) or *Crassostrea virginica* (Eastern oyster). These bivalves are grown in estuarine environments with salinity levels ranging from 15-35 ppt, water temperatures of 10°-25°C, and dissolved oxygen levels above 5 mg/L. Oyster farming can use bottom culture, rack-and-bag systems, or suspended culture techniques. Oysters filter plankton and organic particles, making them efficient biofilters, but they require clean water to avoid contamination from harmful bacteria or pollutants.

Pearl Oyster Culture: Pearl oyster culture involves farming species like *Pinctada fucata* for the production of pearls. This process is carried out in marine environments with clear, clean waters, temperatures between 25°-30°C, and salinity levels of 30-35 ppt. Pearl oysters are suspended in nets or cages, and water quality must be monitored to maintain low pollutant levels, high dissolved oxygen (above 5 mg/L), and a stable pH (7.8-8.2). The implantation of a nucleus inside the oyster triggers pearl formation, and the oysters are carefully tended to for 6-24 months until harvest.

Pearlspot Culture: Pearlspot (*Etroplus suratensis*) is a hardy brackish water fish, cultured in ponds, cages, or pens in India and Southeast Asia. It tolerates salinities from 0-35 ppt and grows well in temperatures of 25°-30°C. Pearlspot is omnivorous, consuming a diet of natural algae, detritus, and supplemental feed. The species is highly resilient to environmental fluctuations, making it ideal for aquaculture in estuarine regions.

Pen Culture Unit: Pen culture units are enclosures built in shallow coastal waters, often using nets or stakes, to confine fish or shrimp while allowing for a semi-natural environment. These pens are designed to keep species like milkfish, seabass, or mullets in brackish or marine waters while maintaining water flow and preventing stock escape. Pen dimensions can vary, but they often cover large areas up to 1 hectare. Site selection for pen culture considers factors such as water depth, tidal range, and protection from strong currents or waves.

Pen Culture: Pen culture is similar to cage culture but uses fixed enclosures in shallow water bodies, where the bottom substrate serves as the natural habitat. Pens are typically used in lagoons, estuaries, or shallow coastal areas. Species like milkfish and shrimp are cultured, and water quality is maintained naturally through tidal exchange. Stocking densities vary between 5-10 fish/m^2, and nutrient enrichment of the water is often needed to support growth.

Pen Design: Pen culture involves enclosing a section of a coastal water body, typically using nets or stakes driven into the seabed, to form a large, semi-natural environment for the cultivation of species like milkfish, seabass, or shrimp. Pen size can vary, but they typically cover areas from 0.1 to 1 hectare, depending on stocking density and species. Construction must be resilient to tidal fluctuations, with materials resistant to corrosion and fouling. Mesh size and height are designed to prevent stock escape and predator entry.

pH Monitoring: pH is a measure of the acidity or alkalinity of water and is a critical parameter in aquaculture, as it affects the solubility and toxicity of chemicals like ammonia and metals. Most marine species thrive in water with a pH range of 7.5 to 8.5. Deviations from this range can cause stress, reduce growth rates, and increase susceptibility to disease. pH fluctuations can be managed through buffering agents, regular water exchanges, and maintaining optimal alkalinity levels (80–120 mg/L as $CaCO_3$).

Pole-Based Culture: Pole-based culture is used primarily for seaweed farming, where long poles are driven into the seabed, and seaweed is attached at various points along ropes or lines stretched between poles. This system is effective in shallow, intertidal zones and allows seaweed to grow submerged during high tide while being accessible at low tide for maintenance. Species like **Gracilaria** or **Kappaphycus** are commonly cultured using this method, with the ideal site featuring stable salinity (25-35 ppt) and nutrient-rich waters.

Pond Aerators: Pond aerators are mechanical devices used in aquaculture ponds to increase dissolved oxygen levels, improve water circulation, and prevent the buildup of toxic gases like ammonia and carbon dioxide. Common types include paddle wheel aerators, diffused air systems, and airlift pumps. Proper use of aerators helps maintain DO levels above 5 mg/L, ensuring optimal conditions for fish and shrimp growth while preventing oxygen depletion and mass mortality events.

Pond Preparation: Pond preparation is a critical step in aquaculture to ensure a healthy environment for fish or shrimp. This process includes drying the pond, liming to adjust soil pH (optimal range 6.5-8.5), and fertilizing to promote the growth of natural food sources such as phytoplankton. Proper preparation helps prevent disease outbreaks and ensures optimal stocking conditions for the cultured species.

Post-Stocking Strategies: Post-stocking strategies are management practices implemented after stocking to ensure optimal growth and survival. These include monitoring water quality (ammonia <0.5 mg/L, dissolved oxygen >5 mg/L), providing appropriate feed, controlling diseases, and maintaining proper aeration. Regular sampling to assess growth rates and adjusting feeding schedules are part of these strategies to maximize production efficiency.

Pre-Stocking Strategies: Pre-stocking strategies involve preparing the culture environment before the introduction of seed or juveniles. This includes managing water quality (pH 6.5-8.5, dissolved oxygen >5 mg/L), ensuring proper pond fertilization, checking for predators, and adjusting the salinity and temperature levels according to species requirements. Pre-stocking management is essential for minimizing stress and maximizing survival rates during stocking.

Rabbitfish Culture: Rabbitfish (*Siganidae*), primarily herbivores, are cultured in both coastal and lagoon environments. They thrive in waters with salinities ranging from 15-35 ppt and temperatures of 25°-30°C. Rabbitfish feed on algae and seagrass, and their culture contributes to ecosystem health by controlling algal blooms in coastal aquaculture settings. They are raised in both ponds and cages with moderate stocking densities.

Rack Culture Unit: Rack culture involves placing racks made from PVC or metal frames in shallow coastal waters for growing species like oysters or clams. The racks are elevated above the seabed, allowing shellfish to filter feed while being protected from bottom-dwelling predators. The racks are often designed to withstand tidal fluctuations, with tray or basket arrangements that facilitate easy access for maintenance, monitoring, and harvesting. This system provides an efficient method for culturing shellfish with minimal environmental impact.

Rack Culture: Rack culture is used for cultivating mollusks like oysters and mussels by suspending them on racks made of wooden or PVC frames. Racks are placed in shallow coastal waters where tidal action ensures a continuous supply of nutrients. Rack systems are effective in areas with salinity levels between 25-35 ppt and water temperatures of 20°-30°C. This method helps improve water circulation and reduce sediment accumulation.

Rack Design: Rack culture is employed for shellfish like oysters or clams, where racks of mesh trays or baskets are elevated above the seabed. The racks, typically made from metal or PVC, are placed in shallow coastal waters, allowing the shellfish to filter feed while being protected from bottom predators. The height of the racks is adjusted based on the tidal range, ensuring that the shellfish are submerged during high tide but accessible for maintenance and harvesting during low tide.

Raft Culture Unit: Raft culture units are floating structures used for the cultivation of shellfish, such as mussels or oysters, where ropes or nets are suspended beneath a raft. The raft is typically made of wood or steel and floats on the water surface, anchored securely in areas with good water flow and nutrient availability. This system allows for high-density cultivation with optimal growth conditions. Rafts are designed to handle varying sea conditions and require periodic cleaning to remove fouling organisms.

Raft Culture: Raft culture involves the suspension of culture organisms, such as mussels or seaweed, on floating rafts anchored in the water. This system allows for high-density stocking and efficient use of space in coastal areas. Raft culture requires areas with stable water flow, salinity of 25-35 ppt, and temperatures between 20°-30°C. Rafts are typically made of bamboo or PVC pipes, with longlines used to suspend the culture materials.

Raft Design: Raft culture is used in shellfish farming, particularly for mussels and oysters, where ropes or nets are suspended from floating rafts. The raft framework is usually constructed of wood or steel, with floats made of plastic barrels. Each raft can cover 100-200 square meters, depending on the density of shellfish lines. Rafts are moored in place using heavy anchors, and they must be designed to withstand strong currents and tides, with rope or line spacing that maximizes water flow to the filter-feeding organisms.

Rope-Based Culture: Rope-based culture is another seaweed farming technique where seaweed fronds are tied to ropes or lines that are either suspended from floating buoys or anchored to the seabed. This system is widely used for species like **Eucheuma** and **Kappaphycus** in tropical coastal waters. The ropes allow the seaweed to remain submerged, ensuring access to light and nutrients for optimal growth. This method is highly scalable and can be integrated with other mariculture systems such as fish or shellfish farming.

Salinity Management: Salinity management is critical in coastal aquaculture, as most marine species require specific salinity ranges for optimal growth and survival. The ideal salinity range varies depending on the species; for example, shrimp thrive in salinity levels of 10–35 ppt, while finfish like sea bass prefer 20–35 ppt. Maintaining stable salinity levels prevents osmotic stress, which can lead to weakened immune responses and poor growth rates. Tidal flow and freshwater inputs are often managed to stabilize salinity in culture systems.

Salmon Culture: Salmon culture refers to the farming of salmon, typically in coastal waters using sea cages or pens. Optimal conditions for salmon farming include water temperatures of 10°-15°C and dissolved oxygen levels above 6 mg/L. Salinity levels should range between 30-35 ppt, reflecting marine conditions. Salmon require a protein-rich diet (30-50% protein content), and environmental factors like light exposure are managed to stimulate growth and control maturation. Disease management and biosecurity are crucial to avoid outbreaks such as sea lice infestations and bacterial infections like furunculosis.

Scallop Culture: Scallop culture typically involves species like *Pecten maximus* or *Argopecten irradians*, farmed in coastal areas using bottom or suspended culture systems. Optimal water conditions include temperatures of 10-20°C and salinity levels of 30-35 ppt. Scallops are filter feeders, requiring high-quality water with abundant plankton. Dissolved oxygen should be maintained above 5 mg/L, and ammonia levels kept low (< 0.1 ppm) to avoid stress. Scallop culture is sensitive to environmental changes, particularly harmful algal blooms and water pollution.

Sea Bream Culture: Sea bream, especially species like *Sparus aurata*, is a high-value marine fish cultured in cages or tanks in coastal areas. The fish require optimal salinity of 25-35 ppt and water temperatures of 22°-28°C. Sea bream are carnivorous, and feeds with 40-50% protein are required for optimal growth. Intensive systems with controlled feeding and water quality management are commonly used for commercial production.

Sea Cucumber Farming: Sea cucumber farming primarily involves species like *Holothuria scabra*, cultured in coastal ponds or pens. Optimal conditions include salinity levels of 25-35 ppt and water temperatures of 26°-30°C. Sea cucumbers are detritivores, feeding on organic matter in sediments, and their farming is often integrated into multi-trophic aquaculture systems to improve sediment quality. Maintaining water quality with dissolved oxygen levels above 4 mg/L and low ammonia (<0.5 ppm) is essential for their growth and survival.

Sea Farming Machinery: Machinery used in sea farming includes a range of equipment designed to support various operations in mariculture, such as automatic feeders, water pumps, aerators, and harvesting systems. These machines ensure efficient feeding, oxygenation, and waste removal in marine cage systems, which maintain optimum growth conditions for species like seabass, sea bream, or shellfish. Automation has been increasingly adopted to reduce labor and improve accuracy in feed distribution and environmental control.

Seabass Culture: Seabass (*Lates calcarifer*) is cultured in both brackish and marine waters, with ideal salinity levels of 10-30 ppt and water temperatures of 26°-30°C. The species is cultivated in ponds, cages, or recirculatory systems. Seabass are carnivorous and require high-protein feeds (35-40% protein content). Nursery rearing is critical for ensuring high survival rates before transferring to grow-out systems.

Seaweed Culture: Seaweed culture, or mariculture of macroalgae, includes farming species like *Gracilaria*, *Eucheuma*, and *Sargassum* in coastal waters. This form of aquaculture can be done using longlines, rafts, or net systems in shallow waters with optimal salinity levels of 30-35 ppt, water temperatures of 20°-30°C, and sufficient sunlight for photosynthesis. Seaweed absorbs nutrients like nitrogen and phosphorus from the water, making it environmentally beneficial as a biofilter. It is harvested for various uses such as food, feed, biofuel, and industrial applications, including the extraction of valuable polysaccharides like agar, alginate, and carrageenan.

Seaweed Farming Systems: Seaweed farming systems vary from simple off-bottom methods to complex floating rafts and longline systems. In floating methods, seaweed such as **Gracilaria** or **Kappaphycus** is attached to ropes suspended between floating buoys, allowing it to grow in nutrient-rich waters. Optimal conditions include salinity levels of 25-35 ppt and temperatures of 20°-30°C. Longline systems allow for vertical growth in the water column, providing better water flow and light penetration for efficient seaweed cultivation.

Seaweed Identification: Seaweed identification is essential for determining species suited for commercial cultivation, such as those producing agar, carrageenan, or alginates. Important species include **Gracilaria**, **Kappaphycus**, and **Sargassum**, each identifiable by its color, frond shape, and growth habit. For instance, **Gracilaria** has thin, branching fronds, while **Kappaphycus** has thicker, fleshy branches. Identifying the right species helps optimize farming systems, as each seaweed type requires specific conditions such as temperature (20°-30°C), salinity (25-35 ppt), and nutrient availability.

Sediment Quality: In coastal aquaculture and mariculture, sediment quality refers to the physical, chemical, and biological characteristics of the seabed or pond bottom. The sediment's organic matter content, grain size, redox potential, and nutrient load (such as nitrogen and phosphorus) influence its suitability for aquaculture. Poor sediment quality, often indicated by high organic loading and low oxygen levels, can lead to anaerobic conditions, harmful gas production (H_2S), and disease outbreaks. Optimal sediment quality ensures proper benthic activity and minimizes stress on cultured species.

Seed Acclimatization: Seed acclimatization in aquaculture refers to the process of gradually adjusting hatchery-reared or wild-caught juvenile organisms (seed) to the environmental conditions of the culture system, such as temperature, salinity, and water quality. This is crucial to minimize stress and mortality. For instance, in shrimp farming, post-larvae are acclimatized by gradually changing water salinity from hatchery levels (often 30-35 ppt) to the farm's salinity (which could be as low as 10-20 ppt). The process is conducted over several hours or days to ensure smooth transition and higher survival rates.

Seed Collection: Seed collection refers to obtaining fry, fingerlings, or post-larvae from either the wild or hatcheries. Wild seed collection often targets natural breeding grounds, while hatchery production ensures a more controlled and reliable supply. Proper handling, transport, and acclimatization of seed are critical to minimize stress and mortality and quality control measures are essential for ensuring healthy stock for aquaculture.

Seed Survival Estimation: Seed survival estimation involves determining the percentage of seed (juveniles or post-larvae) that survive from stocking to a specific time point, often during grow-out stages. It is calculated by comparing the number of individuals stocked with the number surviving after a set period. In shrimp culture, survival rates of 70-80% are typically targeted. Factors influencing survival include water quality, stocking density, feed quality, and disease management. Regular monitoring and health assessments help to optimize conditions and improve survival rates.

Semi-Intensive Aquaculture: Semi-intensive aquaculture involves moderate stocking densities (5-15 fish/m^2) and the supplementation of natural food with formulated feeds. Pond preparation, water quality management, and fertilization are practiced to enhance productivity, which typically ranges from 2-5 tons/ha/year. Water quality parameters such as dissolved oxygen (5-7 mg/L) and pH (6.5-8.0) are maintained to support fish health.

Shellfish Identification: Shellfish identification focuses on recognizing different species of mollusks and crustaceans, such as shrimp, crabs, oysters, mussels, and clams, which are commonly cultured in coastal and marine environments. Each species is identified by its shell structure, size, and external features like antennae, claws, or siphons. For example, mussels have elongated shells with a dark, bluish-black color, while oysters have rough, irregularly shaped shells. Accurate identification helps in selecting species with optimal growth conditions and market value.

Shore-Based Aquaculture: Shore-based aquaculture refers to aquaculture operations located onshore, typically involving tanks, ponds, or raceways. These systems allow for greater control of water quality and are often used for hatchery operations, broodstock conditioning, or early-stage rearing. Salinity, temperature, and dissolved oxygen levels are closely monitored and controlled, often using recirculating aquaculture systems (RAS) to ensure sustainability and high production efficiency.

Shrimp Farming: Shrimp farming primarily focuses on species like *Penaeus vannamei* or *Penaeus monodon*, cultured in brackish water ponds or coastal systems. Optimal conditions include water temperatures of 28°-30°C, salinity levels of 10-35 ppt, and dissolved oxygen levels above 5 mg/L. Shrimp feed consists of 30-40% protein, and farmers manage water quality through regular monitoring of ammonia (below 0.1 ppm), nitrite (below 0.25 ppm), and pH (7.5-8.5). Biosecurity measures are implemented to prevent viral diseases like white spot syndrome.

Site Selection: Site selection in aquaculture is a critical factor that influences the success of farming operations. Key considerations include water quality (pH 6.5-8.5, salinity 10-35 ppt), access to fresh seawater, protection from extreme weather, and minimal exposure to pollutants. The site should be accessible for transportation and have stable environmental conditions that match the biological requirements of the species being cultured.

Soft-Shell Crab Production: Soft-shell crab production capitalizes on the natural molting cycle of crabs, harvesting them immediately after they shed their exoskeleton before it hardens. Crabs are typically held in tanks or cages with water temperatures between 25°-30°C and salinity levels of 20-30 ppt. Optimal management includes providing abundant food, maintaining good water quality (dissolved oxygen above 4 mg/L, ammonia below 0.1 ppm), and closely monitoring the crabs to collect them at the precise time post-molt.

Soil Quality Management: Soil quality management focuses on maintaining a pond or culture system's substrate to support healthy aquatic environments. Key aspects include maintaining a suitable pH (6.5-7.5), managing organic matter levels, and preventing the accumulation of harmful substances like hydrogen sulfide. Soil amendments like lime and organic fertilizers are applied to improve pond productivity and prevent nutrient imbalances.

Stocking Strategies: Stocking strategies involve determining the appropriate species, number, and size of fish or shellfish to introduce into the culture system. Stocking densities vary based on culture intensity, with traditional systems stocking 1-3 fish/m^2 and intensive systems stocking up to 30 fish/m^2. Timing and method of stocking (e.g., gradual acclimatization) are critical to reduce stress and enhance growth performance.

Tidal Influence: Tidal influence plays a crucial role in mariculture, particularly in coastal systems, as it affects water exchange, nutrient replenishment, and waste removal. Tidal fluctuations determine the movement of water in and out of marine enclosures or ponds, helping to regulate salinity, temperature, and dissolved oxygen. Optimal tidal flow ensures the dispersal of waste products, reduces the risk of disease accumulation, and enhances the supply of nutrients for filter feeders like mollusks.

Traditional Aquaculture: Traditional aquaculture refers to farming systems that rely on minimal inputs and mimic natural processes. These systems, such as homestead ponds and coastal paddy-cum-fish culture, use indigenous methods to cultivate fish and other aquatic organisms. Water management is often rudimentary, and fish rely on natural food chains, resulting in low productivity (0.5-1.0 tons/ha/year) but also low environmental impact.

Undrainable Ponds: Undrainable ponds refer to aquaculture systems that do not have natural or artificial drainage outlets. These ponds rely on water exchange through evaporation, seepage, or manual pumping. Maintaining water quality in such ponds requires careful monitoring of dissolved oxygen (above 4 mg/L), ammonia (below 0.5 ppm), and pH levels (6.5-8.5). These ponds are often used for extensive or semi-intensive aquaculture practices, where low stocking densities and natural productivity play a significant role in sustaining growth.

Water Quality Management: Water quality management involves monitoring and maintaining key parameters such as temperature (25°-30°C), salinity (10-35 ppt), dissolved oxygen (>5 mg/L), pH (6.5-8.5), and ammonia levels (<0.5 mg/L). Proper management ensures a healthy environment for cultured species, minimizes disease outbreaks, and promotes optimal growth. Techniques include aeration, water exchange, and the use of water treatments like lime or probiotics.

Water Quality Parameters: Water quality parameters in aquaculture refer to key factors like temperature, salinity, dissolved oxygen, pH, ammonia, nitrite, and nitrate concentrations, which are essential for the growth and health of cultured species. Ideal conditions vary by species, but for many marine organisms, water temperature should range between 20°-30°C, salinity levels between 25-35 ppt, and dissolved oxygen levels above 5 mg/L. pH should be maintained between 7.5-8.5, with ammonia levels below 0.1 ppm, nitrite below 0.5 ppm, and nitrate below 50 ppm. Regular monitoring ensures optimal growth and prevents disease outbreaks.

Yellowtail Culture: Yellowtail (Seriola quinqueradiata) is a popular species for mariculture, particularly in Japan. The fish are cultured in offshore cages with salinity levels of 30-35 ppt and water temperatures of 20°-28°C. Yellowtail require a high-protein diet (45-50%) and grow rapidly under intensive culture conditions, reaching market size in 12-18 months. Stocking densities range from 10-15 kg/m in marine cages.

Zero Water Exchange System: Zero water exchange systems involve no routine water replacement throughout the production cycle, instead relying on biofloc technology or other filtration methods to recycle nutrients and maintain water quality. These systems are used in shrimp and fish farming to conserve water, reduce effluent discharge, and enhance biosecurity. Key parameters include dissolved oxygen above 4 mg/L, ammonia below 1 mg/L, and regular monitoring of total suspended solids to ensure optimal conditions for growth.

CHAPTER 14

BREEDING AND SEED PRODUCTION OF MARINE AND BRACKISH WATER FISHES

Aeration Systems: Aeration systems introduce oxygen into water to maintain adequate dissolved oxygen levels (>5 ppm) for fish survival and growth. In aquaculture, aerators like paddlewheels or diffusers are commonly used to improve oxygenation and prevent hypoxic conditions.

Age at Maturity: Age at maturity is the age at which fish first reach reproductive capability. In marine aquaculture, species like seabass may reach sexual maturity around 2–4 years, while others, like milkfish, mature between 2 and 5 years, depending on environmental factors such as water temperature and nutrition.

Aquaculture Biosecurity: Aquaculture biosecurity refers to practices aimed at preventing the introduction and spread of pathogens within a facility. Measures include disinfection of equipment, quarantine of new stock, regular health checks, and controlling the movement of people and animals within hatcheries.

Aquaculture Equipment: Aquaculture equipment includes tools and machinery used in the cultivation of fish, such as aerators, pumps, tanks, filtration systems, and feeding devices. High-quality equipment is essential for maintaining water quality and ensuring the health and growth of cultured species.

Artemia: Artemia, or brine shrimp, are commonly used as live feed in aquaculture due to their high nutritional content (rich in essential fatty acids) and small size, making them ideal for larval fish. They are hatched from cysts in saline water (salinity: 20-30 ppt).

Artificial Insemination: Artificial insemination is a breeding technique where sperm is manually introduced to eggs, bypassing the need for natural spawning. It allows for better control over fertilization, especially in species with difficult breeding habits or in selective breeding programs.

Bacterial Infections: Bacterial infections in fish, such as those caused by *Vibrio* or *Aeromonas* species, are common in both marine and brackish water systems. They can cause symptoms like ulcers, hemorrhages, and fin rot. Effective treatment involves antibiotics, maintaining optimal water quality, and stress reduction.

Brackishwater Aquaculture: Brackishwater aquaculture is the cultivation of fish and shellfish in waters with intermediate salinity levels (5-30 ppt). Common species include milkfish and shrimp. Management of salinity, water quality, and disease is critical for the success of brackishwater farms.

Brackishwater Fish Seed: Brackishwater fish seed consists of young fish, such as milkfish or mullet, raised in hatcheries or collected from natural brackish environments. These seeds are crucial for sustaining aquaculture production, especially in regions with fluctuating salinity (5–20 ppt).

Brackishwater Fishes: Brackishwater fishes inhabit environments with varying salinity levels between freshwater and seawater, typically ranging from 5 to 20 ppt. Species such as mullet and milkfish thrive in these transitional zones, and aquaculture practices for them focus on managing fluctuating salinity and water quality parameters for optimal growth and reproduction.

Brackishwater Nursery: A brackishwater nursery is a specialized facility where juvenile fish, such as fingerlings, are raised in water with moderate salinity (5-15 ppt) before being transferred to grow-out ponds. These nurseries require careful management of water quality, including salinity and pH, and feeding practices to promote healthy growth.

Breeding Seasonality: Breeding seasonality refers to the specific time of year when fish species reproduce, often influenced by environmental cues like temperature, photoperiod, and food availability. For example, many marine species spawn during warmer months (25°-30°C), and breeders often manipulate light and temperature to extend or induce spawning seasons.

Breeding Techniques: Breeding techniques in aquaculture include natural spawning, induced breeding through hormone treatments, and artificial insemination. Environmental factors like temperature (26°-30°C) and photoperiod are often manipulated to enhance breeding success and synchronize spawning events.

Brooder Disease: Brooder diseases are illnesses that affect broodstock, often leading to reproductive failure. Diseases can be bacterial (e.g., Aeromonas), viral (e.g., viral nervous necrosis), or parasitic. Managing water quality, providing a balanced diet, and ensuring biosecurity measures are key to preventing brooder diseases.

Brooder Maturation Diet: Brooder maturation diets are specially formulated feeds for broodstock, rich in proteins (40-50%), lipids (especially omega-3 fatty acids), and vitamins to support gonadal development and enhance reproductive performance. These diets improve egg quality and larval survival rates.

Brooder Nutrition: Brooder nutrition involves providing a diet rich in proteins, lipids (especially omega-3 fatty acids), and vitamins necessary for reproductive success. For marine brooders, a diet containing 45-55% protein is optimal for gonadal development and healthy egg production.

Brooder Stress Management: Brooder stress management involves minimizing stress factors that affect reproductive performance and health of broodstock, such as poor water quality, overcrowding, and handling. Stress negatively impacts spawning and egg quality. Optimal parameters, such as temperature (26°-30°C), dissolved oxygen (>5 ppm), and gentle handling techniques, are essential to reduce cortisol levels and ensure successful breeding.

Broodstock Management: Broodstock management includes selecting and maintaining healthy, sexually mature fish in optimal conditions (temperature 26°-30°C, diet rich in nutrients) to ensure successful reproduction and seed production. This process involves monitoring fish health, water quality, and reproductive cycles.

Broodstock Quarantine: Broodstock quarantine is a precautionary measure involving the isolation of newly acquired or sick fish for a period (typically 2-4 weeks) to monitor for disease before introducing them into the main population. During quarantine, water quality, health, and behavior are closely observed.

Broodstock: Broodstock refers to mature, sexually capable fish that are maintained in aquaculture systems to produce offspring. These fish are carefully selected based on traits like growth rate, disease resistance, and reproductive fitness. For optimal breeding, broodstock management involves controlled environmental conditions such as temperature (optimal range 24°–28°C for tropical species), salinity (varies by species, often 30–35 ppt for marine fishes), and a well-balanced diet rich in essential nutrients like proteins and lipids.

Captive Breeding: Captive breeding involves breeding fish in controlled environments rather than in the wild. This practice helps conserve endangered species, improve aquaculture production, and reduce the need for wild seed resources.

Cobia: Cobia (*Rachycentron canadum*) is a fast-growing marine fish that thrives in open water or offshore cage systems with salinity levels around 30–35 ppt and temperatures between 24°–30°C. Cobia are valued for their high-quality flesh and rapid growth rates, making them a popular species for aquaculture.

Copepods: Copepods are small crustaceans used as live feed for marine fish larvae, particularly during early development stages. They are nutritionally superior to other live feeds due to their high lipid and fatty acid content, and their natural swimming behavior triggers feeding in larvae.

Disease Management: Disease management in aquaculture involves strategies to prevent, diagnose, and treat fish diseases. This includes biosecurity protocols, water quality management, regular health monitoring, and treatments using antibiotics, vaccines, or disinfection methods, depending on the pathogen involved.

Egg Incubation: Egg incubation in fish is the phase from fertilization to hatching. The incubation period is species-dependent and controlled by water temperature, oxygen levels, and water flow. Optimal temperature ranges for egg incubation in marine species are typically between 24°-28°C, and salinity should be maintained at around 30-35 ppt to ensure healthy embryo development.

Fertilization: Fertilization in fish is the union of sperm and egg cells, usually occurring externally in the water column for most marine species. Fertilization success is influenced by factors such as water quality, temperature (optimal range 24–28°C), and the proximity of spawning individuals.

Fingerlings: Fingerlings are juvenile fish that have reached a size suitable for stocking in grow-out systems or nurseries. They range from 5-10 cm in length, and their survival depends on providing high-quality water (temperature 26°-30°C, dissolved oxygen >5 ppm) and proper nutrition.

Fish Breeding Techniques: Fish breeding techniques involve methods like natural spawning, induced spawning (via hormonal administration), and artificial fertilization. Key environmental parameters like temperature (27°-30°C) and salinity (25-35 ppt) are controlled to mimic natural conditions. Hormones like LHRH or HCG may be administered to trigger spawning.

Fish Brooder Qualities: The qualities of an ideal fish brooder include good health, genetic diversity, appropriate age, and size, as well as proven reproductive success. Maintaining brooders in optimal conditions (temperature 26°-30°C, high-quality nutrition) ensures high fertility rates and healthy offspring.

Fish Brooders: Fish brooders are sexually mature fish that are kept in captivity for breeding purposes. The selection of brooders is based on size, health, and reproductive potential. Key parameters such as temperature (26°-30°C) and diet rich in proteins and essential fatty acids are critical for maintaining their reproductive health.

Fish Disinfection: Fish disinfection involves treating fish or eggs to eliminate pathogens before introduction into hatcheries or nurseries. Common methods include immersion in disinfectants like formalin (25-100 ppm) or iodine solutions, which help prevent the spread of diseases.

Fish Eggs: Fish eggs are the fertilized or unfertilized reproductive cells of fish. They require specific conditions for development, including temperature (26°-28°C), salinity (30-35 ppt), and oxygen levels (>5 ppm). Egg health and fertilization success are influenced by water quality and broodstock nutrition.

Fish Farming: Fish farming is the practice of raising fish in controlled environments such as ponds, cages, or tanks. It includes a variety of systems, from extensive (low input) to intensive (high input), and focuses on producing species like tilapia, catfish, and shrimp for commercial purposes.

Fish Hatcheries: Fish hatcheries are controlled environments where eggs are incubated, and larvae are reared under optimized conditions (temperature 26°-30°C, salinity 30-35 ppt) to produce fry and fingerlings for aquaculture or conservation. Hatcheries utilize advanced technologies to manage water quality, feeding regimes, and health monitoring to ensure high survival rates.

Fish Health Parameters: Fish health parameters include water quality (temperature, dissolved oxygen, ammonia, pH), physical appearance, behavior, and growth rates. Regular monitoring of these parameters helps in early detection of disease or stress in the stock, ensuring optimal growth and reducing mortality.

Fish Hormones: Fish hormones are biochemical messengers, such as gonadotropins and sex steroids, that regulate reproductive processes including gametogenesis, ovulation, and spermatogenesis. These hormones are influenced by external environmental factors like temperature and photoperiod.

Fish Larvae: Fish larvae are the early developmental stage that follows hatching, characterized by rapid growth and morphological changes as they progress through various stages (e.g., yolk sac, post-larval). This stage is critical for survival, and larvae require optimal environmental conditions, including water temperature, food availability, and salinity, for proper growth and development.

Fish Mortality: Fish mortality refers to the death of fish in an aquaculture system, which can result from poor water quality, diseases, inadequate nutrition, or handling stress. Monitoring key parameters like ammonia (<0.5 ppm), temperature (26°-30°C), and dissolved oxygen (>5 ppm) is critical to reducing mortality rates.

Fish Nutrition: Fish nutrition involves providing balanced diets that meet the specific nutritional needs of fish at various life stages. Diets typically contain proteins (30-50%), lipids, carbohydrates, vitamins, and minerals, with specific formulations for larval, juvenile, and broodstock fish.

Fish Parasites: Fish parasites include protozoans, helminths, and crustaceans that infect fish, often leading to stress, poor growth, and mortality. Effective management includes regular health checks, maintaining water quality, and using anti-parasitic treatments or disinfection protocols to prevent infestations.

Fish Seed Identification: Fish seed identification involves distinguishing different species of larvae or juveniles based on morphological characteristics such as body shape, pigmentation, fin structure, and meristic counts (e.g., fin rays). Accurate identification is essential for ensuring the correct species are cultured.

Fish Seed Resources: Fish seed resources refer to the availability of juvenile fish (seed) for stocking in aquaculture systems, whether sourced from hatcheries or natural populations. Sustainable management of seed resources is essential for the long-term viability of aquaculture operations.

Fish Spawning: Fish spawning is the release of eggs and sperm into the water for fertilization. It can occur naturally or be induced through hormonal treatments. Environmental conditions like temperature (25°-30°C) and light cycles are optimized to trigger natural spawning behaviors in captive fish.

Fish Transport Methods: Fish transport methods refer to techniques used to safely move live fish, broodstock, or seed from one location to another. Key factors include controlling temperature (24°-28°C), oxygen levels (dissolved oxygen >5 ppm), and reducing stress through anesthetics or proper packing densities in oxygenated bags or tanks to ensure minimal mortality.

Fish Vaccination: Fish vaccination involves administering vaccines to protect fish against specific bacterial or viral pathogens, such as *Vibrio* spp. or infectious salmon anemia virus. Vaccines can be delivered via injection, immersion, or oral methods and are critical for disease prevention in aquaculture.

Fry: Fry refers to the early juvenile stage in fish, immediately after the larval stage. At this stage, the fry are actively feeding and growing. Proper nutrition, including live feeds such as Artemia or rotifers, along with optimal water conditions (temperature 26°-30°C, salinity 25-35 ppt) are essential for fry survival and growth.

Fungal Infections: Fungal infections, often caused by *Saprolegnia*, commonly affect fish eggs and larvae. These infections thrive in poor water conditions and are typically treated with antifungal agents like malachite green or formalin, while maintaining good water quality (oxygen >5 ppm, temperature 26°-28°C).

Gamete Development: Gamete development refers to the maturation of eggs in females (oogenesis) and sperm in males (spermatogenesis) during the reproductive cycle. This process is regulated by hormones like gonadotropins, which are influenced by environmental conditions such as light cycles and water temperature.

Gametogenesis: Gametogenesis is the process of gamete (egg or sperm) production in fish, involving the development and maturation of germ cells. In females, this process is called oogenesis, and in males, spermatogenesis. Gametogenesis is influenced by hormonal signals, environmental factors like temperature and photoperiod, and can take several weeks to months depending on the species.

Gonad Development: Gonad development refers to the growth and maturation of reproductive organs (ovaries and testes) in fish. This process is regulated by hormones and environmental cues, with fish undergoing various stages of gonadal growth, from immature to fully mature capable of producing viable gametes.

Grouper: Grouper (*Epinephelus spp.*) are high-value marine species that require specific conditions for breeding, including stable water temperatures (26°–30°C) and high salinity (30–35 ppt). Groupers are often raised in cage culture systems and are known for their slow growth but premium market price.

Hatchery Management: Hatchery management involves creating optimal environmental conditions and managing fish health for breeding and seed production. This includes maintaining water quality (temperature: 26°-30°C, salinity: 30-35 ppt, dissolved oxygen >5 ppm), providing proper nutrition, disease management, and biosecurity measures to ensure high survival rates during larval stages.

Hatchery Technology: Hatchery technology encompasses the equipment and methods used in fish hatcheries, including water circulation systems, aeration, filtration, temperature control, and automatic feeders. Advancements in hatchery technology improve survival rates and ensure the efficient production of fry and fingerlings.

Health Monitoring: Health monitoring in aquaculture involves regularly checking fish for signs of disease, stress, or poor growth. This includes visual inspection, water quality testing, and sometimes laboratory diagnostics to detect pathogens. Maintaining optimal conditions (oxygen >5 ppm, ammonia <0.5 ppm) is key to fish health.

Hormonal Administration: Hormonal administration involves injecting or implanting hormones like LHRH or HCG into broodstock to induce maturation and spawning. This technique is used when natural spawning triggers are insufficient, and it helps synchronize breeding cycles.

Hormonal Control: Hormonal control in fish breeding involves the use of exogenous hormones, like gonadotropins or GnRH analogs, to induce or enhance reproductive processes such as gametogenesis and spawning. This technique is often used in induced breeding to synchronize maturation and ovulation.

Hypophysation: Hypophysation is the process of inducing ovulation in fish through the administration of pituitary gland extracts or synthetic hormones. This method is commonly used in aquaculture to control and synchronize spawning, especially in species that do not breed readily in captivity.

Induced Breeding: Induced breeding is the artificial stimulation of fish to spawn by administering hormonal injections (e.g., gonadotropins or GnRH analogs). This technique is widely used in aquaculture to overcome environmental and biological barriers to natural reproduction, ensuring year-round seed production.

Larval Development: The larval development in fish refers to the stages between hatching and becoming juvenile, characterized by metamorphosis through various stages like yolk sac absorption, first feeding, and organogenesis. Optimal water temperatures for marine larvae range between 24°-28°C, with salinity levels maintained at 25-35 ppt. Adequate oxygen levels (5-6 ppm) and controlled light cycles (12:12 hours light) are essential for successful larval growth.

Larval Disease: Larval diseases can be caused by bacteria, viruses, fungi, or parasites and are often a result of poor water quality or overcrowding. Common larval diseases include bacterial infections like vibriosis. Maintaining optimal water parameters (temperature 26°-30°C, salinity 30-35 ppt) and providing proper nutrition can help prevent outbreaks.

Larval Feeding: Larval feeding in aquaculture involves providing appropriate live or formulated feeds to fish larvae, such as rotifers, Artemia, or microdiets. Live feeds are often preferred during early stages as they match the size and nutritional requirements of larvae, ensuring optimal growth.

Larval Stages: Larval stages are the early developmental phases of fish post-hatching, including the yolk-sac stage, pre-feeding, and first-feeding larvae. These stages are crucial for survival and growth, requiring specific feeding regimes (live feed like rotifers) and water quality (temperature: 26°-30°C, salinity: 30-35 ppt).

Life Cycle: The life cycle of fish includes various stages from fertilized egg, larvae, juvenile, and adult, culminating in reproduction. Understanding the complete life cycle, including the timing of metamorphosis and sexual maturity, is essential for successful aquaculture practices.

Sexual Dimorphism: Sexual dimorphism refers to the differences in appearance between male and female fish of the same species, often related to size, coloration, or fin morphology. These differences can become more pronounced during the breeding season and are useful for identifying mature broodstock.

Live Feed: Live feed refers to organisms such as Artemia, rotifers, and copepods used as the first feed for larvae due to their high nutritional value and size. Live feed is essential in early larval stages when fish larvae cannot consume artificial diets.

Marine Aquaculture: Marine aquaculture involves the cultivation of fish, crustaceans, and mollusks in seawater environments, typically with salinities of 30-35 ppt. Marine species like grouper, sea bass, and shrimp are farmed using various systems such as cages, ponds, and recirculating aquaculture systems.

Marine Fish Seed: Marine fish seed refers to the juvenile stage of marine species (fry or fingerlings) produced either in hatcheries or collected from the wild for aquaculture purposes. Seed production requires precise control of environmental parameters such as temperature (24°–28°C) and salinity (30–35 ppt).

Marine Fishes: Marine fishes are species that live and breed in saltwater environments such as oceans and seas, typically at salinities of 30–40 ppt. These species include important aquaculture candidates like seabass, grouper, and cobia, which require specialized systems that simulate marine conditions for breeding and seed production.

Marine Nursery: A marine nursery is designed for rearing juvenile marine fish species in seawater conditions (salinity 30-35 ppt) until they are ready for transfer to grow-out systems. It involves managing water quality, stocking densities, and diet to maximize growth and minimize stress in the developing fish.

Maturation: Maturation in fish refers to the physiological process by which fish reach sexual maturity and are capable of reproduction. Environmental factors like photoperiod and temperature play a crucial role, with optimal temperature ranges varying by species. For instance, 24°-28°C is ideal for many tropical marine fish species to ensure gonadal development.

Milkfish: Milkfish (*Chanos chanos*) is a widely farmed brackishwater fish, thriving in salinities of 10–30 ppt and temperatures between 24°–32°C. Known for their rapid growth and tolerance to various environmental conditions, milkfish are an important species in aquaculture, especially in Southeast Asia.

Mullet: Mullet (*Mugilidae spp.*) are brackishwater fishes commonly found in estuaries and lagoons. They tolerate salinity fluctuations (from 5 to 35 ppt) and are often cultivated in extensive or semi-intensive systems due to their ability to feed on detritus and algae.

Natural Breeding: Natural breeding occurs when fish reproduce in their natural environment or in captivity without human intervention. The environmental cues for spawning, such as changes in temperature, photoperiod, and salinity, mimic natural conditions to trigger gametogenesis and release. For many marine species, successful natural breeding requires optimal parameters such as a water temperature range of 24–28°C and salinity around 30–35 ppt.

Natural Seed Availability: Natural seed availability refers to the abundance of fish larvae or juveniles that can be collected from natural water bodies such as estuaries, coastal areas, or rivers. These seeds are often used in aquaculture for stocking ponds or cages, though over-reliance on natural seed can lead to ecological impacts.

Nursery Techniques: Nursery techniques involve the rearing of juvenile fish from post-larval stages (fry) to fingerlings before stocking in grow-out systems. These techniques include managing water quality, providing proper feed, and ensuring adequate space to reduce stress and cannibalism.

Nutritional Requirements: Nutritional requirements in fish involve the balance of proteins (40-55% for marine species), lipids, carbohydrates, vitamins, and minerals essential for growth, health, and reproduction. Different life stages (larvae, juveniles, broodstock) have varying nutrient demands.

Ovulation: Ovulation is the release of mature eggs from the ovaries into the oviduct in female fish, typically triggered by hormonal changes and environmental factors. In aquaculture, ovulation can be induced using hormonal injections to synchronize spawning for seed production.

Packing of Fish Seeds: Packing of fish seeds involves carefully preparing juvenile fish (fry, fingerlings) for transportation, using oxygenated plastic bags with appropriate water volumes and densities. Maintaining optimal temperature (24°-26°C) and oxygen saturation is crucial to ensure their survival during transit.

Phytoplankton: Phytoplankton are microscopic algae that form the base of the aquatic food chain. In aquaculture, they are used to culture zooplankton (e.g., rotifers, copepods), which in turn serve as live feed for fish larvae. Phytoplankton growth requires specific light, nutrient, and temperature conditions.

Quarantine Procedures: Quarantine procedures involve isolating newly acquired or sick fish from the main stock to prevent disease transmission. Fish are typically kept in separate tanks for observation and treatment, and water quality is closely monitored during this period.

Reproductive Physiology: Reproductive physiology in fish involves the study of the hormonal and environmental factors regulating reproductive processes like gametogenesis, spawning, and sexual maturation. Key hormones include gonadotropins and sex steroids, which are influenced by external stimuli such as photoperiod and temperature.

Reproductive Season: The reproductive season refers to the time of year when fish naturally spawn, influenced by environmental factors like temperature, salinity, and daylight length. For many marine species, this period occurs during the warmer months when conditions are ideal for egg development and larval survival, typically between spring and summer.

Rotifers: Rotifers are microscopic zooplankton used as an initial live feed for marine and freshwater fish larvae. They are cultured in tanks and enriched with nutrients to provide essential fatty acids, proteins, and vitamins for larval growth.

Sea Ranching: Sea ranching is a practice where hatchery-reared juvenile fish are released into the wild to grow and be harvested later. This is often used to replenish depleted fisheries. Survival rates depend on the species and environmental factors like temperature and predators.

Seabass: Seabass (*Lates calcarifer*), a popular species for aquaculture, is a euryhaline fish capable of surviving in both marine and brackish water. It grows best in temperatures between 26°–30°C and salinities of 15–35 ppt. Seabass are known for their high market value and adaptability to various culture systems.

Seed Collection Methods: Seed collection involves gathering fish larvae or juveniles (seed) from wild populations or hatcheries for aquaculture. Methods include plankton nets for larval collection or using traps and nets for juveniles. The collected seed must be handled with care to avoid stress and injury.

Seed Handling: Seed handling includes the care and techniques used in the transfer and management of fish larvae or juveniles, focusing on minimizing physical stress and injury. Gentle handling practices, using scoop nets or submerged containers, and avoiding overcrowding help ensure the seed's health and survival.

Seed Packing Procedures: Seed packing procedures involve preparing juvenile fish for transport by placing them in oxygenated plastic bags or containers with water, ensuring proper density, temperature (24°-26°C), and dissolved oxygen. Effective packing minimizes stress and mortality during transportation

Seed Survival Rate: Seed survival rate is the percentage of fish larvae, fry, or fingerlings that survive to a specified stage of development. Factors affecting survival include water quality (temperature 26°-30°C, salinity 30-35 ppt), feed quality, and effective disease management. High survival rates (>80%) are often a key indicator of hatchery success.

Sexual Maturity: Sexual maturity in fish is the stage at which they develop the ability to reproduce, characterized by the full development of gonads (ovaries in females and testes in males). This is influenced by age, size, and environmental conditions, with marine species often reaching sexual maturity at 2–3 years, depending on species, with conditions such as temperature and diet affecting the timing.

Spawn Collection: Spawn collection refers to the process of collecting eggs and sperm during spawning events, either naturally or through induced breeding techniques. Collected spawn is then fertilized and incubated under controlled conditions in hatcheries.

Spawning Triggers: Spawning triggers are environmental or hormonal cues that stimulate fish to spawn. These can include changes in water temperature, salinity, photoperiod, or the administration of hormones. For instance, increasing the temperature to 27°-30°C often induces spawning in many tropical species.

Spermatogenesis: Spermatogenesis is the process by which male fish produce and mature sperm cells. This process occurs in the testes and is regulated by hormonal signals, including testosterone and other androgens. Optimal conditions such as temperature and photoperiod support spermatogenesis.

Temperature Control: Temperature control in aquaculture involves regulating water temperature to suit the optimal growth and reproduction ranges of cultured species. For marine and brackish water fish, the optimal range is often 26°-30°C. Temperature fluctuations can lead to stress, disease, and reduced growth rates.

Transport of Brooders: The transport of brooders requires minimizing stress and maintaining optimal conditions (e.g., temperature 24°-28°C, oxygenation) to ensure their health during transit. Techniques include using oxygenated water, anesthetics, and controlled packing densities to reduce stress.

Viral Diseases: Viral diseases, such as viral nervous necrosis (VNN) or infectious pancreatic necrosis (IPN), pose significant challenges in aquaculture. Prevention involves strict biosecurity, avoiding the introduction of infected stock, and, where available, the use of vaccines.

Water Circulation Systems: Water circulation systems are used in hatcheries and farms to maintain water quality by promoting water flow, oxygenation, and the removal of waste products. Recirculating aquaculture systems (RAS) are advanced forms that recycle water while maintaining optimal parameters.

Water Quality Management: Water quality management in aquaculture is the process of maintaining suitable physical and chemical parameters like temperature, salinity, pH (7.5-8.5), dissolved oxygen (>5 ppm), and ammonia (<0.5 ppm) to ensure optimal conditions for fish health and growth.

Wild Seed Resources: Wild seed resources refer to naturally occurring larvae, fry, or fingerlings in marine and brackish environments, often collected for aquaculture. Overexploitation of these resources can threaten wild populations, so sustainable collection practices are important.

Zooplankton: Zooplankton, such as rotifers and copepods, are a critical live feed for fish larvae. They provide essential nutrients like fatty acids and proteins, and their size and movement stimulate larval feeding behavior.

CHAPTER 15

CULTURE OF SEAWEEDS

Agar: Agar is a hydrocolloid extracted from red seaweeds, particularly species of *Gracilaria* and *Gelidium*. It is widely used in the food industry as a gelling agent and in microbiological research as a culture medium. Agar extraction involves boiling the seaweed to release the polysaccharides, followed by filtration and gel formation. The quality of agar is influenced by the seaweed species, cultivation conditions, and processing methods. Agar's unique properties, such as its ability to form strong gels at low concentrations, make it a valuable product in various applications.

Algal Blooms: Algal blooms occur when there is a rapid increase in the population of algae in aquatic environments, often driven by excess nutrients (eutrophication) and favorable environmental conditions. While some algal blooms are harmless, others, such as harmful algal blooms (HABs), can produce toxins that affect marine life, human health, and aquaculture operations. Monitoring and managing nutrient levels in water bodies are crucial for preventing and mitigating algal blooms.

Algal Cultivation Techniques: Algal cultivation techniques in seaweed farming involve various methods for growing seaweeds in marine or controlled environments. These techniques range from traditional methods like off-bottom farming and floating raft systems to modern integrated multi-trophic aquaculture (IMTA), where seaweeds are grown alongside other aquaculture species like fish or shellfish to improve nutrient cycling. Successful cultivation depends on optimizing parameters such as water temperature, salinity, light intensity, and nutrient availability. For example, optimal growth for many red seaweeds like *Gracilaria* occurs in water temperatures of 20°-28°C and salinity levels of 25-35 ppt.

Alginate: Alginate is a naturally occurring polysaccharide extracted from the cell walls of brown seaweeds, such as *Macrocystis pyrifera, Ascophyllum nodosum,* and *Laminaria spp.* It is used extensively in the food industry as a thickener, emulsifier, and stabilizer, and in pharmaceuticals for wound dressings and dental molds. The extraction process involves treating the seaweed with alkali to convert alginic acid to sodium alginate. Optimal growth conditions for alginate-producing seaweeds include cool, nutrient-rich waters, typically at temperatures of 10°-20°C, with high wave action and strong currents to stimulate growth and alginate production.

Animal Feed: Seaweed-based animal feed supplements are gaining popularity due to their high nutritional value, including vitamins, minerals, amino acids, and bioactive compounds. Seaweeds like *Ascophyllum nodosum* and *Laminaria digitata* are commonly used in feed for livestock, poultry, and aquaculture species. These supplements improve animal health, boost immune function, and enhance the nutritional quality of animal products such as meat, milk, and eggs. The inclusion rate of seaweed in animal feed typically ranges from 1-5%, depending on the species and desired health benefits.

Aquatic Biodiversity: Aquatic biodiversity refers to the variety of species and ecosystems found in marine and freshwater environments. Seaweed farms can influence aquatic biodiversity by providing habitat for various organisms, such as fish, invertebrates, and microorganisms. However, intensive seaweed farming can also impact local biodiversity through habitat modification, nutrient enrichment, and competition with native species. Balancing the benefits of seaweed farming with the protection of aquatic biodiversity is a key consideration in sustainable aquaculture practices.

Asexual Reproduction: Asexual reproduction in seaweeds occurs without the fusion of gametes and includes methods such as spore formation, vegetative propagation, and fragmentation. Spores can be released into the water to settle and grow into new individuals or vegetative fragments of the thallus can regenerate into complete organisms. Asexual reproduction allows for rapid colonization and expansion of seaweed populations.

Bioactive Compounds: Bioactive compounds are naturally occurring chemicals in seaweeds that have biological effects such as antioxidant, anti-inflammatory, or antimicrobial properties. These compounds, including polyphenols, polysaccharides, and fatty acids, are of interest for use in nutraceuticals, pharmaceuticals, and cosmetics. For instance, fucoidan from brown seaweeds has shown potential in cancer therapy. The concentration of bioactive compounds can vary depending on the species, cultivation conditions, and post-harvest processing, making it essential to optimize farming practices for high-value compound production.

Biofouling: Biofouling refers to the accumulation of unwanted organisms such as barnacles, mussels, and other sessile invertebrates, on seaweed farming structures, ropes, and the seaweeds themselves. Biofouling can impede seaweed growth, reduce light penetration, and increase the weight and drag on farming gear. Effective biofouling management involves regular maintenance, the use of anti-fouling coatings, and optimizing the timing of cultivation to avoid peak fouling seasons. In some cases, rotational farming practices are implemented to minimize biofouling impacts.

Brown Seaweeds: Brown seaweeds (Phaeophyceae) are large, multicellular algae predominantly found in colder ocean waters. Their brown color is due to the presence of the pigment fucoxanthin. Brown seaweeds, such as kelp (*Laminaria* and *Macrocystis*), are harvested for alginates, which are used in food, cosmetics, and pharmaceuticals. Optimal growth conditions include temperatures of 5°-15°C and nutrient-rich waters.

Carrageenan: Carrageenan is a family of sulfated polysaccharides extracted from red seaweeds, particularly from species like *Kappaphycus alvarezii* and *Eucheuma denticulatum*. It is widely used as a gelling, thickening, and stabilizing agent in the food, pharmaceutical, and cosmetic industries. Carrageenan comes in three main types: kappa, iota, and lambda, each differing in their gelling properties. The yield and quality of carrageenan are influenced by factors such as species, cultivation conditions, and post-harvest processing. Optimal water temperatures for carrageenan-producing seaweeds typically range from 25°-30°C, and salinity levels around 30-35 ppt are ideal for maximum production.

Climate Impact on Seaweeds: Climate change affects seaweeds by altering water temperature, salinity, and nutrient availability. Seaweeds are sensitive to changes in their environment; for instance, increases in sea surface temperature can stress some species, leading to reduced growth rates and changes in distribution patterns. Ocean acidification, a result of increased CO_2 absorption by seawater, can also affect calcifying seaweeds like coralline algae, reducing their ability to build calcium carbonate structures. However, some seaweed species may benefit from higher CO_2 levels, enhancing photosynthesis and growth under certain conditions.

Coastal Management: Coastal management involves strategies and actions aimed at preserving coastal ecosystems while balancing the needs of human activities such as fishing, tourism, and aquaculture. Effective coastal management requires integrating environmental conservation with the sustainable use of resources. In seaweed farming, coastal management plays a crucial role in site selection, ensuring that farming activities do not disrupt local ecosystems or interfere with other marine industries. Zoning regulations and environmental impact assessments are often employed to achieve these goals.

Coir Rope: Coir rope, made from coconut husk fibers, is a biodegradable material often used in seaweed aquaculture. It provides a natural and sustainable alternative to synthetic ropes and is particularly favored for use in regions where environmental conservation is a priority. Coir ropes are durable, resistant to saltwater, and decompose naturally over time, reducing waste in marine environments.

Commercial Species: Commercial seaweed species refer to the types of seaweed cultivated for economic purposes, primarily for food, industrial, pharmaceutical, and agricultural products. Key commercial species include *Kappaphycus alvarezii* and *Eucheuma denticulatum* for carrageenan production, *Gracilaria* for agar extraction, and *Laminaria* and *Macrocystis* for alginate. The choice of species depends on factors such as environmental conditions, market demand, and the desired end product. Commercial seaweed farming often focuses on species that grow rapidly and can be harvested multiple times a year, optimizing yield and profitability.

Community Structure: Community structure in seaweed ecosystems refers to the composition and interactions of various organisms, including different seaweed species, herbivores, epiphytes, and other marine life. In seaweed farming, maintaining a balanced community structure is essential for preventing overgrowth by undesirable species such as epiphytes or biofouling organisms, which can impact crop health. Managing community structure through site selection, cultivation techniques, and monitoring can enhance the resilience and productivity of seaweed farms while supporting local biodiversity.

Cultivation Depth: Cultivation depth in seaweed farming is determined by the species being farmed, light availability, and water movement. For example, *Eucheuma* and *Kappaphycus* are typically cultivated at depths of 1-3 meters, where light penetration is sufficient for photosynthesis, and water movement ensures nutrient delivery. Deeper cultivation can be challenging due to reduced light and the potential for decreased growth rates. Managing the cultivation depth is crucial for balancing light, nutrients, and protection from surface disturbances.

Culture Maintenance: Culture maintenance in seaweed farming involves the ongoing care and management of seaweed crops to ensure optimal growth and productivity. This includes regular monitoring of environmental conditions such as water temperature, salinity, and nutrient levels, as well as controlling pests, diseases, and biofouling. Proper maintenance ensures that seaweed grows at the desired rate and reaches harvest size with high-quality biomass. Techniques may also involve periodic thinning of seaweed to reduce competition for light and nutrients and maximize overall farm productivity.

Culture Monitoring: Culture monitoring in seaweed farming involves tracking the health, growth, and environmental conditions of seaweed crops throughout the cultivation cycle. This may include regular sampling of water quality parameters (e.g. temperature, salinity, nutrient levels), visual inspections for signs of disease or pests, and growth measurements. Effective monitoring allows farmers to make timely adjustments to farming practices, such as modifying nutrient inputs or adjusting the depth of cultivation, to optimize yields and prevent crop losses. Monitoring is critical for ensuring sustainable and efficient production.

Curation Techniques: Curation techniques for seaweed refer to the methods used to maintain the quality of seaweed during storage and processing. This may include temperature control, moisture regulation, and the use of preservatives to prevent microbial growth. In the case of high-value seaweed products, such as carrageenan or agar, careful curation ensures that the seaweed retains its functional properties throughout the supply chain. Proper curation techniques are essential for maintaining the economic value and safety of seaweed products, particularly in global markets.

Distribution of Seaweeds: The distribution of seaweeds is influenced by environmental conditions such as light, temperature, salinity, and nutrient availability. Seaweeds are broadly classified into three groups—green (Chlorophyta), brown (Phaeophyta), and red (Rhodophyta)—each with different habitat preferences. For example, red seaweeds such as *Gracilaria* and *Kappaphycus*, are typically found in warmer, tropical waters, while brown seaweeds, such as kelp, dominate cooler, nutrient-rich environments. Understanding the natural distribution of seaweeds aids in site selection and species-specific cultivation strategies.

Economic Importance: Seaweed has substantial economic importance due to its diverse applications in food, pharmaceuticals, cosmetics, and biofuel industries. The global seaweed market was valued at over $15 billion, with continuous growth projected. Seaweed farming provides employment and income for coastal communities, particularly in Asia, and plays a vital role in the sustainable blue economy by supporting food security, climate mitigation, and the production of bio-based materials.

Ecosystem Management: Ecosystem management in seaweed farming focuses on maintaining the health and productivity of marine ecosystems while supporting sustainable aquaculture practices. This includes monitoring and regulating the impacts of farming on local biodiversity, water quality, and habitat structure. Ecosystem-based management practices may involve integrated multi-trophic aquaculture (IMTA), where seaweed farming is combined with the cultivation of other species, such as fish or shellfish, to create a balanced and sustainable system that reduces environmental impacts.

Environmental Factors: Environmental factors, such as light, temperature, salinity, water movement, and nutrient availability, are critical determinants of seaweed growth and productivity. These factors influence photosynthesis, metabolism, and reproductive success. For example, high light availability promotes photosynthesis, while stable temperatures in the optimal range (20°-30°C) enhance growth rates for most commercially cultivated seaweeds. Understanding and managing environmental factors are essential for optimizing seaweed farming, preventing crop losses, and ensuring sustainability.

Epiphytes: Epiphytes are organisms, such as algae, fungi, or small invertebrates, that grow on the surface of seaweed, often competing for light, nutrients, and space. The presence of epiphytes can reduce the photosynthetic efficiency of seaweeds and lead to stunted growth or increased susceptibility to disease. In seaweed farming, regular monitoring and cleaning of the crop, as well as the selection of species with natural resistance to epiphytic growth, are common practices to minimize the impact of epiphytes.

Eucheuma Cultivation: Eucheuma species, such as *Eucheuma denticulatum*, are cultivated primarily for the extraction of carrageenan, a gelatinous substance used in various industries. Eucheuma thrives in tropical, coastal environments with clear water, temperatures between 25°-30°C, and salinity levels of 30-35 ppt. It is often farmed using the longline or raft method, with ropes or nets suspended in shallow waters. Eucheuma cultivation requires careful monitoring of environmental conditions to prevent issues like epiphyte overgrowth, disease outbreaks, and nutrient depletion, all of which can impact crop yields.

Fertilizers: Seaweed-based fertilizers are organic products derived from seaweed biomass, rich in essential nutrients like nitrogen, potassium, phosphorus, and trace elements. They promote plant growth, enhance soil health, and improve crop resistance to stress. Seaweed extracts, particularly from brown algae like *Ascophyllum nodosum*, are used as liquid fertilizers and biostimulants due to their content of plant hormones, such as auxins and cytokinins. Fertilizers derived from seaweed can be applied to various crops, offering a sustainable alternative to chemical fertilizers, and are particularly beneficial in organic farming systems.

Fixed Bottom Long-Line: The fixed bottom long-line system is a seaweed cultivation method where lines are anchored to the seafloor at regular intervals. This system is suitable for shallow waters where the seaweed can be grown near the bottom, receiving sufficient light and nutrients. The fixed nature of the lines makes this method resilient to strong currents and wave action, making it ideal for regions with higher water movement.

Fragmentation: Fragmentation is a form of asexual reproduction where parts of the seaweed break off due to physical forces such as waves, and these fragments grow into new individuals. This method is common in species like *Gracilaria* and is crucial for their propagation in natural and aquaculture settings. Fragmentation can contribute to the resilience and spread of seaweed populations in dynamic environments.

Global Seaweed Market: The global seaweed market is valued at over USD 17 billion and is growing rapidly due to increased demand for seaweed products in food, cosmetics, pharmaceuticals, and biofuels. Asia-Pacific dominates this market, driven by strong production in countries like China, Japan, and Indonesia. Key market segments include red seaweeds (used for carrageenan and agar) and brown seaweeds (for alginates). Growth factors include the rising interest in sustainable, plant-based products and seaweed's use as a carbon sink to combat climate change.

Gracilaria Cultivation: *Gracilaria* species are commonly cultivated for agar production and are farmed in a variety of environments, from coastal waters to brackish lagoons. These seaweeds prefer warm temperatures (20°-30°C) and salinity levels of 15-30 ppt, making them adaptable to a range of aquaculture systems. *Gracilaria* is often cultivated using floating rafts or longline systems, where it benefits from moderate water movement and nutrient-rich environments. Proper site selection and nutrient management are key to optimizing *Gracilaria* yields, and this species is often integrated into multi-trophic aquaculture systems.

Grazing Pressure: Grazing pressure refers to the impact of herbivorous organisms, such as fish, sea urchins, and mollusks, on seaweed crops. Grazing can significantly reduce the biomass and productivity of cultivated seaweeds. In intensive farming systems, protective measures like predator exclusion devices or the use of species-resistant to grazing are employed to mitigate this pressure. The level of grazing pressure depends on the density and type of herbivores present in the farming area, and managing this factor is essential for maintaining healthy seaweed crops.

Green Seaweeds: Green seaweeds (Chlorophyta) contain chlorophyll a and b, giving them their characteristic green color. They are commonly found in shallow coastal waters and estuaries. Some species, like *Ulva* (sea lettuce), are cultivated for human consumption and biofuel production. Green seaweeds prefer nutrient-rich environments with moderate light exposure, and optimal growth occurs in temperatures between 10°-25°C.

Harvesting Guidelines: Harvesting guidelines for seaweed farming are protocols ensuring sustainable extraction that minimizes environmental impact and promotes regrowth. These guidelines typically include the optimal timing of harvest, often aligned with the reproductive cycles of seaweed species, and the correct harvesting technique, such as cutting the plant at a specific height to allow regeneration. For example, in *Eucheuma* and *Kappaphycus* farming, cutting should occur 15-20 cm above the holdfast. Harvesting frequency may vary, but a 45-60 day cycle is common, depending on species and environmental conditions like water temperature, salinity, and nutrient availability.

Harvesting Methods: Harvesting methods in seaweed farming depend on the species and cultivation techniques used. Common methods include manual harvesting, where farmers cut seaweed from ropes or nets, or mechanical harvesting, using specialized equipment to remove seaweed from large farms. Timing is crucial to maximize yield and ensure the quality of the biomass. For instance, *Kappaphycus* is typically harvested 45-60 days after planting, when it reaches optimum size and carrageenan content. Post-harvest handling is equally important to prevent spoilage and maintain product quality, particularly for high-value seaweed products.

Human Food: Seaweeds have been consumed as food for centuries, particularly in Asia, where species like *Porphyra* (nori), *Undaria pinnatifida* (wakame), and *Laminaria* (kombu) are staples in diets. Seaweeds are rich in essential nutrients, including iodine, iron, calcium, and vitamins A, C, and B12. They also contain dietary fiber and bioactive compounds that have potential health benefits, such as reducing blood pressure and cholesterol levels. Seaweed is consumed in various forms, including fresh, dried, and as an ingredient in soups, salads, and sushi.

Hydrocolloids: Hydrocolloids are water-soluble polysaccharides extracted from seaweeds, such as agar, carrageenan, and alginate, which are widely used as gelling, thickening, and stabilizing agents in the food and pharmaceutical industries. For example, agar is extracted from red seaweeds like *Gracilaria* and used in food products, microbiological media, and biotechnology applications. Hydrocolloid production requires specific processing techniques to extract and purify the polysaccharides while maintaining their functional properties.

International Scenario: Internationally, seaweed aquaculture is a major industry in countries like China, Indonesia, and South Korea, which dominate global production. Asia accounts for over 80% of the world's seaweed production, with China leading in kelp cultivation. The global market is driven by the demand for seaweed as food, hydrocolloids (agar, carrageenan, alginates), and biofuels. The industry is also expanding in Europe and North America, focusing on sustainability, ecosystem services, and bioremediation.

Kappaphycus Cultivation: Kappaphycus cultivation focuses on farming species like *Kappaphycus alvarezii*, which is widely used for producing carrageenan, a hydrocolloid used in the food and pharmaceutical industries. This seaweed is typically cultivated in tropical waters with temperatures between 25°-30°C, salinity levels of 30-35 ppt, and moderate water movement to ensure nutrient uptake. Kappaphycus is grown on ropes or nets in shallow waters, and its rapid growth rate makes it one of the most commercially viable seaweed species. Regular monitoring for pests, diseases, and environmental changes is crucial to maintaining high yields.

Laboratory-Scale Culture: Laboratory-scale culture of seaweeds involves controlled cultivation of seaweed species in laboratory settings to study their growth, reproduction, and biochemical properties. This is essential for selecting strains with desirable traits for large-scale farming and for researching the effects of environmental variables like light, temperature, and nutrients. Optimal conditions vary by species but typically involve maintaining water temperatures between 10°-20°C, salinity levels of 30-35 ppt, and ensuring adequate light intensity for photosynthesis, usually around 50-150 μmol photons m^2/s.

Land-Based Cultivation: Land-based seaweed cultivation involves growing seaweed in controlled environments, such as tanks or ponds, rather than in open coastal waters. This method allows for greater control over water quality, temperature, and nutrient levels, leading to optimized growth and higher yields. Land-based systems are often used for high-value species or in regions where open-water farming is impractical due to environmental conditions or regulations.

Life Cycle of Seaweeds: The life cycle of seaweeds typically involves an alternation of generations between a diploid sporophyte and a haploid gametophyte. In many species, the sporophyte produces spores that develop into gametophytes, which in turn produce gametes for sexual reproduction. Some species exhibit a direct life cycle, where the organism only has one generation, either sexual or asexual. The life cycle dynamics influence cultivation practices and management in aquaculture.

Long-Line Cultivation: Long-line cultivation is a method of seaweed farming where ropes or lines are suspended horizontally in the water column, and seaweed is attached to them. This method allows for efficient use of space and is commonly used for species like kelp. The lines are typically anchored at both ends, and the seaweed grows as it receives nutrients and light from the surrounding water. The depth of the lines is adjusted to optimize growth conditions.

Marine Biotechnology: Marine biotechnology harnesses the biological resources of the ocean, including seaweeds, for applications in medicine, industry, and environmental management. It includes the extraction of bioactive compounds from marine organisms, genetic engineering of seaweeds for enhanced production, and the use of seaweed-based bioproducts in pharmaceuticals and nutraceuticals. The high polysaccharide content and unique metabolites in seaweeds make them valuable for developing new materials and drugs, while biotechnological advances in cultivation and processing can increase yield and sustainability.

Marine Botany: Marine botany focuses on the study of marine plants, including seaweeds, seagrasses, and mangroves. It examines their physiology, distribution, and ecological roles in marine environments. Marine botanists are interested in how these plants adapt to saline conditions, their contribution to primary production, and their interactions with other marine organisms. This field is essential for understanding and conserving marine ecosystems.

Marine Pollution Impact: Marine pollution, including oil spills, plastic waste, and chemical contaminants, can severely affect seaweed growth and the overall health of marine ecosystems. Pollutants can disrupt photosynthesis, nutrient absorption, and reproduction in seaweeds, leading to decreased biomass and biodiversity loss. Heavy metals, in particular, can accumulate in seaweed tissues, making them unsafe for consumption. Monitoring and mitigating marine pollution is crucial for maintaining the sustainability of seaweed farming and preserving coastal ecosystems.

Marine Resources: Marine resources encompass all living and non-living materials derived from the ocean that are used by humans. This includes fish, shellfish, seaweed, and minerals, as well as genetic resources for pharmaceuticals and biotechnology. Sustainable management of marine resources is crucial to prevent overexploitation and degradation of marine ecosystems. Seaweed farming is increasingly recognized as a sustainable use of marine resources due to its low environmental impact and ability to sequester carbon.

Marine Sustainability: Marine sustainability in the context of seaweed farming focuses on balancing the economic benefits of cultivation with the protection of marine ecosystems. Sustainable practices include minimizing the use of chemicals, reducing the environmental footprint of farming operations, and promoting the health of surrounding marine life. Seaweed farming is often considered a sustainable form of aquaculture because it requires no freshwater or fertilizers and can help mitigate climate change by absorbing carbon dioxide. Integrating seaweed farming with other sustainable marine practices, such as integrated multi-trophic aquaculture (IMTA), can enhance overall sustainability.

Marine Vegetation: Marine vegetation includes various aquatic plants and algae, such as seagrasses, mangroves, and seaweeds, which play vital roles in marine ecosystems by providing habitat, stabilizing sediments, and contributing to nutrient cycling. Seaweeds, in particular, are critical primary producers in coastal ecosystems, supporting marine food webs and contributing to oxygen production. They also help mitigate coastal erosion and can act as natural water filters by absorbing excess nutrients from runoff.

National Scenario: In the national context, seaweed aquaculture has gained prominence as a sustainable industry in many coastal countries. For example, in India, seaweed farming is encouraged by government initiatives aiming to boost blue economy sectors. Coastal regions like Tamil Nadu and Gujarat are emerging hubs for the cultivation of species like *Kappaphycus alvarezii* and *Gracilaria*, supported by initiatives to improve coastal livelihoods and the introduction of modern farming techniques.

Net-Style Farm Systems: Net-style farm systems involve using nets or mesh structures to cultivate seaweeds in marine environments. Seaweed is either directly attached to the nets or grown within enclosures. This method provides stability in rough waters and allows for high-density cultivation. Net-style systems are often used for smaller, fast-growing species and can be combined with other aquaculture activities, such as shellfish farming.

Nursery Methods: Nursery methods in seaweed culture involve the initial growing phase where seed stock is carefully nurtured under controlled conditions before being transplanted to the main farming area. These methods may include land-based tanks, sea-based enclosures, or floating systems where parameters such as light, temperature, and nutrient availability are optimized for early growth. The nursery phase is crucial as it determines the strength and viability of the seaweed before it is exposed to the open water environment. Successful nursery practices can significantly increase the survival and productivity of the crop.

Nutraceuticals: Nutraceuticals are products derived from seaweeds that provide health benefits beyond basic nutrition, often containing bioactive compounds such as vitamins, minerals, antioxidants, and omega-3 fatty acids. Seaweeds like Spirulina and Chlorella are popular nutraceuticals due to their high nutrient content and potential health benefits, including boosting the immune system and reducing inflammation. The production of seaweed-based nutraceuticals involves careful cultivation, harvesting, and processing to preserve the active compounds and ensure product quality.

Nutrient Availability: Nutrient availability, particularly nitrogen, phosphorus, and potassium, plays a critical role in seaweed cultivation. These nutrients are absorbed from the surrounding water and are essential for photosynthesis, growth, and reproduction. Seaweeds in nutrient-rich environments typically exhibit faster growth rates and higher biomass yields. For instance, *Gracilaria* species thrive in waters with nitrate levels ranging from 1-3 mg/L. In areas with low nutrient availability, fertilizers or nutrient enrichment techniques may be employed to enhance growth.

Nylon Rope: Nylon rope is widely used in seaweed aquaculture due to its strength, durability, and resistance to marine conditions. Nylon ropes are used in various cultivation systems, such as long-lines and nets, where they provide support for the growing seaweed. However, concerns about plastic pollution have led to increased interest in more sustainable alternatives like coir or biodegradable ropes.

Overharvesting Risks: Overharvesting of natural seaweed beds can lead to depletion of local populations, disruption of marine ecosystems, and loss of biodiversity. Sustainable harvesting practices involve regulating the amount and timing of harvests to allow seaweed populations to regenerate. In some regions, overharvesting has prompted the development of seaweed farming as an alternative, reducing pressure on wild stocks while meeting the growing demand for seaweed products. Managing overharvesting risks is crucial for preserving the ecological and economic benefits of seaweed resources.

Pharmaceuticals: Pharmaceuticals derived from seaweeds include bioactive compounds that have medicinal properties, such as antiviral, antibacterial, and anticancer effects. For example, carrageenan from red seaweeds has been explored for its potential in drug delivery systems, while compounds like fucoxanthin from brown seaweeds are being studied for their anti-obesity and anti-inflammatory effects. Seaweed farming for pharmaceutical purposes requires precise control over cultivation and processing to ensure the consistency and efficacy of the bioactive compounds.

Phycology: Phycology is the scientific study of algae, encompassing both microalgae and macroalgae (seaweeds). This field investigates the biology, taxonomy, ecology, and economic uses of algae. Phycologists study the role of algae in marine and freshwater ecosystems, their physiological processes, and their potential in industries such as biofuels, food, and pharmaceuticals.

Phytochemicals: Phytochemicals in seaweeds are bioactive compounds that have beneficial health effects, including antioxidants, anti-inflammatory agents, and anti-cancer properties. These include polyphenols, flavonoids, carotenoids, and polysaccharides. For instance, fucoidan, a sulfated polysaccharide found in brown seaweeds, has been studied for its potential in cancer therapy and immune modulation. The concentration of phytochemicals in seaweeds can vary significantly depending on environmental factors, species, and post-harvest processing techniques.

Polysaccharides: Polysaccharides in seaweeds are complex carbohydrates that serve as energy reserves and structural components. They are classified into storage polysaccharides, such as laminarin in brown algae, and structural polysaccharides, such as cellulose and alginates in the cell walls. Polysaccharides like carrageenan, agar, and alginates are commercially valuable for their gelling, thickening, and stabilizing properties, widely used in the food, pharmaceutical, and cosmetic industries. Optimum extraction of these polysaccharides often involves temperature-controlled processing, with conditions tailored to the specific type of seaweed and desired product.

Post-Harvest Handling: Post-harvest handling includes all activities carried out after harvesting to maintain the quality and safety of seaweed before it is processed or sold. This involves washing to remove sand, salt, and impurities, proper drying to reduce moisture content, and storage under controlled conditions to prevent spoilage. In some cases, pre-processing steps like blanching or chemical treatments may be applied to enhance the preservation of certain bioactive compounds. Effective post-harvest handling is essential to maintaining the commercial value and safety of seaweed products.

Pre-Harvest Activities: Pre-harvest activities in seaweed farming involve preparing the crop for optimal yield and quality at the time of harvest. This includes monitoring environmental conditions, such as water temperature, salinity, and nutrient levels, and managing pests or diseases that could impact growth. Adjusting the density of seaweed on ropes or nets to ensure sufficient space for growth and light penetration is also critical. Pre-harvest activities help ensure that the seaweed reaches its maximum biomass and desired biochemical composition at the time of harvest.

Propagation Methods: Seaweed propagation methods include vegetative propagation (e.g., using fragments or cuttings) and spore-based methods. In aquaculture, vegetative propagation is often preferred for its simplicity and reliability. Propagation techniques vary by species, with some requiring specific conditions such as light, temperature, and nutrient levels to optimize growth and survival rates during cultivation.

Raft Culture: Raft culture is a method used in seaweed aquaculture where seaweed species, like Kappaphycus or Gracilaria, are attached to ropes or nets suspended in the water from floating rafts. This method allows for optimal water movement around the seaweed, ensuring good nutrient exchange and light exposure, which are critical for growth. The rafts are usually anchored in protected coastal waters with moderate wave action and a depth ranging from 2 to 10 meters. Raft culture is commonly used in large-scale seaweed farming due to its efficiency in maximizing space and ease of maintenance.

Red Seaweeds: Red seaweeds (Rhodophyta) are primarily found in marine environments and are characterized by their reddish color due to the presence of phycoerythrin. They are economically important for producing agar and carrageenan, which are used as gelling agents in the food and pharmaceutical industries. Red seaweeds thrive in tropical and subtropical waters, with optimal growth temperatures ranging from 15° to 30°C.

Reproduction Modes: Seaweeds exhibit diverse reproductive strategies, including sexual and asexual reproduction. Sexual reproduction involves the fusion of gametes, while asexual reproduction occurs through mechanisms such as fragmentation, spore production, or vegetative propagation. The mode of reproduction varies by species and environmental conditions, influencing the growth dynamics and population structure of seaweed beds.

Salinity Levels: Salinity levels are a crucial factor in seaweed cultivation, influencing species selection and growth rates. Most commercially cultivated seaweeds, such as Kappaphycus and Eucheuma, thrive in salinity ranges between 30-35 parts per thousand (ppt). Variations in salinity, due to freshwater influx from rivers or heavy rains, can lead to stress and reduced growth. Maintaining stable salinity levels within the optimal range is essential for the healthy development of seaweed and to avoid osmotic stress or other physiological disturbances.

Sea Cultivation: Sea cultivation refers to the farming of seaweeds in open coastal waters, using various methods such as long-line systems, rafts, or nets. This form of cultivation takes advantage of natural water movement and nutrient availability, making it more cost-effective for large-scale production. Site selection is crucial, with optimal locations offering suitable water depth, current, and nutrient concentrations.

Seaweed Aquaculture: Seaweed aquaculture, or seaweed farming, is the cultivation of various species of marine macroalgae, such as kelp, red, green, and brown seaweeds, in controlled or semi-controlled environments, both in coastal waters and land-based systems. This practice is aimed at optimizing growth conditions such as water temperature (10°-20°C for many species), salinity (around 30-35 ppt), nutrient availability (particularly nitrogen and phosphorus), and light exposure. Seaweed aquaculture contributes to sustainable food production, carbon sequestration, and biofuel generation.

Seaweed Biomass: Seaweed biomass refers to the total mass of seaweed produced in a given area or volume. Biomass production is a critical measure of the success of seaweed farming operations and is influenced by factors such as species selection, environmental conditions, and farming techniques. High biomass yields are typically associated with optimal water quality, nutrient availability, and light conditions. Seaweed biomass is harvested for various uses, including food, feed, biofuels, and pharmaceuticals, making it a valuable resource in the global economy.

Seaweed Biostimulants: Seaweed biostimulants are natural products derived from seaweeds, used to enhance plant growth, yield, and stress tolerance. They contain a range of bioactive compounds, including plant hormones, polysaccharides, and trace elements that stimulate plant metabolic processes. Brown seaweeds such as *Ascophyllum nodosum* and *Ecklonia maxima*, are particularly effective as biostimulants. These products are applied as foliar sprays, seed treatments, or soil drenches and are widely used in agriculture to promote sustainable farming practices and reduce the need for chemical inputs.

Seaweed Characteristics: Seaweeds are multicellular marine algae that exhibit a wide range of morphological diversity. They vary from microscopic filamentous forms to large, complex structures like kelp forests. They lack true roots, stems, and leaves but have analogous structures such as holdfasts, stipes, and blades. Seaweeds absorb nutrients directly from the surrounding water and are classified into red (Rhodophyta), green (Chlorophyta), and brown (Phaeophyceae) groups based on their pigmentation and biochemical composition.

Seaweed Classification: Seaweed classification organizes these marine algae into hierarchical categories based on their evolutionary relationships. This includes divisions (Chlorophyta, Phaeophyceae, Rhodophyta), classes, orders, families, genera, and species. Classification helps in understanding the functional roles of different seaweeds in ecosystems and their potential uses in industries like biotechnology and aquaculture.

Seaweed Conservation: Seaweed conservation involves efforts to protect natural seaweed populations and their habitats from overexploitation, pollution, and climate change. Conservation strategies may include establishing marine protected areas, regulating seaweed harvesting, and promoting sustainable farming practices that reduce the pressure on wild seaweed stocks. Additionally, conservation efforts may focus on preserving genetic diversity among seaweed species, which is essential for the resilience of both natural and farmed populations in the face of environmental changes.

Seaweed Drying: Seaweed drying is a crucial post-harvest step to reduce the moisture content of seaweed, preventing spoilage and extending its shelf life. Traditionally, seaweed is sun-dried by spreading it on mats or racks, but controlled drying methods, such as hot air drying, are also used to ensure consistent quality. Drying should be done as quickly as possible to avoid degradation of valuable compounds, like carrageenan or agar. Optimal drying typically reduces the moisture content to below 15%, making the seaweed suitable for long-term storage or processing.

Seaweed Ecology: Seaweed ecology studies the interactions between seaweeds and their surrounding environment, including factors such as light availability, nutrient levels, water movement, and interactions with other marine organisms. Seaweeds play a vital role in marine ecosystems as primary producers, providing food and habitat for various marine species. Understanding seaweed ecology is crucial for the development of sustainable farming practices, as it helps to minimize the ecological impact of cultivation and ensures that seaweed farming can coexist with natural marine ecosystems.

Seaweed Ecosystems: Seaweed ecosystems, such as kelp forests and seagrass beds, are vital components of coastal marine environments. These ecosystems provide habitat, food, and shelter for numerous marine species, contribute to carbon sequestration, and protect shorelines from erosion. Seaweed ecosystems are dynamic, with their structure and function influenced by factors such as light availability, water quality, and species interactions.

Seaweed Extraction: Seaweed extraction involves the process of isolating valuable compounds, such as polysaccharides, proteins, and lipids, from seaweed biomass. Techniques for seaweed extraction vary depending on the target compound and may include mechanical methods (e.g., pressing or centrifugation), solvent extraction, or enzymatic hydrolysis. Extraction processes must be optimized to maximize yield while preserving the functional properties of the extracted compounds. The extracted materials are used in various industries, including food, cosmetics, pharmaceuticals, and biofuels.

Seaweed Farming Economics: The economics of seaweed farming involves the assessment of production costs, market prices, and profitability. Factors influencing profitability include species selection, farming techniques, labor, and post-harvest processing. The initial investment in equipment and farm setup can be significant, but operational costs are generally low compared to other forms of aquaculture. Seaweed farming is seen as a low-input, high-output industry with potential for providing income to coastal communities. However, market fluctuations and environmental challenges such as climate change can impact economic viability.

Seaweed Farming Equipment: Seaweed farming equipment includes essential tools for cultivation, such as longlines, nets, ropes, anchors, and buoys, which are used to create an underwater farm structure. Nets and ropes provide surfaces for seaweed attachment, while buoys maintain the structure's position in the water column. Other tools include harvesting knives, drying racks, and boats for transportation. The selection of equipment depends on farming techniques (e.g., rope vs. raft method) and the species being cultivated. The use of biodegradable materials in equipment is gaining importance to reduce marine pollution.

Seaweed Farming: Seaweed farming is the practice of cultivating seaweed species for commercial purposes, including food, pharmaceuticals, cosmetics, and biofuels. It involves selecting appropriate sites, species, and cultivation methods, such as raft, longline, or net culture. Seaweed farming is considered one of the most sustainable forms of aquaculture, as it requires no fertilizers or freshwater inputs and can help mitigate climate change by absorbing carbon dioxide. Key factors in successful seaweed farming include optimizing water quality, nutrient availability, and managing environmental risks.

Seaweed Harvesting Season: The seaweed harvesting season varies depending on the species and local environmental conditions. Harvesting typically occurs when the seaweed reaches its peak biomass and biochemical composition. For example, Kappaphycus and Eucheuma species are often harvested every 45-60 days in tropical regions, while temperate species like kelp may have a more defined seasonal cycle, with harvesting occurring in the spring and summer months. Timing the harvest correctly is critical to maximize yield and ensure the quality of the seaweed, particularly for commercial products like carrageenan and alginate.

Seaweed Industry: The seaweed industry encompasses the cultivation, harvesting, processing, and commercialization of various seaweed species for use in food, pharmaceuticals, cosmetics, and agriculture. Major players include countries like China, Indonesia, and South Korea, contributing to over 80% of global seaweed production. The industry has seen rapid growth due to the demand for seaweed-derived products like carrageenan, agar, and alginates. Sustainability is a key concern, as overharvesting and environmental degradation pose risks to long-term industry viability.

Seaweed Metabolites: Seaweed metabolites include primary metabolites, such as carbohydrates, proteins, and lipids, which are essential for growth and reproduction, and secondary metabolites, such as phenolic compounds, terpenoids, and alkaloids, which have ecological functions like defense against herbivores and pathogens. These secondary metabolites are also of interest for their antioxidant, anti-inflammatory, and antimicrobial properties, making them valuable for pharmaceutical and cosmetic applications. Seaweed metabolites vary by species and environmental conditions, with brown seaweeds typically producing high levels of phenolics and red seaweeds being rich in sulfated polysaccharides.

Seaweed Powder: Seaweed powder is produced by drying and grinding seaweed biomass into a fine powder. It is used in food products, cosmetics, and as a dietary supplement due to its high concentration of nutrients and bioactive compounds. Seaweed powder can be derived from various species, each offering different benefits; for example, kelp powder is high in iodine, while spirulina powder is rich in protein and antioxidants. The drying process is crucial to preserve the nutritional value of the seaweed, with low-temperature drying being preferred to prevent degradation of sensitive compounds.

Seaweed Preservation: Seaweed preservation involves methods to extend the shelf life of harvested seaweed and maintain its quality for processing or consumption. Common preservation techniques include drying, freezing, and chemical preservation (e.g., using salt or acids). The choice of preservation method depends on the intended use of the seaweed; for instance, dried seaweed is often used in food products, while frozen seaweed may be processed further for pharmaceutical or cosmetic applications. Effective preservation prevents spoilage, microbial contamination, and loss of valuable compounds.

Seaweed Processing: Seaweed processing transforms raw seaweed into various products, such as food, hydrocolloids (agar, carrageenan), bioactive compounds, and biofuels. Processing techniques depend on the end product and include mechanical extraction, chemical treatments, drying, and purification processes. For example, agar is extracted from Gracilaria through a series of steps, including boiling, filtration, and gel formation. Seaweed processing requires careful control of parameters, such as temperature, pH, and drying time, to maintain the functionality and quality of the final product.

Seaweed Production: Seaweed production involves the cultivation, harvesting, and processing of marine macroalgae. Production systems vary from small-scale artisanal farms to large commercial operations. Optimum conditions for production include high nutrient concentrations, moderate water currents, and temperatures ranging between 15°-20°C for temperate species. Global seaweed production reached approximately 35 million metric tons in recent years, driven by increasing demand in food and non-food sectors.

Seaweed Taxonomy: Seaweed taxonomy classifies marine macroalgae into three primary groups: Chlorophyta (green algae), Phaeophyceae (brown algae), and Rhodophyta (red algae). This classification is based on their pigment composition, cell wall structure, and reproductive modes. Taxonomic studies of seaweeds are essential for species identification, ecological research, and understanding evolutionary relationships among algae.

Seaweed-Derived Products: Seaweed-derived products include a wide range of goods such as food products (nori, kelp), hydrocolloids (agar, carrageenan, alginates), fertilizers, animal feed, biofuels, and cosmetics. These products leverage the unique properties of seaweeds, such as their high polysaccharide content, bioactive compounds, and nutrient density. For example, alginates from brown seaweed are used as thickening agents in the food industry, while carrageenan from red seaweed is used in dairy products and pharmaceuticals.

Seed Stock: Seed stock in seaweed aquaculture refers to the juvenile or propagative material used to initiate the cultivation process. These may include spores, seedlings, or vegetative fragments that are cultivated in nurseries before being transferred to grow-out sites. Seed stock quality is critical for ensuring successful crop establishment, and proper selection is based on traits such as growth rate, resilience to environmental conditions, and disease resistance. Maintaining healthy seed stock is fundamental to achieving high yields in seaweed farming.

Sexual Reproduction: Sexual reproduction in seaweeds involves the production of gametes (sperm and eggs) that fuse to form a zygote, which develops into a new individual. This process may involve alternation of generations, where a diploid sporophyte alternates with a haploid gametophyte. Fertilization can occur in the water column or within the seaweed thallus, depending on the species.

Single Rope Raft: The single rope raft method is a traditional technique for seaweed farming, where a single rope is used to suspend seaweed in the water column. The rope is typically anchored at both ends and can be raised or lowered to optimize light and nutrient exposure. This method is commonly used for red and green seaweeds in regions with calm waters.

Site Selection: Site selection for seaweed farming involves evaluating environmental factors such as water quality, temperature, salinity, and nutrient availability, along with logistical considerations like access to markets and protection from storms. Ideal sites typically have steady water flow, clean water with minimal pollutants, and suitable depths (usually between 2 to 10 meters depending on the species). Salinity levels of 30-35 ppt, temperatures ranging from 20°-30°C, and high nutrient availability are often critical for maximizing growth in seaweed culture. Proper site selection is essential to ensure sustainable yields and minimal environmental impact.

Soil Conditioner: Seaweed-based soil conditioners are products that improve soil structure, water retention, and nutrient availability, enhancing plant growth and productivity. These conditioners are often made from brown seaweeds, which contain polysaccharides like alginate, that improve soil aggregation and moisture retention. They also introduce beneficial microbes and micronutrients into the soil, promoting healthy plant root development. Seaweed soil conditioners are especially effective in sandy or degraded soils, helping to restore fertility and support sustainable agricultural practices.

Species Identification: Species identification in seaweeds involves the use of morphological characteristics, such as color, shape, and size, along with molecular tools like DNA barcoding. Accurate species identification is crucial for biodiversity assessments, conservation efforts, and optimizing aquaculture practices. Traditional methods rely on visual inspection, while modern techniques use genetic markers to differentiate closely related species.

Sustainable Aquaculture: Sustainable aquaculture refers to farming practices that ensure environmental, social, and economic sustainability, minimizing the ecological footprint while producing healthy food. In the context of seaweed farming, sustainability involves practices such as avoiding overharvesting, reducing pollution, and maintaining biodiversity. Seaweed farming is often seen as a highly sustainable form of aquaculture due to its ability to sequester carbon, reduce nutrient pollution in coastal waters, and provide habitat for marine organisms.

Sustainable Farming: Sustainable seaweed farming involves practices that ensure long-term viability and minimal environmental impact. This includes selecting appropriate sites, minimizing the use of chemicals, and integrating seaweed farming with other forms of aquaculture (e.g., multi-trophic aquaculture) to create balanced ecosystems. Sustainable farming practices also focus on protecting local biodiversity, avoiding overexploitation of natural seaweed beds, and contributing to climate change mitigation by sequestering carbon. Adopting sustainability standards ensures the resilience of seaweed farming operations and the protection of marine ecosystems.

Temperature Range: The temperature range is one of the most important factors affecting seaweed growth, reproduction, and metabolism. Different species of seaweed have specific temperature requirements; for instance, *Kappaphycus alvarezii* thrives in tropical waters with temperatures ranging from 25°-30°C. Temperatures outside this range can lead to reduced growth rates, reproductive failures, or even die-offs. Regular monitoring and selection of sites with consistent temperature profiles are essential for optimizing seaweed production.

Water Quality: Water quality is a fundamental parameter in seaweed aquaculture, as it directly influences growth, health, and productivity. Key factors include clarity, temperature, salinity, nutrient levels (such as nitrogen and phosphorus), and dissolved oxygen. Optimal water quality for seaweed farming typically involves clear waters with minimal sedimentation, salinity ranging from 30-35 ppt, temperatures between 20°-30°C, and a good supply of macronutrients like nitrates and phosphates. Continuous monitoring is essential to avoid issues like nutrient depletion, contamination, or harmful algal blooms.

CHAPTER 16

ADVANCES IN AQUACULTURAL PRACTICES

Advanced Aquacultural Practices: Advanced aquacultural practices involve modern, innovative techniques aimed at increasing productivity, sustainability, and resource efficiency in aquaculture. These practices include biofloc technology, recirculatory aquaculture systems (RAS), integrated multi-trophic aquaculture (IMTA), and aquaponics. Emphasis is placed on reducing water use, enhancing biosecurity, and optimizing feed conversion ratios (FCR) to improve growth rates and reduce environmental impacts. Advanced methods typically involve precise control of water quality, automated feeding, and the integration of renewable energy sources.

Advanced Aquaculture Equipment: Advanced aquaculture equipment includes cutting-edge technologies designed to optimize production efficiency, water quality, and fish health in modern aquaculture systems. This includes automated feeders, high-efficiency aerators, sensors for real-time water quality monitoring, advanced filtration systems, and robotic systems for fish sorting and grading. Innovations such as intelligent feeding systems, which adjust feeding rates based on fish behavior and biomass, contribute to improved feed conversion ratios (FCR) and reduced environmental impact. The integration of IoT (Internet of Things) and AI technologies into aquaculture equipment further enhances operational precision and resource management.

Aeration Systems: Aeration systems in aquaculture provide the necessary oxygenation of water, ensuring that cultured species have adequate dissolved oxygen levels to support metabolic processes. Aeration methods include mechanical surface aerators, air diffusers, and paddle wheels. The optimal dissolved oxygen level for most aquaculture systems is 5-7 mg/L, with aeration capacity adjusted based on stocking density, biomass, and system type to prevent hypoxia and promote healthy growth.

Alkalinity: Alkalinity in aquaculture measures the water's capacity to neutralize acids, mainly influenced by carbonate and bicarbonate ions. It is critical for buffering pH fluctuations and providing a stable environment for both cultured species and microbial communities. Optimal alkalinity levels for most freshwater systems range from 50 to 200 mg/L as $CaCO_3$. Regular monitoring and supplementation with alkaline compounds, like sodium bicarbonate, ensure stable alkalinity levels, supporting healthy aquatic life.

Ammonia Control: Ammonia control is critical in aquaculture as ammonia, produced from the decomposition of organic matter and excretions, is toxic to aquatic species. Maintaining low ammonia concentrations is achieved through biological filtration, particularly nitrification, where ammonia is converted to nitrite and then nitrate by beneficial bacteria. Optimal ammonia levels should be kept below 0.05 mg/L in aquaculture systems to avoid toxicity, with water temperature, pH, and aeration playing crucial roles in ammonia dynamics.

Aquaculture Effluents: Aquaculture effluents are wastewater discharged from aquaculture systems containing organic matter, nutrients, and chemicals such as nitrogen, phosphorus, and uneaten feed. Proper effluent management is crucial to minimize environmental impacts, including eutrophication and water pollution. Effluent treatment methods include sedimentation, filtration, and nutrient recovery processes. Regulatory guidelines often set limits on nutrient discharge, with nitrogen concentrations typically kept below 10 mg/L and phosphorus below 1 mg/L in treated effluents.

Aquaculture Equipment: Aquaculture equipment encompasses all the tools, machinery, and technology used in the production and management of aquatic organisms. This includes aerators, feeders, pumps, filtration systems, water quality monitoring devices, and harvesting tools. Advanced equipment plays a critical role in optimizing production efficiency, maintaining water quality, and ensuring biosecurity. Proper equipment selection and maintenance are crucial for sustainability, energy efficiency, and cost-effectiveness in both extensive and intensive systems.

Aquaculture Monitoring Systems: Aquaculture monitoring systems use sensors and software to continuously track water quality parameters such as dissolved oxygen, pH, temperature, ammonia, and nitrate levels. These systems provide real-time data and alerts, enabling farmers to make informed decisions and promptly address issues that could affect the health and growth of cultured species. Advanced monitoring systems integrate with automated feeding and aeration systems, improving overall farm efficiency and reducing labor costs.

Fish Growth Monitoring: Fish growth monitoring involves the regular measurement of fish size and weight to assess growth rates, feed efficiency, and overall health. Growth is influenced by factors such as water quality, temperature, stocking density, and feed quality. Modern systems use automated imaging and sensor technologies to monitor growth in real-time, reducing the need for manual sampling. Optimal growth rates are species-specific, with regular monitoring ensuring that feeding and environmental conditions are adjusted accordingly.

Aquaculture Nutrition: Aquaculture nutrition focuses on the dietary requirements of aquatic species to ensure optimal growth, health, and production. Nutritional formulations are designed to meet specific species' needs, including protein, lipids, carbohydrates, vitamins, and minerals. Protein content, which is crucial for growth, typically ranges from 30% to 50% depending on the species (e.g., 35-40% for tilapia and 40-45% for shrimp). Proper nutrition management enhances feed efficiency, reduces waste, and improves overall system sustainability.

Aquaculture Ponds: Aquaculture ponds are controlled water bodies designed for the cultivation of fish, crustaceans, and other aquatic organisms. Ponds are managed for optimal environmental conditions, such as temperature, dissolved oxygen, and pH, to maximize growth and productivity. Proper pond management involves regular monitoring of water quality parameters, with temperature typically kept between 25°-30°C, DO above 5 mg/L, and pH between 6.5 and 9, depending on the species cultured.

Aquaculture System Modeling: Aquaculture system modeling involves using computational tools and simulations to predict the performance of aquaculture systems under different conditions. Models can simulate various aspects of aquaculture, such as water quality dynamics, growth rates, and nutrient cycling, helping optimize system design and management. For example, RAS models can predict how different filtration setups or stocking densities affect ammonia levels and fish growth. Modeling is an essential tool for decision-making in advanced aquaculture practices, as it helps identify potential issues, reduce costs, and improve system efficiency before implementation.

Aquaculture System Resilience: Aquaculture system resilience refers to the ability of an aquaculture operation to withstand and recover from environmental, economic, and biological stresses, such as disease outbreaks, climate change, and market fluctuations. Resilient systems are designed with adaptability in mind, incorporating elements like diversified species, biosecurity measures, and flexible water management strategies. In recirculating systems, resilience may involve redundancy in filtration systems and backup power supplies to mitigate risks from mechanical failures. Enhancing resilience is crucial for ensuring the long-term sustainability and profitability of aquaculture operations in an increasingly unpredictable global environment.

Aquaculture Tanks: Aquaculture tanks are enclosed systems used to culture aquatic species in a controlled environment, allowing precise management of water quality, feeding, and disease control. Tanks vary in size and design, ranging from small-scale hatcheries to large-scale commercial systems. Key parameters include maintaining water temperature within species-specific ranges (e.g., 20°-28°C for tilapia), DO levels above 4 mg/L, and continuous filtration to remove waste products.

Aquaculture Waste Reuse: Aquaculture waste reuse involves repurposing organic waste products such as uneaten feed, feces, and nutrient-rich water. This waste can be used in integrated aquaculture systems, such as aquaponics, or as fertilizers for agricultural crops. The reuse of aquaculture waste not only reduces environmental impact but also creates additional revenue streams. For instance, nutrient-rich effluents can be applied to crops at rates of 50-100 m /ha to enhance soil fertility and reduce reliance on chemical fertilizers.

Aquaculture: Aquaculture, or aquatic farming, refers to the cultivation of aquatic organisms such as fish, crustaceans, mollusks, and plants in controlled environments. It includes both freshwater and marine systems, with various methods like pond culture, cage culture, and recirculatory systems. Optimal water quality parameters are critical, including dissolved oxygen levels above 5 mg/L, temperature control depending on species (e.g., 20°-28°C for tilapia), and pH levels ranging from 6.5 to 8.5. Nutrient management, disease control, and sustainable practices are essential for minimizing environmental impacts.

Aquaponics: Aquaponics is an integrated farming system that combines aquaculture with hydroponics, where fish waste provides nutrients for plant growth, and plants help filter and purify the water for the fish. The system relies on a balance between fish stocking density and plant production, with nutrient levels carefully managed to ensure both fish and plants thrive. Water pH is generally maintained between 6.8 and 7.2, while nitrate levels are kept below 100 mg/L to prevent toxicity. Aquaponics is considered a sustainable farming method, as it recycles water and nutrients, reducing the need for external inputs.

Aquatic Ecosystem Health: Aquatic ecosystem health in aquaculture refers to the overall condition and balance of the water body, including its biological, chemical, and physical components. A healthy ecosystem supports sustainable production by maintaining high water quality, biodiversity, and resilience against diseases and environmental stressors. Regular monitoring of water parameters, species diversity, and sediment quality is necessary to ensure ecosystem health. Preventative measures like controlled stocking densities and eco-friendly practices contribute to long-term sustainability.

Aquatic Species Health: Aquatic species health in aquaculture is a critical factor that influences growth, survival, and overall productivity. Health management includes disease prevention through biosecurity measures, vaccination, proper nutrition, and maintaining optimal water quality. Key health indicators include the behavior, growth rate, and physical condition of the cultured species. Regular health monitoring helps detect early signs of disease, such as changes in behavior or appearance, allowing for timely intervention. For example, maintaining dissolved oxygen levels above 5 mg/L is crucial for most fish species to reduce stress and support a healthy immune system.

Bioaugmentation: Bioaugmentation is the introduction of beneficial microbial strains into aquaculture systems to enhance biological processes such as nutrient cycling, organic matter degradation, and disease control. These microorganisms, including nitrifying bacteria and probiotics, improve water quality by accelerating the breakdown of waste products like ammonia and nitrite. Bioaugmentation is commonly used in biofloc systems and RAS to maintain a stable microbial community, promoting system health and resilience. Successful bioaugmentation requires regular monitoring and may involve dosing with specific microbial products tailored to the system's needs.

Biofilters: Biofilters are critical components of Recirculatory Aquaculture systems (RAS) and aquaponics that facilitate biological filtration by hosting nitrifying bacteria that convert toxic ammonia (NH_3) into less harmful nitrate (NO_{3-}). These filters work through the nitrification process, where ammonia is oxidized to nitrite and then to nitrate. Optimal biofilter performance requires a large surface area for bacterial colonization, with effective operation depending on maintaining water flow rates and oxygen levels that support bacterial growth. Biofilters help maintain water quality and reduce the need for frequent water exchanges.

Biofloc Harvesting: Biofloc harvesting involves the removal and collection of microbial aggregates (Biofloc) that form in aquaculture systems rich in organic matter and nutrients. These bioflocs contain a mixture of microorganisms, uneaten feed, and waste products that can be used as a protein-rich feed supplement. Harvesting biofloc requires balancing aeration, water flow, and flocculant addition to ensure efficient recovery while maintaining water quality. Optimal biofloc concentration is generally 10-30 mL/L in the system water.

Biofloc Technology (BFT): Biofloc technology (BFT) is an advanced aquaculture practice that promotes the growth of beneficial microbial communities (bioflocs) in the water, which enhance water quality by converting waste products like ammonia into microbial biomass. This biomass is then consumed by the cultured species as an additional protein source, reducing feed costs. BFT systems typically maintain a carbon-to-nitrogen (C

Biosecurity: Biosecurity in aquaculture involves implementing practices and protocols to prevent the introduction and spread of pathogens, diseases, and invasive species. This includes measures such as quarantine of new stock, disinfection of equipment, and strict hygiene protocols. Biosecurity is essential for protecting both farmed species and the surrounding environment. Key practices include maintaining high water quality, controlling access to facilities, and regular health monitoring to detect and mitigate potential outbreaks early.

Carbon Dioxide Management: Carbon dioxide (CO_2) management in aquaculture is vital for maintaining water quality and preventing CO_2 toxicity, which can impair respiration and cause stress in fish. In Recirculatory Aquaculture systems (RAS), CO_2 levels can build up due to fish respiration and bacterial activity. Efficient gas exchange systems, including aeration, degassing towers, or CO_2 strippers, are used to maintain safe CO_2 concentrations, generally below 20 mg/L for most species. Proper CO_2 management also helps maintain the pH balance in aquaculture systems, which is crucial for optimal fish health.

Carbon Sources: In aquaculture, carbon sources are materials added to systems like Biofloc Technology (BFT) to balance the carbon-to-nitrogen (C) ratio, promoting microbial growth and enhancing water quality. Common carbon sources include molasses, wheat bran, and rice bran. The ideal C ratio for biofloc systems is between 10-15:1, which helps in converting toxic ammonia into microbial biomass. Proper management of carbon sources helps reduce the need for water exchange and improves the overall sustainability of the aquaculture system.

Carrying Capacity: Carrying capacity in aquaculture is the maximum biomass of organisms that a system can support without compromising water quality, growth rates, and overall ecosystem health. It depends on factors such as water exchange rates, aeration, nutrient input, and waste removal efficiency. For example, in intensive aquaculture systems, carrying capacity might range from 20 to 50 kg/m , depending on species and system design. Exceeding carrying capacity can lead to stress, disease outbreaks, and environmental degradation.

Chemical Application: Chemical application in aquaculture refers to the use of chemicals, such as disinfectants, therapeutants, and water conditioners, to manage water quality, prevent disease, and enhance production. Common chemicals include chlorine, formalin, and potassium permanganate, which are used for disinfection and parasite control. Proper dosage and application methods are critical to avoid toxicity to cultured species and minimize environmental impact. For instance, chlorine should be used at concentrations below 10 mg/L to disinfect water without harming fish. Regulations and best practices guide the safe use of chemicals in aquaculture.

Climate Impact on Aquaculture: Climate impact on aquaculture includes the effects of changing weather patterns, ocean acidification, and temperature fluctuations on the productivity and sustainability of aquaculture operations. Warmer water temperatures can accelerate growth rates but may also increase disease outbreaks and stress in cultured species. Rising sea levels and extreme weather events pose additional risks to coastal aquaculture. Adaptive strategies, such as selective breeding for climate resilience and the development of climate-smart aquaculture practices, are becoming increasingly important.

Coupled Aquaponics: Coupled aquaponics systems integrate aquaculture and hydroponics in a single-loop system where water flows continuously between the fish tanks and plant beds. Nutrient-rich water from the aquaculture component provides essential nutrients to the plants, which, in turn, help filter and clean the water for reuse in the fish tanks. Optimal system performance depends on balancing fish stocking densities, plant biomass, and water quality parameters, with ammonia levels ideally kept below 0.5 mg/L and pH between 6.5 and 7.0 for most plants.

Decoupled Aquaponics: Decoupled aquaponics is an advanced system where the aquaculture and hydroponics components operate independently, allowing for better control over water quality and nutrient levels in each section. This separation enables optimized conditions for both fish and plants, improving system resilience and reducing risks associated with waterborne diseases. Decoupled systems allow for higher water recirculation and reduced waste discharge, enhancing the sustainability of integrated aquaculture systems.

Deep Water Culture (DWC): Deep Water Culture (DWC) is a hydroponic technique used in aquaponics where plants are suspended with their roots submerged in nutrient-rich water. This system is suitable for growing leafy greens and other fast-growing plants. In DWC systems, aeration is critical to provide dissolved oxygen to the roots, with DO levels typically maintained above 5 mg/L. DWC systems are simple and efficient, relying on the nutrient content from the aquaculture component to sustain plant growth.

Denitrification: Denitrification is the biological process by which nitrate (NO_3^-) is reduced to nitrogen gas (N_2) under anoxic conditions, effectively removing nitrogen from the aquaculture system. This process is carried out by denitrifying bacteria, which thrive in low-oxygen environments and use nitrate as an electron acceptor. Optimal conditions for denitrification include maintaining a dissolved oxygen level below 0.5 mg/L and providing a sufficient organic carbon source to fuel bacterial activity.

Disease Control: Disease control in aquaculture encompasses strategies to prevent, manage, and treat diseases in cultured species. This includes maintaining water quality, optimizing stocking densities, administering vaccines or medications, and using biological control agents like probiotics. Disease control is critical for reducing mortality and ensuring sustainable production, with an emphasis on early detection and minimizing stress to prevent outbreaks. Common diseases such as viral, bacterial, and parasitic infections are addressed through integrated health management practices.

Disinfection: Disinfection in aquaculture refers to the application of chemical or physical agents to destroy pathogens in water, equipment, and facilities. Disinfection techniques include the use of chlorine, iodine, and hydrogen peroxide, as well as physical methods like heat treatment. For example, chlorine can be used at concentrations of 5-10 mg/L for effective disinfection, with careful dechlorination to protect aquatic life. Disinfection is essential for biosecurity, particularly in Recirculatory Aquaculture Systems (RAS) and hatcheries, where disease prevention is critical to maintaining healthy stock and preventing economic losses.

Dissolved Oxygen (DO): Dissolved oxygen (DO) is one of the most critical water quality parameters in aquaculture, as aquatic species rely on adequate oxygen levels for respiration. DO levels are influenced by factors such as temperature, stocking density, and aeration. Most fish species require DO levels above 5 mg/L for optimal growth and health, while lower levels can lead to stress, reduced growth rates, and mortality. Regular monitoring and aeration adjustments are necessary to maintain proper oxygenation.

Eco-friendly Aquaculture Practices: Eco-friendly aquaculture practices involve the adoption of sustainable farming methods that minimize environmental impact while promoting the health of both cultured species and surrounding ecosystems. This includes the use of organic feeds, reduced chemical inputs, water recycling, and integrated multi-trophic aquaculture (IMTA). By focusing on energy efficiency, waste reduction, and resource optimization, eco-friendly practices help ensure that aquaculture can meet the growing demand for seafood while preserving natural resources.

Economic Analysis: Economic analysis in aquaculture involves evaluating the financial feasibility and profitability of aquaculture operations. This includes calculating costs such as feed, energy, labor, and capital investments, as well as revenues from the sale of cultured species. Factors like Feed Conversion Ratio (FCR), stocking density, and system efficiency play key roles in determining profitability. Economic analysis also considers external factors such as market prices, regulatory compliance, and environmental sustainability. Conducting thorough economic analysis helps aquaculture businesses optimize operations and make informed decisions to improve financial outcomes.

Energy Efficiency in Aquaculture: Energy efficiency in aquaculture involves minimizing energy consumption while maintaining optimal production levels. This can be achieved through the use of energy-efficient pumps, aeration systems, and lighting, as well as implementing renewable energy sources like solar or wind power. Energy-efficient systems reduce operational costs and environmental impact. For example, Recirculating Aquaculture Systems (RAS) can consume as much as 50% less energy compared to traditional pond systems when optimized for energy use, through heat recovery and efficient water circulation.

Environmental Impact Assessment (EIA): Environmental Impact Assessment (EIA) in aquaculture is a systematic process used to evaluate the potential environmental effects of proposed aquaculture projects before they are implemented. EIA focuses on identifying, predicting, and mitigating negative environmental impacts on water quality, biodiversity, local ecosystems, and surrounding communities. It involves analyzing parameters such as nutrient loading, chemical use, and waste discharge, often under regulatory frameworks. For sustainable aquaculture, EIA ensures that practices minimize ecological damage, align with conservation efforts, and comply with national and international environmental standards.

Environmental Sustainability: Environmental sustainability in aquaculture refers to practices that minimize the negative impact on natural ecosystems while ensuring long-term productivity. This involves reducing resource use (e.g., water, feed), controlling waste outputs, and adopting renewable energy sources. Advanced aquaculture practices like RAS, IMTA, and BFT are designed to enhance sustainability by improving resource efficiency, reducing pollution, and promoting biodiversity. Environmental sustainability is also supported by regulations, certifications, and best practices that encourage responsible production and conservation of aquatic ecosystems.

Feed Conversion Ratio (FCR): The Feed Conversion Ratio (FCR) is a measure of the efficiency of feed utilization in aquaculture, calculated as the amount of feed required to produce a unit of body weight gain in cultured species. An optimal FCR varies by species and feed quality but typically ranges from 1.0 to 1.8 for efficient species like tilapia and shrimp. Lower FCR values indicate better feed efficiency, reducing costs and minimizing waste accumulation in the system.

Filter Types: Filter types in aquaculture systems include mechanical filters, biological filters, and chemical filters, each serving a distinct function in maintaining water quality. Mechanical filters remove particulate matter and uneaten feed through screening or sedimentation, while biological filters host nitrifying bacteria that convert toxic ammonia into less harmful nitrates through the nitrification process. Chemical filters, such as activated carbon, remove dissolved organic compounds and other pollutants. The choice of filter type depends on the system (e.g., RAS, BFT), species being cultured, and water quality requirements. Proper filter selection and maintenance are crucial for sustaining healthy aquatic environments.

Fish Feed Formulation: Fish feed formulation is the process of developing balanced diets for aquaculture species, tailored to their specific nutritional needs. It involves selecting and blending ingredients such as fishmeal, plant proteins, oils, vitamins, and minerals. Optimal protein content for carnivorous fish like salmon ranges from 35-45%, while omnivorous species like tilapia require 25-35%. The goal is to achieve a high Feed Conversion Ratio (FCR), typically between 1.0 and 1.5, while promoting growth, health, and water quality by minimizing nutrient discharge.

Fish Species Compatibility: Fish species compatibility in aquaculture refers to the selection of species that can co-exist in the same system without causing undue stress, competition, or predation. Factors influencing compatibility include water quality preferences (e.g., temperature, salinity, pH), diet, behavior, and growth rates. For instance, tilapia and catfish are often considered compatible in polyculture systems due to their similar environmental requirements and differing feeding habits. Ensuring compatibility is crucial for maximizing growth and survival rates in multi-species systems like IMTA, polyculture, or aquaponics.

Hardness: Water hardness in aquaculture refers to the concentration of dissolved calcium and magnesium ions, which are essential for osmoregulation and overall fish health. Optimal hardness levels vary by species; freshwater species generally require hardness levels between 50 to 150 mg/L as $CaCO_3$, while marine species thrive in harder water. Hardness also influences the availability of trace elements and the efficiency of biofiltration, making it an important water quality parameter to monitor and adjust.

Harvesting Techniques: Harvesting techniques in aquaculture refer to the methods used to collect fish, shrimp, or other aquatic species from culture systems for processing and sale. Techniques vary depending on the species and system, including netting, draining ponds, or using pumps. In RAS, careful harvesting methods are used to minimize stress and injury to the fish. Harvest timing is critical and is often based on market size, species growth rates, and seasonal demand. Post-harvest handling, including chilling and transportation, also plays a significant role in maintaining product quality and ensuring profitability.

Health Management: Health management in aquaculture focuses on maintaining the well-being of cultured species through disease prevention, early diagnosis, and treatment. Key components include water quality monitoring, proper nutrition, vaccination, and the use of prophylactics such as probiotics or immunostimulants. Effective health management requires regular surveillance, biosecurity measures, and timely intervention to prevent the spread of diseases, ensuring high survival rates and optimal production.

Heavy Metal Removal: Heavy metal removal in aquaculture systems is necessary to prevent the accumulation of toxic metals like lead, cadmium, mercury, and arsenic, which can be harmful to both cultured species and human consumers. Advanced techniques for heavy metal removal include phytoremediation using plants, biofilters with specialized bacteria, and chemical precipitation. Activated carbon and ion-exchange resins are also employed in RAS and other systems to reduce heavy metal concentrations. Effective removal processes aim to keep heavy metal levels below safe thresholds, such as lead concentrations below 0.005 mg/L, in accordance with water quality guidelines.

Heterotrophic Bacteria: Heterotrophic bacteria in aquaculture play a crucial role in breaking down organic matter, including uneaten feed, fish excreta, and other detritus. In biofloc systems, these bacteria form bioflocs that help recycle waste into biomass that can be consumed by fish and shrimp, contributing to improved feed conversion ratios. Heterotrophic bacteria thrive in environments with high organic loads and require a balanced carbon-to-nitrogen ratio, typically maintained at around 15:1 to 20:1, to function effectively. Proper management of heterotrophic bacteria is essential for maintaining water quality and preventing the accumulation of toxic byproducts.

Indoor BFT: Indoor Biofloc Technology (BFT) systems are controlled aquaculture environments where microbial flocs develop and convert waste products into microbial biomass, which can be consumed by cultured species. These systems are typically implemented in closed environments to reduce the impact of external factors like temperature fluctuations and contaminants. Indoor BFT systems require careful management of carbon-to-nitrogen ratios (10-15:1), dissolved oxygen (above 4 mg/L), and light exposure to optimize microbial growth and water quality, leading to enhanced production efficiency and reduced water exchange.

Integrated Multi-Trophic Aquaculture (IMTA): IMTA is a sustainable aquaculture practice that involves the cultivation of multiple trophic levels of organisms, such as fish, shellfish, and seaweeds, in the same system. The waste products of one species serve as nutrients for others, creating a balanced ecosystem. For example, finfish excreta provides nutrients for seaweeds, while shellfish help filter particulates from the water. IMTA systems require careful management of water flow and nutrient cycling to optimize the growth of all species. This approach enhances environmental sustainability by reducing nutrient pollution and increasing overall productivity.

Integrated Pest Management (IPM): Integrated Pest Management (IPM) in aquaculture is an environmentally responsible approach to controlling pests, such as parasites, predatory species, or invasive organisms, that can threaten cultured species. IPM combines biological, mechanical, and chemical methods to minimize pest populations while reducing the use of harmful pesticides. For example, in shrimp farming, natural predators like fish or crabs can be introduced to control pest populations, while regular monitoring and water management help prevent infestations. IPM promotes sustainable production by minimizing environmental impact and reducing the risk of pesticide resistance.

Life Cycle Assessment (LCA): Life Cycle Assessment (LCA) in aquaculture is a methodology used to evaluate the environmental impacts of aquaculture operations throughout their entire lifecycle, from feed production and water use to waste disposal and product processing. LCA helps identify areas where resource use can be minimized and sustainability improved. For example, LCA can reveal the carbon footprint of producing a kilogram of fish in a RAS compared to a traditional pond system, helping guide decisions on energy use, feed ingredients, and waste management. LCA is a key tool in promoting environmentally sustainable aquaculture practices

Microbial Communities: Microbial communities in aquaculture systems are composed of bacteria, fungi, protozoa, and other microorganisms that play a vital role in nutrient cycling, organic matter decomposition, and disease prevention.

In systems like biofloc and recirculating aquaculture systems (RAS), managing the diversity and balance of microbial populations is crucial for maintaining water quality and promoting the health of cultured species. Optimal conditions include maintaining stable temperature, pH, and DO levels to support beneficial microbial activity.

Nitrification: Nitrification is a two-step biological process in aquaculture systems where ammonia (NH_3) is converted to nitrite (NO_2^-) by ammonia-oxidizing bacteria (AOB), and then nitrite is converted to nitrate (NO_3^-) by nitrite-oxidizing bacteria (NOB). This process is essential for controlling ammonia toxicity. Optimal conditions for nitrification include a temperature range of 25°-30°C, a pH of 7.0-8.5, and dissolved oxygen levels above 4 mg/L to ensure efficient bacterial activity.

Nitrogen Cycle: The nitrogen cycle in aquaculture involves the conversion of toxic nitrogenous compounds, like ammonia and nitrite, into less harmful forms through biological processes. Ammonia is oxidized to nitrite by ammonia-oxidizing bacteria (AOB), and nitrite is further oxidized to nitrate by nitrite-oxidizing bacteria (NOB). Maintaining optimal conditions for nitrification, such as a temperature range of 25°-30°C, dissolved oxygen (DO) levels above 4 mg/L, and a pH of 7.5-8.5, is essential for efficient nitrogen removal.

Nutrient Composition: Nutrient composition in aquaculture refers to the balance of essential nutrients required for the growth and health of both cultured species and supporting organisms like plants in aquaponics. This includes macronutrients like nitrogen, phosphorus, and potassium, as well as micronutrients like iron, manganese, and zinc. Optimal nutrient levels depend on species; for example, nitrate levels should be kept below 100 mg/L in aquaponics systems, while phosphorus should range from 1-10 mg/L. Regular testing and supplementation ensure that all organisms receive adequate nutrition.

Nutrient Cycling: Nutrient cycling in aquaculture refers to the movement and transformation of nutrients, such as nitrogen and phosphorus, within a system. In integrated systems like aquaponics and IMTA, nutrient cycling plays a crucial role in supporting both plant and animal growth. For instance, fish excreta provides nutrients for plants, while plants help absorb excess nutrients, preventing buildup and water quality degradation. Efficient nutrient cycling minimizes waste, reduces the need for external inputs, and enhances system sustainability. Key processes involved include nitrification, mineralization, and nutrient uptake by plants or filter feeders.

Nutrient Dosing: Nutrient dosing in aquaculture refers to the controlled addition of nutrients, such as nitrogen, phosphorus, and trace elements, to support the growth of cultured species or plants in aquaponics systems. Proper dosing ensures that nutrients are available at optimum levels without causing harmful excesses. For instance, in aquaponics, maintaining a nitrate concentration of 5-30 mg/L and a phosphorus concentration of 0.5-3 mg/L is critical for plant health while preventing nutrient buildup that could harm aquatic organisms.

Nutrient Film Technique (NFT): The Nutrient Film Technique (NFT) is a hydroponic growing method often used in aquaponics, where a thin film of nutrient-rich water flows continuously over the roots of plants. This technique provides plants with water, nutrients, and oxygen simultaneously, making it highly efficient. In aquaponics, the water is enriched with nutrients from fish waste, and flow rates are typically maintained between 1 to 2 liters per minute to optimize nutrient uptake and prevent root rot.

Nutrient Flow: Nutrient flow in aquaculture systems refers to the movement and recycling of essential nutrients, such as nitrogen and phosphorus, throughout the system. In integrated systems like aquaponics and IMTA, nutrient flow is crucial for supporting both fish and plant growth. For example, fish excreta provides nitrates for plant growth in aquaponics, while seaweeds in IMTA absorb dissolved nutrients from fish waste. Efficient nutrient flow management ensures that nutrient levels are optimized for all organisms, preventing nutrient buildup and promoting system sustainability.

Nutrient Supplementation: Nutrient supplementation in aquaculture involves providing essential vitamins, minerals, and trace elements to ensure the optimal growth and health of cultured species. This can include adding specific nutrients to feeds, such as phosphorus, calcium, or vitamin C, which are necessary for bone development, immune function, and overall metabolism. Supplementation is especially important in intensive systems like RAS, where natural nutrient sources may be limited. Ensuring proper nutrient supplementation helps prevent deficiencies that could lead to poor growth, deformities, or disease.

Operational Measures: Operational measures in aquaculture refer to the day-to-day management practices that ensure the smooth functioning of aquaculture systems. These include feeding schedules, water quality monitoring, disease control, and system maintenance. Key operational measures vary depending on the system type (e.g., RAS, BFT, IMTA) but generally involve monitoring parameters like temperature, dissolved oxygen, and nutrient levels, as well as adjusting stocking densities and feeding rates to optimize growth and minimize waste. Efficient operational measures are essential for maximizing productivity, ensuring biosecurity, and maintaining environmental sustainability.

Organic Carbon Sources: In aquaculture, organic carbon sources, such as molasses, sugar, and starch, are used to stimulate microbial growth, which aids in the conversion of nitrogenous wastes like ammonia into microbial biomass. This is critical in biofloc systems, where maintaining a carbon-to-nitrogen (C) ratio between 10:1 and 20:1 promotes the proliferation of beneficial bacteria, enhancing water quality and providing an additional protein source for cultured species. Optimal carbon dosing is crucial for balancing water chemistry and preventing oxygen depletion.

Outdoor BFT: Outdoor biofloc technology (BFT) systems utilize natural sunlight and environmental conditions to support the growth of bioflocs in open or semi-enclosed environments, such as ponds or tanks. These systems are more susceptible to environmental fluctuations like temperature and rainfall, which require careful monitoring. Outdoor BFT systems can support various species, particularly in regions with warm climates, where temperature management is less of an issue. Similar to indoor systems, maintaining an optimal carbon-to-nitrogen ratio (10-15:1) and dissolved oxygen levels above 4 mg/L is crucial for system stability and fish health.

Ozone Treatment: Ozone treatment in aquaculture involves the use of ozone (O_3) as a powerful oxidizing agent to disinfect water and break down organic compounds. Ozone effectively eliminates pathogens, improves water clarity, and reduces Biochemical Oxygen Demand (BOD). However, its application requires careful control to avoid toxicity, with optimal residual ozone levels typically maintained below 0.1 mg/L to prevent harm to aquatic organisms.

Periphyton: Periphyton is a complex community of microorganisms, including algae, bacteria, and invertebrates, that grow on submerged surfaces in aquatic environments. In aquaculture, periphyton serves as a natural food source for fish and shrimp, enhancing growth and survival rates. Periphyton-based aquaculture systems, such as periphyton aquaculture, promote biodiversity and nutrient recycling, improving water quality and system sustainability. The growth of periphyton depends on light availability, water quality, and substrate type, and it can be encouraged by installing artificial substrates like bamboo or nets in ponds or tanks.

pH Stabilization: pH stabilization in aquaculture is crucial for maintaining a healthy environment for aquatic species and beneficial bacteria. The optimal pH range varies by species, but most freshwater species thrive in a pH range of 6.5 to 8.5. Sudden pH fluctuations can stress fish and inhibit nitrification. Buffering agents like calcium carbonate are commonly used to stabilize pH, especially in systems with soft water or high organic loads.

Phosphorus Cycling: Phosphorus cycling in aquaculture refers to the movement and transformation of phosphorus within an aquaculture system, from feed inputs to waste outputs. Phosphorus is a critical nutrient for fish and plant growth, but excess phosphorus can lead to eutrophication and water quality issues. In advanced systems like Recirculatory Aquaculture Systems (RAS) and integrated multi-trophic aquaculture (IMTA), phosphorus is carefully managed to optimize growth while minimizing environmental impact. Techniques such as biofiltration, phytoremediation, and careful feed formulation help control phosphorus levels, typically maintaining concentrations below 0.1 mg/L in aquaculture effluent.

Phototrophic Organisms: Phototrophic organisms, such as algae and cyanobacteria, utilize light energy to convert carbon dioxide into organic compounds through photosynthesis. In aquaculture systems, phototrophs contribute to nutrient cycling by absorbing excess nutrients like nitrogen and phosphorus, thereby improving water quality. In biofloc and pond systems, they can serve as a natural food source for fish and shrimp. The growth of phototrophic organisms depends on factors such as light intensity, nutrient availability, and water temperature, with optimal conditions for many algae species including light levels of 30-100 µmol photons/m^2/s and temperatures between 20°-30°C.

Phytoremediation: Phytoremediation is a biological method for improving water quality in aquaculture systems by using plants to absorb, filter, and remove contaminants such as heavy metals, excess nutrients, and organic pollutants from the water. Commonly used plants for phytoremediation include water hyacinth, duckweed, and various species of aquatic macrophytes. These plants can reduce nutrient levels, particularly nitrogen and phosphorus, thereby mitigating the risk of eutrophication. Phytoremediation is often incorporated into integrated aquaculture systems to enhance sustainability and reduce the need for chemical treatments.

Plankton: Plankton, including phytoplankton and zooplankton, are microscopic organisms that form the base of aquatic food chains. In aquaculture, plankton plays a crucial role as a natural food source for filter-feeding species like shellfish and larval fish. Phytoplankton contributes to oxygen production through photosynthesis, while zooplankton serves as prey for various cultured species. Plankton populations need to be monitored, as blooms can lead to oxygen depletion (hypoxia) or the release of harmful toxins, particularly in intensive aquaculture systems.

Plant Species in Aquaponics: Plant species selection in aquaponics is critical for maintaining a balanced system where fish waste can be efficiently used as a nutrient source for plant growth. Common plant species include leafy greens (e.g., lettuce, spinach), herbs (e.g., basil, mint), and fruiting plants (e.g., tomatoes, cucumbers). These plants thrive in nutrient-rich water with pH levels between 6.8 and 7.2 and nitrate concentrations below 100 mg/L. Proper plant selection ensures efficient nutrient uptake, helps maintain water quality, and enhances the productivity and sustainability of aquaponics systems.

Polyculture: Polyculture in aquaculture involves the cultivation of multiple species in the same water body or system, usually from different trophic levels, to maximize resource use and improve system productivity. For example, a typical polyculture system might include a mix of herbivorous, omnivorous, and carnivorous species, or species that occupy different niches, such as fish, shrimp, and mollusks. Polyculture systems can help improve feed conversion ratios, reduce waste, and provide environmental benefits by mimicking natural ecosystems. Effective polyculture requires careful species selection and management to avoid competition and ensure balanced growth.

Pond Lining: Pond lining involves the installation of impermeable materials, such as polyethylene or PVC liners, at the bottom of aquaculture ponds to prevent seepage, erosion, and contamination of groundwater. Liners help in maintaining water quality and ease of pond management, especially in areas with poor soil conditions. Pond liners also facilitate easier sludge removal and reduce water usage by minimizing seepage losses, making them essential for intensive aquaculture systems. Properly installed liners can last over 10 years, depending on material quality.

Probiotics in Aquaculture: Probiotics are beneficial microorganisms introduced into aquaculture systems to improve the health and growth of cultured species. They enhance water quality by outcompeting harmful pathogens, aiding digestion, and promoting nutrient absorption. Effective probiotics include species of *Bacillus, Lactobacillus,* and *Pseudomonas,* which are added to water or feed. Optimal probiotic use requires careful dosing and monitoring to ensure beneficial effects, typically at concentrations between 10^6 and 10^9 CFU/g of feed.

Protein Skimmers: Protein skimmers, also known as foam fractionators, are devices used in aquaculture, particularly in marine systems, to remove organic compounds from water by creating foam. These systems enhance water quality by eliminating proteins, lipids, and other dissolved organic matter before they decompose into harmful substances like ammonia. Protein skimmers are most effective at salinity levels above 30 ppt and require precise air-to-water ratio control to maximize foam production and efficiency.

Recirculatory Aquaculture Systems (RAS): Recirculatory Aquaculture Systems (RAS) are land-based, closed-loop systems where water is continuously filtered and reused, reducing water consumption and improving environmental sustainability. Key components include mechanical filters, biofilters, and disinfection units. RAS systems aim to maintain optimal water quality with ammonia levels below 0.05 mg/L, nitrite below 1 mg/L, and stable pH (6.8-7.5). Temperature control and dissolved oxygen above 6 mg/L are vital for species health. RAS allows for intensive production of species such as salmon and tilapia in a controlled environment, minimizing disease risks and environmental impact.

Salinity Control: Salinity control is vital in aquaculture, particularly for species that require specific salinity ranges for optimal growth and health. For instance, marine species such as shrimp and sea bass require salinities between 15-35 ppt, while brackish species like tilapia can tolerate a broader range of 0-25 ppt. Managing salinity involves regular monitoring and adjustments through water exchange or the addition of salts, ensuring that the salinity levels remain within species-specific tolerances.

Sediment Management: Sediment management in aquaculture involves the control and removal of organic and inorganic sediments that accumulate at the bottom of ponds, tanks, and raceways. Excessive sediment buildup can lead to poor water quality, oxygen depletion, and the release of toxic substances like hydrogen sulfide. Regular dredging, siphoning, and the use of sediment traps are common practices. Effective sediment management ensures that ponds remain productive, with optimal sediment depth typically kept below 10 cm to minimize negative impacts on water quality.

Sludge Management: Sludge management in aquaculture deals with the collection, treatment, and disposal of solid waste that accumulates in the form of organic sludge at the bottom of ponds or tanks. If not properly managed, sludge can deteriorate water quality by releasing ammonia, hydrogen sulfide, and other harmful compounds. Techniques like dewatering, composting, and anaerobic digestion are employed to treat and repurpose sludge. Regular removal and proper treatment ensure a healthy production environment and prevent system clogging.

Solids Removal: Solids removal in aquaculture involves the extraction of particulate matter, including uneaten feed and feces, to maintain water quality and prevent the buildup of harmful waste products. Techniques include sedimentation, screening, and filtration. Optimal removal of solids is essential for preventing oxygen depletion and nutrient overloads, with systems often designed to maintain total suspended solids (TSS) levels below 50 mg/L for recirculating systems and below 100 mg/L for biofloc systems.

Species Selection: Species selection in aquaculture involves choosing appropriate aquatic species for farming based on factors such as environmental conditions, market demand, and compatibility with farming systems like RAS, BFT, or IMTA. Important considerations include growth rates, feed conversion efficiency, disease resistance, and tolerance to water quality parameters like temperature and salinity. For example, tilapia is favored in tropical freshwater systems due to its hardiness and fast growth, while salmon is commonly farmed in colder marine environments. Proper species selection is essential for optimizing production and sustainability.

Suspended Solids: Suspended solids in aquaculture systems consist of organic and inorganic particles, including uneaten feed, fecal matter, and detritus. High concentrations can reduce water quality by increasing turbidity, depleting dissolved oxygen, and clogging filtration systems. The optimal level of suspended solids varies depending on the system but typically should be kept below 50 mg/L in recirculating systems to ensure proper water clarity and efficient filtration.

System Design: System design in aquaculture refers to the layout, engineering, and construction of aquaculture facilities, including tanks, ponds, filtration systems, and water circulation mechanisms. Proper system design is critical to ensuring efficient production, water quality management, and biosecurity. Design considerations include species-specific needs, such as tank depth and water flow rates, as well as system type (e.g., RAS, BFT, or IMTA). Effective system design optimizes resource use, minimizes environmental impact, and facilitates ease of operation and maintenance, contributing to sustainable aquaculture practices.

Tank Maintenance: Tank maintenance in aquaculture involves regular cleaning, monitoring, and upkeep of tanks to ensure optimal conditions for the cultured species. Key aspects of tank maintenance include removing uneaten feed and waste, checking water quality parameters (e.g., ammonia, nitrate, pH), and inspecting filtration and aeration systems. Regular maintenance helps prevent disease outbreaks, improves water quality, and supports healthy growth. For example, in recirculatory aquaculture systems (RAS), tanks are typically cleaned weekly, and water quality parameters are monitored daily to maintain a stable environment.

Temperature Regulation: Temperature regulation in aquaculture is essential to ensure the optimal growth and metabolism of cultured species. Different species have specific temperature ranges for optimal growth; for example, tilapia thrives in water temperatures between 25°-30°C, while shrimp prefer 26°-30°C. Temperature fluctuations can cause stress and impact immune response, so systems must be equipped with heating or cooling mechanisms to maintain stable temperatures within species-specific ranges.

Trace Elements in Aquaculture: Trace elements, or micronutrients, are essential minerals required in small amounts for the optimal growth, reproduction, and health of aquatic species. Key trace elements include zinc, copper, iron, manganese, and selenium, which play roles in enzymatic reactions, oxygen transport, and immune function. Deficiencies or excesses of these elements can lead to poor growth, deformities, or even mortality. For example, optimal copper concentrations in aquaculture systems typically range from 0.005 to 0.03 mg/L, depending on species, to support enzyme function and disease resistance without causing toxicity.

Turbidity: Turbidity in aquaculture refers to the cloudiness or haziness of water caused by suspended particles such as algae, sediment, and organic matter. High turbidity can reduce light penetration, inhibit photosynthesis in phytoplankton-based systems, and interfere with fish feeding behavior. Maintaining turbidity at optimal levels, typically below 30 NTU (Nephelometric Turbidity Units) in most systems, is crucial for ensuring water clarity, promoting healthy fish growth, and preventing excessive sedimentation.

Ultraviolet (UV) Sterilization: UV sterilization in aquaculture uses ultraviolet light to kill or inactivate harmful microorganisms such as bacteria, viruses, and algae in water. UV systems are often integrated into recirculating aquaculture systems (RAS) to control pathogen loads and improve water quality. The effectiveness of UV sterilization depends on factors such as water clarity, flow rate, and UV dosage, with a common dosage range being 30,000 to 45,000 $\mu Ws/cm^2$ for effective pathogen control without damaging beneficial bacteria.

Waste Management: Waste management in aquaculture is the process of handling and treating solid and liquid waste generated by cultured species, including uneaten feed, feces, and metabolic byproducts. Effective waste management practices include mechanical filtration, sedimentation, and biofiltration, as well as the use of waste recycling methods like nutrient recovery in IMTA systems or composting for agricultural use. Waste management is crucial for preventing water quality deterioration, minimizing environmental impact, and maintaining system sustainability. For example, in BFT systems, microbial communities help convert waste into usable biomass, reducing waste accumulation.

Water Chemistry: Water chemistry in aquaculture refers to the chemical characteristics of water, including parameters such as pH, alkalinity, hardness, dissolved oxygen, ammonia, nitrite, nitrate, and salinity. Maintaining optimal water chemistry is essential for the health and growth of cultured species, with acceptable ranges varying depending on species and system type. For example, a stable pH of 7.5-8.5, dissolved oxygen levels above 5 mg/L, and ammonia levels below 0.05 mg/L are crucial for most freshwater aquaculture systems.

Water Exchange Rates: Water exchange rates refer to the frequency and volume of water replaced in aquaculture systems to maintain water quality and remove excess nutrients and waste. The optimal exchange rate depends on system type, species, and stocking density. In flow-through systems, water exchange rates might range from 5-10% per day, while recirculating aquaculture systems (RAS) are designed to minimize water exchange to 1-5% per day, relying on filtration and treatment to maintain water quality.

Water Filtration Systems: Water filtration systems in aquaculture are critical for maintaining water quality by removing suspended solids, organic matter, and dissolved pollutants. Filtration methods include mechanical filters, biological filters, and chemical filters, each targeting specific contaminants. For example, sand filters effectively remove particles larger than 20 microns, while activated carbon filters absorb dissolved organic compounds. Proper filtration sizing and maintenance are essential to avoid clogging and ensure optimal water turnover rates, typically between 0.5 to 1 system volume exchange per hour.

Water Quality Management: Water quality management in aquaculture is the practice of maintaining optimal physical, chemical, and biological conditions for the health and growth of cultured species. Key parameters include temperature (e.g., 20°-30°C depending on species), dissolved oxygen (above 5 mg/L), pH (6.5-8.5), ammonia (NH_3) levels below 0.05 mg/L, and nitrite (NO_{2-}) levels below 1 mg/L. Effective management involves continuous monitoring and adjustments through aeration, filtration, and water exchanges to prevent stress and disease in the cultured organisms.

Water Recirculation: Water recirculation in aquaculture refers to the continuous reuse of water within a closed-loop system, typically in recirculatory aquaculture systems (RAS). The water is filtered and treated to remove waste, and then recirculated back into the system. Water recirculation reduces water consumption by up to 90% compared to traditional flow-through systems, making it an environmentally sustainable option. Key components include mechanical filtration, biofiltration, and water sterilization. Effective water recirculation requires maintaining optimal water quality parameters, such as low ammonia and nitrate levels, stable pH, and adequate dissolved oxygen.

Water Reuse Systems: Water reuse systems in aquaculture refer to systems designed to recycle and treat water for continuous use, reducing the need for fresh water and minimizing waste discharge. Recirculating aquaculture systems (RAS) are a common type of water reuse system, employing mechanical and biological filtration, as well as UV sterilization or ozone treatment, to maintain water quality. RAS systems typically require minimal water exchange, often as low as 1-5% per day, while keeping parameters such as ammonia below 0.5 mg/L and dissolved oxygen above 5 mg/L. These systems support sustainability and resource efficiency in intensive aquaculture operations.

Water Sterilization: Water sterilization in aquaculture involves removing or inactivating harmful pathogens, parasites, and microorganisms to ensure a disease-free environment for cultured species. Common methods include ultraviolet (UV) sterilization, ozonation, and chlorination. UV sterilization is often preferred due to its effectiveness at wavelengths around 254 nm, which can inactivate most pathogens without altering water chemistry. Ozone treatment is also effective but requires careful monitoring to avoid toxicity. Effective sterilization reduces the risk of disease outbreaks, contributing to higher survival rates and better overall production.

Yield Optimization: Yield optimization in aquaculture involves adjusting production practices to maximize the quantity and quality of harvested species. This can include improving feed efficiency, managing stocking densities, enhancing water quality, and reducing disease risks. For example, optimizing Feed Conversion Ratios (FCR) through balanced diets and careful feeding schedules can significantly improve yields. Advanced technologies, such as automated feeders and sensors for water quality monitoring, contribute to yield optimization by providing precise control over the aquaculture environment. Successful yield optimization balances economic returns with environmental sustainability.

CHAPTER 17

INTEGRATED AQUACULTURE

Agro-Ecosystem: An agro-ecosystem refers to a farm system where both agricultural crops and livestock are integrated with aquaculture to create a sustainable, self-sufficient environment. It focuses on the flow of energy and nutrients between plants, animals, and aquatic organisms, enhancing productivity while minimizing waste and environmental impacts.

Ammonification: Ammonification is the process by which organic nitrogen compounds in animal waste or dead organisms are converted into ammonia (NH_3) by bacteria. This process is critical in integrated aquaculture systems as ammonia serves as a key nutrient for phytoplankton growth. However, ammonia levels above 0.5 mg/L can become toxic to fish, requiring proper management of organic inputs.

Animal Manure: Animal manure, such as cow dung, poultry litter, or pig waste, is a valuable organic fertilizer in integrated aquaculture systems. It provides essential nutrients like nitrogen, phosphorus, and potassium, promoting primary productivity in ponds and reducing the need for artificial feeds. Application rates vary but typically range between 100-300 kg/ha depending on the manure type and pond size.

Animal Waste Recycling: In integrated aquaculture systems, animal waste recycling involves using livestock manure to fertilize ponds, promoting the growth of natural fish food like phytoplankton and zooplankton. This approach reduces feed costs and recycles nutrients within the system, supporting both fish and plant growth.

Aquaculture Economics: Aquaculture economics focuses on the financial viability, cost structure, and profitability of fish farming operations. Key factors include input costs (seed, feed, labor), capital investment, revenue generation, and market trends. A typical break-even point for small-scale aquaculture projects is within 1-2 years, depending on species and system intensity.

Aquaculture Inputs: Aquaculture inputs encompass all resources required for the farming process, including seed (fingerlings), feed, fertilizers, chemicals, and equipment for pond preparation, water quality management, and disease control. Proper management of these inputs ensures optimal fish growth, health, and production efficiency.

Aquaponics: Aquaponics is a closed-loop system that combines aquaculture (fish farming) with hydroponics (soilless plant cultivation). In this system, nutrient-rich fish waste is used to fertilize plants, while plants help filter and purify the water, which is recirculated back to the fish tanks. Optimal fish-to-plant ratios, water pH (6.8–7.2), and dissolved oxygen levels (>5 mg/L) are essential for balancing the needs of both organisms. Aquaponics systems are highly sustainable, requiring minimal external inputs like fertilizers and ensuring efficient water and nutrient use.

Aquatic Biofilters: Aquatic biofilters are systems that use natural biological processes to remove waste and toxins from water in aquaculture systems. These filters rely on beneficial bacteria to convert ammonia into less harmful nitrate (nitrification), which can then be absorbed by plants. Biofilters are essential for maintaining water quality, particularly in recirculating aquaculture systems (RAS).

Aquatic Plants: Aquatic plants play a crucial role in aquaculture by providing oxygen, shelter, and food for various fish species. In integrated systems, species like duckweed, Azolla, and water spinach are cultivated alongside fish to absorb nutrients from the water, improve water quality, and provide an additional food source for herbivorous fish.

Aquatic Weeds: Aquatic weeds are undesirable plant species that grow excessively in water bodies, often competing with fish for nutrients and oxygen. Common aquatic weeds include water hyacinth, hydrilla, and duckweed. If left uncontrolled, they can impede fish movement, reduce water quality, and cause oxygen depletion in ponds. Effective control measures, such as manual removal or biological control using herbivorous fish, help maintain balanced ecosystems.

Azolla: Azolla is a nitrogen-fixing aquatic fern that is often cultivated in integrated aquaculture systems due to its rapid growth and high protein content. It is used as a supplemental feed for fish and livestock, as well as a biofertilizer in rice-fish farming systems, where it enriches water with nutrients and improves productivity.

Bhasabandha: Bhasabandha is a traditional aquaculture system practiced in Odisha, India, where fish are cultured in natural depressions or dug-out ponds created in floodplains. These ponds receive water from nearby rivers or rainfall, and fish like rohu, catla, and mrigal are reared in these seasonal water bodies. Bhasabandha systems are low-input, relying primarily on natural productivity for fish feed.

Bheries: Bheries are traditional brackish water aquaculture systems found in West Bengal, India, where fish are cultured in large, shallow water bodies formed by bunding natural tidal creeks. These systems primarily raise shrimp and carp using natural tidal water exchange to regulate salinity and nutrient levels. Bheries have become a vital component of coastal aquaculture, with water salinity levels typically ranging between 5-15 ppt.

Biogas Production: Biogas production is a process where organic waste, such as animal manure or crop residues, is anaerobically digested by microorganisms to produce methane gas, which can be used as a renewable energy source. In integrated aquaculture systems, livestock waste can be processed in biogas digesters to generate energy while producing nutrient-rich slurry that can fertilize fish ponds or crops.

Biomass Production: Biomass production in aquaculture refers to the total weight of fish or aquatic organisms grown in a system over a specific period. It depends on stocking density, feed quality, water management, and species growth rates. In well-managed integrated systems, biomass production can reach 10-15 tons per hectare annually. Maximizing biomass requires efficient nutrient cycling, with feed conversion ratios (FCR) ideally kept at 1.2 to 1.5 for efficient growth.

Capital Investment: Capital investment in aquaculture includes the initial costs for land, pond construction, water supply systems, and equipment such as aerators, feeders, and monitoring systems. Typical capital investments for small-scale aquaculture projects range from 2 to 5 lakh per hectare, depending on the intensity and type of system used (e.g., extensive, semi-intensive, or intensive).

Carbon Cycle: The carbon cycle in integrated aquaculture involves the exchange of carbon between the atmosphere, water, plants, and animals. Carbon enters the system through respiration, photosynthesis, and decomposition of organic matter. The balance of dissolved carbon dioxide (CO_2) is critical for maintaining optimal pH levels (6.5-8.5) and supporting photosynthetic organisms like algae, which form the base of the aquatic food chain.

Carbon Dioxide: Carbon dioxide (CO_2) is a by-product of fish respiration and organic matter decomposition. In aquaculture systems, high CO_2 levels (>20 mg/L) can lower pH, leading to stress and reduced growth rates. Managing CO_2 levels involves aeration to increase oxygen diffusion and promote gas exchange, ensuring a healthy environment for fish and aquatic organisms.

Carrying Capacity: Carrying capacity refers to the maximum biomass or population that an ecosystem can support without degrading environmental quality. In aquaculture, this is determined by factors like water quality, available nutrients, and oxygen levels. For example, intensive systems typically maintain 50-100 kg of fish per cubic meter, while semi-intensive systems support 5-10 kg per cubic meter. Overloading the system beyond its carrying capacity leads to poor growth, increased disease risk, and water quality deterioration.

Catla: Catla (*Catla catla*) is a surface-feeding, fast-growing Indian major carp species popular in polyculture. It thrives in temperatures between 25°–32°C and feeds on zooplankton, making it an excellent candidate for integrated systems with species that feed at different trophic levels.

Cattle-Fish Integration: In cattle-fish integration, cattle are raised adjacent to fish ponds, and their manure is used to fertilize the pond, promoting the growth of phytoplankton and zooplankton, which serve as natural feed for the fish. This system reduces the cost of fish feed and improves nutrient cycling. Optimum stocking density for cattle is about 2-3 cows per hectare of fish pond, with manure application rates of around 100-150 kg/ha to maintain proper nutrient balance.

Chemical Composition of Animal Waste: Animal waste used in integrated aquaculture contains essential nutrients such as nitrogen (1-2%), phosphorus (0.2-0.5%), and potassium (0.5-1%), along with trace minerals. These nutrients enhance primary productivity in fish ponds by stimulating the growth of phytoplankton and zooplankton, which serve as natural food for fish. Manure quality varies depending on the type of livestock, diet, and handling methods.

Common Carp: Common carp (*Cyprinus carpio*) is a widely cultured fish due to its hardiness and adaptability to different environments. It prefers water temperatures between 25°–30°C and is typically omnivorous, consuming detritus, plankton, and supplemental feed in aquaculture systems. Common carp is often integrated into polyculture systems alongside species like rohu and catla.

Cost-Benefit Analysis: Cost-benefit analysis (CBA) in aquaculture evaluates the economic feasibility of a project by comparing the costs of inputs, operations, and maintenance with the expected financial returns. A positive cost-benefit ratio (greater than 1) indicates profitability, guiding decision-making on project investments, expansion, or diversification.

Crop Rotation: Crop rotation is the agricultural practice of growing different types of crops sequentially on the same land to enhance soil health, improve nutrient availability, and reduce pest and disease buildup. In integrated aquaculture systems, crops like rice, vegetables, or legumes are rotated with fish farming, where fish effluents enrich the soil with nutrients. This cyclical process optimizes land use and ensures sustainability by enhancing soil fertility and preventing nutrient depletion, with crop rotation cycles typically ranging from 3 to 5 years.

Dissolved Oxygen: Dissolved oxygen (DO) refers to the amount of oxygen present in water, essential for fish respiration and maintaining aerobic microbial processes. Optimal DO levels for most freshwater fish species range from 5-8 mg/L. Low oxygen levels (<3 mg/L) can lead to stress, reduced growth, and even mortality, requiring aeration or water exchange to maintain adequate oxygen levels.

Duck-Fish Integration: Duck-fish integration involves rearing ducks directly on fish ponds, where their droppings provide organic fertilizer to stimulate algal and plankton growth, which serve as natural fish food. Ducks also help control weeds and insects in ponds, reducing the need for chemical inputs. The recommended stocking density is around 200-300 ducks per hectare of pond, and they provide an effective nutrient input for fish, especially carp species.

Ecological Interactions: Ecological interactions in aquaculture systems refer to the relationships between organisms such as fish, plants, and microorganisms. In integrated systems, positive interactions like nutrient cycling between fish and aquatic plants, symbiotic relationships with beneficial bacteria, and natural predator-prey dynamics maintain ecosystem balance. Optimal ecological interactions lead to efficient nutrient utilization and high productivity, with water quality parameters like dissolved oxygen (>6 mg/L) and ammonia (<0.1 mg/L) being carefully managed.

Egg Identification: Egg identification in aquaculture involves the recognition of specific morphological features like size, shape, color, and presence of oil globules that differentiate the eggs of one species from another. Correct identification is crucial in hatcheries to prevent mismanagement of species, as environmental conditions such as temperature (28°–30°C for warm-water species) and salinity (for marine species, 28–35 ppt) must be optimized for successful hatching.

Eutrophication: Eutrophication is the excessive enrichment of water bodies with nutrients, particularly nitrogen and phosphorus, leading to the overgrowth of algae and other aquatic plants. In aquaculture, it can result from over-fertilization or improper waste management. Eutrophication can cause oxygen depletion (hypoxia), fish kills, and poor water quality. Monitoring nutrient levels and maintaining nitrogen below 3 mg/L and phosphorus below 0.1 mg/L is essential to prevent eutrophication.

Farm Effluents: Farm effluents are waste products, including water discharged from aquaculture ponds, which may contain nutrients, organic matter, and chemicals used in fish farming. Effective effluent management is critical to prevent water pollution and maintain the health of surrounding ecosystems. Effluent treatment systems often involve biofiltration, sedimentation, and nutrient recycling.

Farm Labor Management: Farm labor management in aquaculture involves organizing, training, and supervising workers to ensure efficient farm operations. Effective labor management includes assigning tasks like feeding, pond maintenance, and harvesting, as well as ensuring safety and productivity. Labor costs in small-scale aquaculture farms typically account for 15-25% of total operational costs.

Feed Conversion Ratio: Feed conversion ratio (FCR) is a measure of the efficiency with which fish convert feed into body mass. A lower FCR indicates better feed utilization. Optimal FCR values in aquaculture range from 1.2 to 1.5 for species like tilapia, indicating that 1.2-1.5 kg of feed is needed to produce 1 kg of fish. Efficient FCRs reduce input costs and environmental impacts.

Fish Farming: Fish farming, or aquaculture, is the controlled cultivation of aquatic organisms, primarily fish, in tanks, ponds, or enclosures. This method is designed to optimize conditions for the rapid growth of species like tilapia, carp, or catfish through regulated feeding, water quality management, and disease control. Water parameters like dissolved oxygen (4-8 mg/L), pH (6.5-9.0), and temperature (25°-30°C) are critical for ensuring fish health and productivity. Fish farming helps meet the growing demand for seafood and reduces pressure on wild fish stocks.

Fish Feed Optimization: Fish feed optimization involves balancing the nutritional needs of fish with efficient feed usage to promote growth while minimizing waste. Key nutrients include proteins, lipids, vitamins, and minerals, with Feed Conversion Ratios (FCR) ideally ranging from 1.0 to 1.5 in aquaculture. Overfeeding or poor-quality feed can lead to nutrient buildup, water quality deterioration, and higher production costs.

Fish Growth Rate: Fish growth rate is the rate at which a fish increases in size or weight over time, influenced by factors like species, feed quality, water temperature, and stocking density. Growth rates are measured using specific growth rate (SGR), calculated as the percentage increase in weight per day. In aquaculture, fast-growing species like tilapia achieve SGRs of 1.5-2.5% per day, while water temperatures of 25°–30°C typically optimize growth.

Fish Pond Fertilization: Fish pond fertilization involves the addition of organic or inorganic fertilizers to enhance the natural productivity of the pond ecosystem. Organic fertilizers like animal manure or compost stimulate the growth of algae and plankton, which serve as food for fish. Inorganic fertilizers provide essential nutrients like nitrogen and phosphorus, with application rates of 2-5 kg N/ha and 1-2 kg P/ha recommended for optimal pond productivity.

Fish Productivity: Fish productivity refers to the output of fish biomass per unit area or volume in aquaculture systems. It is influenced by factors like stocking density, feed quality, water management, and system design. In well-managed intensive systems, fish productivity can reach 10-15 tons per hectare per year, with feed conversion ratios (FCR) of 1.2–1.5.

Fish Species Selection: Fish species selection is the process of choosing appropriate species for aquaculture based on factors like growth rate, market demand, tolerance to local environmental conditions, and compatibility with other species in integrated systems. Common species for polyculture include carp species (e.g., rohu, catla), tilapia, and shrimp. The selection process ensures economic viability and environmental sustainability.

Food Chain: A food chain in aquaculture refers to the linear sequence of organisms through which energy and nutrients flow, beginning with primary producers (e.g., algae) and ending with apex predators (e.g., carnivorous fish). Integrated aquaculture systems optimize food chains by maintaining species that feed at different levels, enhancing nutrient recycling and productivity.

Goat-Fish Integration: Goat-fish integration combines goat farming with aquaculture, where goat manure is used to fertilize fish ponds. Goats are generally housed near or above ponds, allowing their droppings to naturally enrich pond nutrients. This system can improve fish production by enhancing plankton growth. A density of 10-15 goats per hectare of fish pond is recommended, with careful monitoring of manure application to avoid excessive nutrient loading.

Grass Carp: Grass carp (*Ctenopharyngodon idella*) is a herbivorous fish species used in aquaculture to control aquatic weeds. It grows best in temperatures between 20°–30°C and requires large amounts of aquatic plants, such as hydrilla and water hyacinth, as part of its diet, making it a valuable species for integrated aquaculture systems.

Harvesting Techniques: Harvesting techniques in aquaculture involve the methods used to collect fish from ponds, tanks, or cages. Techniques include seine netting, draining ponds, or mechanical harvesting. The timing and method of harvesting impact fish quality, with optimal harvesting occurring when fish reach marketable size (e.g., 500-800 g for tilapia), ensuring minimal stress and damage.

Input Costs: Input costs in aquaculture refer to the expenses incurred for feed, seed, labor, equipment, and water management. Feed costs alone often account for 50-70% of total operational costs in fish farming. Efficient resource management, like optimizing feed conversion ratios (FCR) and reducing water use, can help lower input costs and increase profitability.

Insect Control: Insect control in integrated aquaculture focuses on managing insect pests that may harm fish or crops in the system. Biological control methods, such as introducing predatory fish species or using beneficial insects, are preferred to reduce the reliance on chemical pesticides. Insects like mosquitoes, which can breed in stagnant water, can be managed through proper pond aeration and water flow regulation to prevent larvae development

Integrated Aquaculture: Integrated aquaculture refers to the practice of combining aquaculture with other agricultural or livestock systems to create a symbiotic environment where waste from one system serves as a resource for another. This approach maximizes resource efficiency by utilizing animal waste as fertilizer for ponds or integrating crop residues into fish farming systems. It reduces the need for external inputs, enhances nutrient recycling, and improves overall farm productivity. For instance, nitrogen and phosphorus from animal waste can enhance primary productivity in fish ponds, promoting fish growth.

Integrated Farm Models: Integrated farm models are holistic approaches that combine fish farming with agriculture, livestock rearing, and other production systems. These models aim to maximize resource efficiency by recycling nutrients and water between the various components of the farm. They help achieve environmental sustainability, reduce input costs, and improve farm productivity.

Integrated Farming Design: Integrated farming design refers to the development of aquaculture systems that combine fish farming with agriculture, livestock, or other farming practices to optimize resource use and enhance sustainability. These designs focus on nutrient recycling, where fish waste can be used as fertilizer for crops, and livestock waste can support aquatic plants or be used in fish feed.

Integrated Farming Systems: Integrated farming systems combine aquaculture with other agricultural or livestock activities to maximize resource use efficiency. These systems recycle nutrients between components (e.g., fish waste as fertilizer for crops), reduce input costs, and improve productivity. Typical models include rice-fish farming or fish-livestock integration, with productivity increasing by up to 30-50% compared to standalone farming.

Kharlands: Kharlands are low-lying coastal lands in Maharashtra, India, that are used for integrated fish and paddy farming. These lands are subject to tidal inundation and have been modified with bunds to control saline water intrusion. Kharlands are typically used for growing rice during the monsoon season and for shrimp farming during the dry season, with salinity levels ranging between 10-25 ppt.

Khazan Lands: Khazan lands refer to reclaimed coastal lands in Goa, India, used for integrated paddy-cum-fish farming. These lands are protected from tidal floods by embankments and sluice gates, which regulate the flow of brackish water for shrimp or fish farming during the off-season. Salinity levels in Khazan lands can vary from 5-30 ppt, depending on the season and tidal influence.

Larval Stages: Larval stages in fish refer to the developmental phases post-hatching, during which the fish undergoes critical growth and morphological changes. This includes the yolk-sac stage, pre-flexion, flexion, and post-flexion stages. Larval rearing requires optimal water quality, temperature (25°–30°C), and feeding regimes to ensure survival and growth. Larvae are often fed live feed like rotifers and Artemia during early stages.

Livestock Grazing: Livestock grazing in integrated farming systems refers to the management of animals like cattle, goats or poultry that graze on pastures or feed from by-products of aquaculture systems. In integrated aquaculture, livestock grazing complements fish farming by providing manure that enhances pond fertility. Controlled grazing prevents overgrazing, which could otherwise lead to soil degradation. Optimal livestock densities depend on the carrying capacity of the land but typically range from 1 to 3 animals per hectare in mixed farming systems.

Livestock Management: Livestock management in integrated systems refers to the practices involved in rearing animals such as poultry, cattle, or ducks in conjunction with fish farming. Livestock waste is often used as fertilizer for aquatic plants or ponds, enhancing the productivity of the entire system. Effective livestock management ensures the sustainability of both animal and fish production.

Livestock Production: Livestock production in integrated aquaculture involves raising animals such as poultry, cattle or goats alongside fish farming, where manure from livestock is used to fertilize ponds, enhancing primary productivity. Proper integration improves nutrient cycling and reduces waste, contributing to overall farm productivity. In these systems, livestock yields can increase by 10-20% when combined with fish farming.

Livestock Waste Management: Livestock waste management in integrated aquaculture systems involves the proper collection, storage, and utilization of animal waste to prevent nutrient overloads and water pollution. Techniques like composting, biogas production, or direct use as pond fertilizer can enhance nutrient recycling while minimizing environmental impacts. Effective management requires balancing waste inputs to maintain optimal water quality.

Market Trends: Market trends in aquaculture refer to the changing demands and pricing for fish and aquatic products. Factors like consumer preferences for sustainable seafood, supply chain disruptions or environmental conditions can influence market trends. Monitoring these trends helps farmers make informed decisions on species selection, harvesting timing, and marketing strategies.

Monoculture: Monoculture is the practice of cultivating a single species in an aquaculture system, which allows for optimized management of feeding, water quality, and disease control. While more straightforward, monoculture systems are prone to higher disease risks and resource inefficiencies, compared to polyculture.

Nitrification: Nitrification is the two-step biological process where ammonia is first oxidized to nitrite (NO_2^-) and then to nitrate (NO_3^-) by nitrifying bacteria such as Nitrosomonas and Nitrobacter. This process is essential in integrated aquaculture for converting toxic ammonia into less harmful nitrate, which can be absorbed by plants or algae. Nitrate levels should be kept below 50 mg/L to prevent excessive algal growth and poor water quality.

Nitrogen Cycle: The nitrogen cycle in integrated aquaculture refers to the movement of nitrogen between the atmosphere, water, plants, animals, and soil. Nitrogen enters the system through animal waste, feed, and fertilizers, and is transformed through processes like ammonification, nitrification, and denitrification. Proper management of nitrogen levels (optimal ammonia <0.5

mg/L, nitrate <50 mg/L) ensures fish health and ecosystem productivity.

Nutrient Absorption: Nutrient absorption in aquaculture refers to the uptake of essential nutrients like nitrogen, phosphorus, and potassium by aquatic plants, algae, and microorganisms in the system. Proper management of nutrient levels, through balanced feeding and waste recycling, is critical to maintaining water quality and ensuring the health and growth of cultured species.

Nutrient Dynamics: Nutrient dynamics in integrated aquaculture refer to the flow and cycling of essential nutrients such as nitrogen, phosphorus, and carbon within the system. Nutrient inputs from animal manure, feed, and crop residues are transformed into organic matter that supports fish growth and plant production. Monitoring parameters like total nitrogen (1-5 mg/L) and phosphorus (0.05-0.1 mg/L) is critical for maintaining the balance between nutrient availability and water quality in aquaculture systems.

Nutrient Enrichment: Nutrient enrichment in aquaculture refers to the process of adding nutrients like nitrogen, phosphorus, and potassium to water bodies to promote the growth of phytoplankton and zooplankton. This process increases the natural food availability for fish but must be carefully managed to avoid excessive nutrient loading, which can lead to eutrophication. Ideal nutrient levels include total nitrogen of 1-3 mg/L and phosphorus of 0.05-0.1 mg/L.

Nutrient Recycling: Nutrient recycling in integrated aquaculture involves the reuse of organic matter and nutrients within the system to reduce waste and improve productivity. Animal waste, uneaten feed, and plant residues decompose, releasing essential nutrients back into the water or soil, where they are absorbed by plants, algae, or fish. This process minimizes the need for external inputs and enhances the sustainability of the system.

Nutrient Uptake: Nutrient uptake refers to the absorption of essential nutrients like nitrogen, phosphorus, and potassium by aquatic organisms, including fish, plants, and microorganisms. Efficient nutrient uptake in integrated systems ensures optimal growth and reduces waste. Fish feed containing 25-35% protein optimizes nitrogen uptake, while aquatic plants like Azolla and duckweed absorb excess nutrients, improving water quality.

Organic Matter Decomposition: Organic matter decomposition is the biological breakdown of plant and animal residues in fish ponds, releasing nutrients like nitrogen, phosphorus, and carbon into the water. Decomposition is carried out by microorganisms and is essential for recycling nutrients that support algal and fish growth. Optimal conditions for decomposition include temperatures between 20°-30°C and dissolved oxygen levels above 2 mg/L.

Paddy-Cum-Fish System: Paddy-cum-fish system is a form of integrated aquaculture where rice and fish are cultivated together in the same field. Fish like carp are introduced into flooded paddy fields, where they feed on weeds, pests, and rice residues, while their waste enriches the soil. This system enhances rice productivity, reduces pest populations, and provides an additional source of income from fish production. Water depth in paddy fields is maintained at 10-15 cm for optimal fish growth.

pH Levels: pH levels indicate the acidity or alkalinity of water in aquaculture systems. Most freshwater fish thrive in slightly alkaline water with a pH of 7.0–8.5, while marine species prefer a pH range of 8.0–8.4. Extreme pH values can stress fish and disrupt biological processes like respiration and growth. Lime ($CaCO_3$) or other buffering agents are used to stabilize pH levels in ponds.

Phosphorus Cycle: The phosphorus cycle in aquaculture involves the movement of phosphorus between organic and inorganic forms in the water and sediment. Phosphorus is a vital nutrient for plant and phytoplankton growth but can lead to eutrophication if over-applied. Animal waste and feed are primary phosphorus sources in integrated systems, with ideal concentrations in fish ponds maintained below 0.05 mg/L to prevent algal blooms.

Pig-Fish Integration: Pig-fish integration involves rearing pigs near fish ponds, where pig waste is directly or indirectly used as fertilizer to enhance primary productivity in the water. This system reduces feed costs and increases the nutrient input for fish growth. Pigs are typically housed on slatted platforms above or near fish ponds, with an ideal pig stocking density of 5-6 pigs per hectare of pond. The manure provides nitrogen and phosphorus, boosting algal growth, which in turn feeds the fish.

Plant-Fish Integration: Plant-fish integration refers to the cultivation of fish alongside aquatic or terrestrial plants in a symbiotic system where fish waste provides nutrients for plant growth, and plants help filter and clean the water. Aquaponics is a well-known example of this integration, where leafy greens or herbs are grown hydroponically with fish like tilapia. The ideal ammonia concentration in the water should be below 0.5 mg/L to prevent toxicity to fish while supporting plant growth.

Pokali: Pokali is a traditional paddy-cum-fish farming system practiced in the coastal areas of Kerala, India, where paddy fields are alternated with shrimp or fish culture during the off-season. After the paddy harvest, the fields are flooded with brackish water for shrimp farming. Salinity in Pokali fields typically ranges from 5-20 ppt, depending on tidal influences. This system utilizes natural nutrient recycling between rice and shrimp/fish culture.

Polyculture: Polyculture involves cultivating multiple fish species together in the same pond or system to utilize different ecological niches and trophic levels. This practice maximizes resource use efficiency and minimizes waste. Common polyculture species include Indian major carps, tilapia, and shrimp, with stocking ratios adjusted based on their feeding habits and growth rates.

Pond Ecosystem: A pond ecosystem in aquaculture consists of biotic (fish, plants, microorganisms) and abiotic (water, nutrients, sunlight) components that interact to sustain life. It is a complex system where water quality, nutrient cycling, and ecological balance are crucial for productivity. Proper management ensures stable dissolved oxygen levels (>5 mg/L), pH (7.0–8.5), and nutrient availability to maintain a healthy ecosystem.

Pond Fertility: Pond fertility is the ability of an aquaculture pond to support aquatic life, influenced by nutrient levels and primary productivity. Fertile ponds have balanced levels of nitrogen and phosphorus (N)

Pond Management: Pond management involves the continuous monitoring and regulation of water quality, nutrient levels, and stocking densities to maintain a productive culture environment. Parameters like dissolved oxygen (>5 mg/L), pH (7.0–8.5), ammonia (<0.1 mg/L), and temperature (25°–30°C) are regularly monitored. Managing feed inputs, waste, and water exchange are critical components.

Pond Preparation: Pond preparation is a crucial step in aquaculture that involves cleaning, drying, and liming ponds to remove pathogens, unwanted species, and organic matter. Liming adjusts pH levels to an optimal range of 7.0–8.0, and organic and inorganic fertilizers are added to enhance primary productivity. This step is fundamental to ensuring a healthy culture environment.

Poultry-Fish Integration: Poultry-fish integration involves rearing poultry, such as chickens, above or near fish ponds, with poultry droppings used to fertilize the pond. Poultry waste provides essential nutrients like nitrogen and phosphorus, promoting the growth of natural fish food. This integration reduces the need for commercial feed. Optimum stocking rates are around 500-600 birds per hectare of pond, with regular monitoring to maintain water quality.

Predatory Fishes: Predatory fishes are species that feed on other aquatic organisms, including smaller fish, invertebrates, and zooplankton. Examples include catfish, snakeheads, and sea bass. In aquaculture, their role can be critical for maintaining ecosystem balance by controlling populations of smaller, unwanted species. In integrated systems, predatory fish need to be carefully managed to prevent excessive predation on stock. Optimal stocking density varies but is typically kept lower (1-2 fish per square meter) to balance their role as both stock and controllers of other populations.

Primary Productivity: Primary productivity is the rate at which autotrophic organisms, like algae and aquatic plants, produce organic matter from inorganic substances using sunlight. In aquaculture, it serves as the base of the food chain, supporting fish growth.

Profitability: Profitability in aquaculture refers to the net income generated from fish farming after deducting operational costs like feed, seed, and labor. Profitability is influenced by fish growth rates, feed efficiency, market prices, and input costs. Small-scale aquaculture farms generally aim for profit margins of 20-30%, with faster-growing species offering quicker returns.

Rabbit-Fish Integration: Rabbit-fish integration is a system where rabbits are reared in enclosures above or near fish ponds. Rabbit manure, rich in nitrogen, phosphorus, and potassium, is used to fertilize fish ponds, promoting the growth of natural fish food like algae and plankton. This integration reduces feed costs while providing high-quality protein sources (rabbits and fish). For optimal nutrient input, about 100-150 rabbits per hectare of pond are maintained, with manure application rates of 50-100 kg/ha.ratio of 10:1 and adequate sunlight (photosynthetically active radiation, 400–700 nm), contributing to plankton bloom and higher fish yields. ratio of 7:1 to 10:1, supporting high algal and plankton productivity, which provides natural food for fish. Fertility can be managed through organic (e.g., livestock manure) or inorganic fertilizers (e.g., urea, superphosphate) applied at 50-100 kg/ha.

Revenue Generation: Revenue generation in aquaculture is the income derived from the sale of fish, shellfish, or other aquatic products. It depends on factors like fish growth rates, market demand, and sales channels. Profit margins typically range from 20-40% in well-managed systems, with faster-growing species (e.g., tilapia) contributing to quicker returns on investment.

Risk Management: Risk management in aquaculture involves identifying, assessing, and mitigating potential risks, such as disease outbreaks, market fluctuations, and environmental changes. Strategies include diversifying species, ensuring biosecurity, and using insurance. Effective risk management minimizes losses and ensures stable production and income.

Rohu: Rohu (*Labeo rohita*) is a freshwater carp species native to South Asia and is a key species in polyculture systems. It prefers temperatures between 25°–32°C and grows well in ponds with moderate water flow and a pH of 7.0–8.0. Rohu primarily consumes phytoplankton and zooplankton, making it compatible with other carp species in polyculture setups.

Silver Carp: Silver carp (*Hypophthalmichthys molitrix*) is a filter-feeding species commonly used in polyculture systems. It consumes phytoplankton and thrives in nutrient-rich waters with temperatures of 18°–28°C. Silver carp helps maintain water quality by reducing excess plankton in ponds, making it an essential species in integrated aquaculture.

Site Selection: Site selection for aquaculture is the process of choosing a location that provides optimal conditions for fish farming, such as water availability, soil quality, and proximity to markets. Factors like water quality (pH 6.5-8.5, DO >5 mg/L), topography, and access to infrastructure play a key role in site suitability. Proper site selection is critical to ensuring long-term farm productivity and sustainability.

Soil Fertility: Soil fertility in integrated aquaculture refers to the ability of pond soils to provide essential nutrients like nitrogen, phosphorus, and potassium to support fish and plant production. Animal waste, crop residues, and organic matter decomposition enhance soil fertility, while proper water management ensures nutrient availability without causing eutrophication. Ideal soil pH for fish ponds ranges from 6.5 to 7.5, promoting nutrient cycling.

Stocking Density: Stocking density refers to the number of fish or biomass introduced per unit volume of water, and it varies based on species, culture system, and intended output. In semi-intensive systems, stocking densities of 10,000–20,000 fingerlings per hectare are typical, while intensive systems may hold up to 100,000 per hectare. Optimizing stocking density helps maximize yield while maintaining water quality and fish health.

Sustainable Farming: Sustainable farming refers to agricultural practices that maintain productivity while conserving resources and minimizing environmental impact. In aquaculture, sustainable methods include using renewable resources, minimizing waste, and adopting practices like polyculture and integrated multi-trophic aquaculture (IMTA). Key sustainability indicators include water use efficiency, low carbon footprint, and maintaining biodiversity. Sustainable systems optimize resource inputs, typically using 20-30% less feed and energy compared to conventional systems.

Tilapia: Tilapia is a fast-growing, hardy species often cultivated in aquaculture systems due to its adaptability to various environmental conditions. It thrives in water temperatures between 25°–30°C, tolerates salinity levels up to 35 ppt, and can be integrated into polyculture systems with minimal feeding requirements due to its omnivorous diet.

Total Alkalinity: Total alkalinity refers to the water's ability to neutralize acids, primarily due to the presence of bicarbonates, carbonates, and hydroxides. In aquaculture, alkalinity levels between 75-200 mg/L (as $CaCO_3$) are considered optimal for buffering against pH fluctuations and supporting healthy aquatic life. Regular monitoring and the addition of lime can help maintain adequate alkalinity levels.

Trophic Levels: Trophic levels represent the hierarchical positions in an ecosystem based on energy flow from primary producers to top predators. In aquaculture, primary producers (phytoplankton) are consumed by herbivores (e.g., tilapia), which are then preyed upon by carnivorous species (e.g., catfish). Understanding trophic levels helps design integrated systems where species at different levels coexist, maximizing resource use efficiency and reducing competition.

Water Hardness: Water hardness is determined by the concentration of calcium and magnesium ions in water, which affects fish osmoregulation and overall water chemistry. Optimal hardness levels for freshwater aquaculture systems typically range from 50-200 mg/L (as $CaCO_3$). Hard water provides essential minerals, but excessive hardness can lead to scaling and poor water quality, requiring dilution or treatment.

Water Hyacinth: Water hyacinth (*Eichhornia crassipes*) is a fast-growing, free-floating aquatic plant that can be both a boon and a bane in aquaculture. While it absorbs excess nutrients and improves water quality, it can also outcompete other species for space and oxygen if not properly managed. In integrated systems, water hyacinth can be used for bioremediation and nutrient recycling.

Water Quality Management: Water quality management involves maintaining optimal physical, chemical, and biological conditions for the cultured species. Key parameters include dissolved oxygen (5–7 mg/L), temperature (25°–30°C), pH (7.0–8.5), ammonia (<0.05 mg/L), and salinity (depending on species). Effective water exchange, aeration, and use of biofilters help manage these parameters.

Water Spinach: Water spinach (*Ipomoea aquatica*), also known as kangkong, is an aquatic plant that is widely grown in integrated aquaculture systems. It absorbs excess nutrients from water, reducing nitrogen and phosphorus levels, and provides an additional source of income or food. It thrives in nutrient-rich waters and helps maintain water quality in aquaculture ponds.

Water Temperature: Water temperature is a critical parameter in aquaculture that influences fish metabolism, growth, and reproduction. Different species have optimal temperature ranges; for example, warm-water species like tilapia thrive at 25°–30°C, while cold-water species like trout prefer 10°–18°C. Maintaining appropriate temperature levels through shading or water circulation helps ensure optimal fish health and productivity.

Weed Control: Weed control in aquaculture involves strategies to manage the growth of aquatic plants and weeds in fish ponds. Methods include manual removal, mechanical harvesting, biological control using fish species like grass carp, or the use of herbicides. Maintaining proper nutrient levels (e.g., phosphorus below 0.1 mg/L) and shading through plant-fish integration can also help prevent excessive weed growth.

Weed Fishes: Weed fishes are species considered undesirable in aquaculture systems due to their ability to compete with cultured species for food and space or negatively impact water quality. They often include species such as Puntius and Gambusia, which may thrive in eutrophic conditions. Management involves controlling their population through predation, netting, or selective breeding to exclude these species from culture systems.

CHAPTER 18

FRESHWATER FISH SPECIES

Family: Cyprinidae (Carps and Minnows)

Indian Major Carps:

Catla (*Catla catla*): Belonging to the Cyprinidae family, Catla is a fast-growing freshwater fish native to South Asia. It has a deep, compressed body with a large head and upturned mouth, which makes it a surface feeder. Catla primarily feeds on zooplankton, though it may also consume phytoplankton in nutrient-rich waters. Its optimal growth occurs in water temperatures between 25°-32°C and in water with a pH range of 6.5-8.5. It is one of the major species used in Indian carp polyculture systems due to its fast growth and high demand.

Rohu (*Labeo rohita*): Rohu is a column-feeding fish, with a streamlined body and silver scales. It is native to the rivers of India and Bangladesh. Rohu primarily feeds on phytoplankton and detritus in its juvenile stages but becomes omnivorous in adulthood. It thrives in water temperatures of 25°-32°C, with a pH range of 7.0-8.0. Rohu is one of the most commonly cultured carps in aquaculture systems across South Asia due to its popularity and high market value.

Mrigal (*Cirrhinus cirrhosus*): A bottom-dwelling freshwater species, Mrigal is a key species in Indian aquaculture. It has an elongated body and feeds primarily on detritus and benthic microorganisms. Mrigal grows optimally at temperatures of 26°-30°C and in slightly alkaline water with a pH range of 7.0-8.5. Like Catla and Rohu, Mrigal is commonly cultured in polyculture systems alongside other carp species.

Minor Carps:

Calbasu (***Labeo calbasu***): A bottom feeder with a robust body and distinctive dark coloration, Calbasu is native to the Indian subcontinent. It feeds on detritus, algae, and small invertebrates. Its optimal water temperature for growth is 25°-30°C, with a preferred pH range of 7.0-8.5. Though less popular than Indian major carps, it is often farmed in regions with available water resources.

Bata (***Labeo bata***): Bata is a small, fast-growing fish with a silver body and reddish fins. It feeds on plankton, aquatic plants, and small insects, making it an omnivorous species. It thrives in freshwater bodies with a temperature range of 22°-30°C and prefers water with a pH of 6.5-8.0. Bata is cultivated in various parts of India, though not as widely as major carps.

Labeo gonius (***Labeo gonius***): Native to slow-moving rivers and reservoirs, Labeo gonius is a bottom feeder that prefers a temperature range of 24°-30°C and water with a neutral to slightly alkaline pH. It feeds mainly on detritus and plant matter. Though not widely farmed, it is an important species in certain local aquaculture systems.

Exotic Carps:

Common Carp (***Cyprinus carpio***): Common carp is one of the most widespread and adaptable freshwater fish in global aquaculture. It has a broad tolerance for environmental conditions, thriving in temperatures of 20°-30°C and pH levels of 6.5-9.0. Common carp is an omnivorous species, feeding on detritus, benthic organisms, and plant material. It is often used in monoculture and polyculture systems worldwide due to its adaptability and fast growth.

Grass Carp (***Ctenopharyngodon idella***): Grass carp is a herbivorous fish native to Eastern Asia but widely introduced worldwide for weed control in aquaculture systems. It feeds on aquatic plants, making it a valuable species for controlling excessive vegetation in ponds and lakes. It grows best in warm waters between 20°-30°C, with a pH range of 7.0-8.5. Grass carp has high growth rates and can reach large sizes in suitable environments.

Silver Carp (***Hypophthalmichthys molitrix***): Silver carp is a filter-feeding fish that consumes phytoplankton and small suspended particles in the water. It is native to East Asia but has been introduced in many countries for aquaculture and water quality management. It grows optimally in water temperatures between 20°-28°C and in water with a pH range of 6.5-8.5. Silver carp is highly efficient at converting primary productivity into fish biomass, making it popular in polyculture systems.

Bighead Carp (***Hypophthalmichthys nobilis***): Bighead carp is another filter feeder, specializing in consuming zooplankton. It is native to China but is farmed worldwide for its rapid growth and large size. Optimal growth occurs in water temperatures of 25°-30°C, with a pH range of 6.8-8.5. Bighead carp is often used in integrated aquaculture systems, where it helps maintain water quality by controlling plankton populations.

Mahseer:

Golden Mahseer (***Tor putitora***): Golden mahseer is an iconic freshwater fish found in rivers and streams in the Himalayan region. It has a streamlined body and golden scales. It is an omnivorous species, feeding on small fish, insects, and plant material. The species prefers clear, fast-flowing waters with a temperature range of 18°-26°C and a pH of 6.5-8.0. Due to its conservation importance, it is both cultured and released into the wild as part of fishery enhancement programs.

Chocolate Mahseer (***Neolissochilus hexagonolepis***): This species is found in rivers and streams of South Asia, including the foothills of the Himalayas. It is omnivorous, feeding on aquatic plants, algae, and invertebrates. Chocolate Mahseer thrives in cool, well-oxygenated water with a temperature range of 18°-24°C and neutral to slightly alkaline pH. It is increasingly important in conservation-based aquaculture due to its declining wild populations.

Red-Finned Mahseer (***Tor tor***): Known for its distinctive red fins, this species is native to fast-flowing rivers in South Asia. It is an omnivorous feeder, consuming small fish, insects, and plant matter. It prefers waters with temperatures of 20°-28°C and a pH of 6.5-7.5. Red-finned mahseer is valued for both sport fishing and conservation aquaculture efforts.

Barbs and Minnows:

Olive Barb (***Systomus sarana***): A medium-sized barb, this species is a bottom and mid-water feeder, primarily consuming plant material, algae, and small invertebrates. It prefers temperatures of 22-28°C and water with a pH range of 6.5-7.5.

Barbs and Minnows:

Pool Barb (***Puntius sophore***): Commonly found in freshwater streams and rivers, Pool Barb has a silver body with black spots near the tail. It is an omnivorous species, feeding on algae, detritus, and small invertebrates. The species prefers water temperatures of 22°-28°C and a pH range of 6.5-7.5. Though small in size, Pool Barb is important in local aquaculture and wild fisheries.

Ticto Barb (*Pethia ticto*): The Ticto Barb is a small, peaceful species native to South Asia. It has distinctive red and black markings on its body. It feeds primarily on small invertebrates and plant material and thrives in water temperatures of 20°-26°C, with a pH of 6.5-7.5. It is often kept in ornamental aquaculture but also contributes to local fisheries.

Spotted Barb (***Puntius sophore***): Similar to the Pool Barb, the Spotted Barb is an omnivorous species that feeds on small invertebrates, algae, and organic matter. It thrives in a temperature range of 22°-28°C, with a neutral pH. The species is often found in shallow rivers and ponds and is a common part of the minor carp fisheries.

Dwarf Barb (*Pethia phutunio*): A small-sized barb, Dwarf Barb is known for its adaptability to various freshwater environments. It is an omnivore, feeding on both plant material and small invertebrates. The optimal water temperature for Dwarf Barb is between 22°-28°C, and it prefers water with a pH of 6.5-7.5. It is used in ornamental aquaculture due to its vibrant appearance.

Rosy Barb (*Pethia conchonius*): The Rosy Barb is a popular ornamental species due to its bright coloration. It is omnivorous, feeding on small insects, plants, and algae. Rosy Barb thrives in water temperatures of 18°-28°C, with a pH of 6.0-7.5. While primarily used in the ornamental trade, it is also found in wild fisheries in South Asia.

Firefin Barb (*Pethia ticto*): Known for its vibrant red fins, Firefin Barb is a small fish species that is primarily omnivorous, feeding on plant material and small aquatic organisms. It prefers water temperatures of 20°-28°C and a pH range of 6.5-7.5. Its bright appearance makes it popular in ornamental fishkeeping, though it is also present in local fisheries.

Deccan Barb (*Puntius deccanensis*): Endemic to the freshwater streams and rivers of the Deccan Plateau in India, the Deccan Barb is a small species that primarily feeds on algae, detritus, and small invertebrates. It grows best in water temperatures of 22°-28°C, with a pH range of 6.5-7.5. Though not widely cultivated, it is significant in local ecosystems and fisheries.

Coldwater fishes

Rainbow Trout (*Oncorhynchus mykiss*): Rainbow trout is a cold-water species native to North American rivers and lakes, especially west of the Rocky Mountains. It is highly adaptable, capable of surviving in a wide range of aquatic environments from lakes to streams, but prefers clear, cool, well-oxygenated waters. Rainbow trout are anadromous, meaning they can live in both freshwater and saltwater, with some populations migrating to the ocean. Their diet consists primarily of invertebrates, but they also consume small fish. They are renowned in aquaculture due to their rapid growth rate, with an optimal temperature range of 12°C to 16°C for their cultivation.

Brown Trout (*Salmo trutta*): Brown trout is a cold-water fish species native to Europe and western Asia but has been introduced to various regions worldwide. They thrive in cool, well-oxygenated rivers and lakes with a temperature range of 4°C to 19°C. Brown trout are adaptable to different habitats, from small streams to large lakes and rivers. Their diet is diverse, feeding on invertebrates, small fish, and even amphibians. Brown trout are known for their elusive behavior, making them a popular species among anglers. Their capacity to adapt to varying environmental conditions makes them a valuable species in aquaculture and fisheries management.

Brook Trout (*Salvelinus fontinalis*): Brook trout, also known as speckled trout, is native to eastern North America and thrives in cold, clear, well-oxygenated streams and lakes, often with temperatures between 10°C to 16°C. They prefer habitats with abundant cover, such as undercut banks, fallen trees, and aquatic vegetation. Brook trout are sensitive to environmental changes, particularly water pollution and habitat destruction. They have a varied diet that includes insects, crustaceans, and small fish. In aquaculture, brook trout are prized for their rapid growth and ability to thrive in colder water conditions, though they require high water quality and careful management.

Arctic Char (*Salvelinus alpinus*): Arctic char is a cold-water species found in Arctic and subarctic regions, inhabiting both freshwater and marine environments. It is closely related to both salmon and trout and is highly adaptable to cold water, thriving in temperatures ranging from 4°C to 12°C. Arctic char exhibits both anadromous (migratory) and resident (non-migratory) life cycles. They are known for their tolerance to extreme environmental conditions, including low temperatures and high salinity. Their diet consists mainly of zooplankton, insects, and small fish. In aquaculture, Arctic char is valued for its firm, pink flesh and is considered a delicacy in many regions.

Lake Trout (*Salvelinus namaycush*): Lake trout is native to the cold, deep lakes of North America, especially in the Great Lakes region. They prefer water temperatures between 4°C to 12°C and are often found at depths of up to 200 meters. Lake trout have a slow growth rate compared to other trout species, and they can live for several decades. They primarily feed on other fish, especially smaller species like smelt and cisco. Lake trout are sensitive to changes in water temperature and quality, making them indicators of ecosystem health. In aquaculture, they are raised for both commercial and recreational purposes but require careful management to mimic their natural deep-water habitat.

Coho Salmon (*Oncorhynchus kisutch*): Coho salmon is a cold-water species native to the North Pacific Ocean and is widely distributed along the coasts of North America and Asia. Coho salmon are anadromous, meaning they are born in freshwater rivers and streams, migrate to the ocean, and return to freshwater to spawn. They prefer water temperatures ranging from 10°C to 15°C. Coho salmon are known for their aggressive behavior and powerful swimming ability, making them a popular target for sport fishing. Their diet consists of plankton, insects, and smaller fish. In aquaculture, Coho salmon are highly valued for their flesh quality and are cultivated in both freshwater and marine environments.

Himalayan Snow Trout (***Schizothorax richardsonii***): Native to the cold, fast-flowing rivers of the Himalayas, this species is well-adapted to low temperatures (8-18°C). It is an omnivore, feeding on algae, aquatic plants, and small invertebrates. Himalayan Snow Trout prefers well-oxygenated water with a neutral to slightly alkaline pH. It is increasingly cultured in colder regions for its economic value.

Flying Barb (***Esomus danricus***): This small barb is native to slow-moving streams and ponds in South Asia. It has a unique ability to leap out of the water, hence its common name. Flying Barb is an omnivorous species, feeding on algae, insects, and detritus. It prefers temperatures between 22°-28°C and water with a neutral pH. It is commonly used in both ornamental aquaculture and local fisheries.

Hill Trout (***Barilius spp.***): Hill Trout is a small, fast-swimming species commonly found in fast-flowing rivers and streams. It is primarily insectivorous, feeding on aquatic and terrestrial insects, though it may consume plant material. Hill Trout thrives in cool, well-oxygenated water with temperatures of 18-24°C and a pH range of 7.0-8.0. It is valued for both its game fishing potential and conservation.

Atlantic Salmon (*Salmo salar*): Atlantic salmon is a cold-water fish species native to the North Atlantic Ocean and rivers that flow into it. They are anadromous, spending most of their lives in the ocean before returning to freshwater to spawn. Atlantic salmon prefer water temperatures between 8°C and 14°C and are often found in fast-flowing, well-oxygenated rivers during their spawning runs. Their diet primarily consists of small fish and crustaceans. Atlantic salmon are one of the most commercially important fish species, with a significant portion of the global supply coming from aquaculture operations. They are valued for their high growth rate and adaptability to farming conditions, although they require stringent water quality and temperature control in captivity.

Family: Siluridae (Catfish)

Indian Catfish:

Walking Catfish (***Clarias batrachus***): Walking Catfish is a hardy, air-breathing species known for its ability to "walk" over land using its pectoral fins. Native to Southeast Asia, it is an omnivore, feeding on detritus, insects, small fish, and plant matter. It thrives in water temperatures of 25°-30°C and can tolerate low dissolved oxygen levels, making it suitable for aquaculture in stagnant or low-oxygen environments. Walking Catfish is farmed across Asia, although its invasive potential in non-native regions is of concern.

Asian Catfish (***Pangasius pangasius***): This species is a fast-growing, omnivorous catfish that is widely cultured in Southeast Asia for its high economic value. Asian Catfish prefers warm water temperatures of 26°-30°C and neutral to slightly alkaline water with a pH of 7.0-8.0. It primarily feeds on algae, plant material, and small invertebrates. Due to its adaptability and high growth rate, it is a key species in the aquaculture industry, especially in large-scale pond and cage systems.

Mystus Catfish:

Dark Mystus vittatus (***Mystus vittatus***): Known for its distinctive banding pattern, Mystus vittatus is a small, omnivorous catfish native to South Asia.

It feeds on small insects, crustaceans, and detritus. Mystus prefers water temperatures between 24°-30°C and can tolerate a wide range of water conditions, including pH levels from 6.5 to 8.0. It is often found in rivers, lakes, and ponds and is valued in local aquaculture for its adaptability.

Mystus cavasius (***Mystus cavasius***): A bottom-dwelling species, *Mystus cavasius* feeds on detritus, insects, and other small aquatic organisms. It thrives in warm water temperatures of 22°-30°C and prefers slightly alkaline water with a pH of 7.0-8.5. The species is important in local fisheries but not widely farmed on a commercial scale.

Striped Catfish (***Mystus tengara***): Striped Catfish is a small species distinguished by its horizontal stripes along its body. It is an omnivore, feeding on algae, small invertebrates, and detritus. It prefers water temperatures of 24°-30°C and slightly alkaline water with a pH range of 7.0-8.0. Striped Catfish is often found in rivers and ponds and is important in local fisheries.

Magur (Clarias species):

Asian Catfish (***Clarias gariepinus***): Often referred to as African Catfish, this species is a highly efficient predator and omnivore that consumes insects, small fish, and plant material. It can tolerate a wide range of environmental conditions, including low oxygen levels, making it a popular species in intensive aquaculture. Optimal water temperatures range between 25°-30°C, with a pH of 6.5-8.0. It is widely farmed across Africa, Asia, and parts of Europe due to its fast growth and high market value.

Other Silurids:

Rita Catfish (***Rita rita***): This species is a medium-sized bottom feeder found in rivers and large reservoirs. It feeds on detritus, small crustaceans, and insects. Rita Catfish prefers warm waters with temperatures of 24°-30°C and slightly alkaline water with a pH of 7.0-8.5. It is important in local fisheries but not commonly used in large-scale aquaculture.

Wallago (***Wallago attu***): Wallago is a large, carnivorous catfish that preys on small fish and invertebrates. It thrives in rivers and lakes with water temperatures of 24°-30°C and a pH range of 6.5-8.0. Wallago attu is an important species in local fisheries and is sometimes cultured in ponds and reservoirs.

Ompok pabda (***Ompok pabda***): Ompok pabda is a small catfish species native to the rivers and lakes of South Asia. It is carnivorous, feeding on small fish, insects, and crustaceans. Optimal water temperatures for this species range from 24°-30°C, with a pH of 7.0-8.5. It is often included in traditional polyculture systems in South Asia.

Singhi (***Heteropneustes fossilis***): Singhi is a freshwater catfish with a high tolerance for low-oxygen environments, due to its ability to breathe air. It is an omnivorous feeder, consuming detritus, small invertebrates, and plant matter.

The species thrives in warm water with temperatures of 24°-30°C and a pH of 6.5-7.5. Singhi is a valuable species in traditional and commercial aquaculture systems in India and Bangladesh.

Family: Bagridae (Bagrid Catfishes)

Sperata aor (***Sperata aor***): Commonly known as "Long-whiskered Catfish," this large freshwater species inhabits rivers and reservoirs of South Asia. It is carnivorous, feeding primarily on small fish and invertebrates. Optimal water conditions include temperatures between 24°-30°C and a pH of 7.0-8.5. Sperata aor is important in local fisheries and aquaculture, particularly in India and Bangladesh, where it is valued for its large size and flesh quality.

Sperata seenghala (***Sperata seenghala***): Known as the "Seenghala Catfish" or "Giant River Catfish," this species is a popular food fish in South Asia. It is a carnivore, consuming fish, crustaceans, and other aquatic organisms. It prefers a temperature range of 24°-30°C and slightly alkaline water with a pH of 7.0-8.5. *Sperata seenghala* is an important species in inland fisheries, where it is often cultured in ponds and reservoirs for its economic value.

Family: Channidae (Snakeheads)

Murrel:

Striped Snakehead (***Channa striata***): This species is a highly predatory fish, known for its ability to survive in low-oxygen environments due to its accessory breathing organs. It feeds on fish, insects, and small crustaceans. Striped Snakehead thrives in water temperatures of 24-32°C and prefers slightly acidic to neutral water (pH 6.5-7.5). It is widely farmed in Southeast Asia and is valued for its fast growth, high protein content, and ability to survive in various aquatic environments.

Spotted Snakehead (***Channa punctata***): Spotted Snakehead is a smaller, carnivorous species found in freshwater rivers, ponds, and swamps. It feeds on fish, insects, and small invertebrates. Optimal water conditions include temperatures of 22°-30°C and a neutral pH. Spotted Snakehead is important in local fisheries and is sometimes cultured in mixed or polyculture systems for its economic value and ability to adapt to diverse environments.

Rainbow Snakehead (***Channa bleheri***): This species is primarily an ornamental fish, prized for its bright colors and small size. Rainbow Snakehead is carnivorous, feeding on small fish and invertebrates, and prefers water temperatures of 22°-28°C with a pH of 6.5-7.5. Though not commonly used in large-scale aquaculture, it is bred for the aquarium trade.

Giant Snakehead (***Channa marulius***): One of the largest snakehead species, Giant Snakehead is a highly aggressive predator, feeding on fish, amphibians, and even small mammals. It prefers warm water conditions of 25°-32°C and slightly alkaline water with a pH of 7.0-8.5. Giant Snakehead is farmed in some regions, though its predatory nature and potential environmental impact limit its widespread cultivation.

Dwarf Snakehead (*Channa gachua*): A small, adaptable species, Dwarf Snakehead is found in freshwater ponds, rivers, and swamps across Asia. It is a carnivore, feeding on insects, small fish, and other invertebrates. The species thrives in water temperatures of 22°-28°C and a pH range of 6.5-7.5. It is used in ornamental aquaculture and is occasionally farmed for local consumption.

Family: Cichlidae (Tilapia)

Tilapia:

Nile Tilapia (**Oreochromis niloticus**): Nile Tilapia is one of the most widely farmed fish species globally, known for its rapid growth, adaptability, and resistance to disease. It is an omnivore, feeding on algae, plankton, and detritus. Nile Tilapia prefers water temperatures of 24°-30°C and a pH range of 6.5-8.0, making it suitable for both freshwater and brackish water aquaculture systems. Its high tolerance for different environmental conditions has made it a cornerstone of commercial aquaculture in Africa, Asia, and Latin America.

Mozambique Tilapia (**Oreochromis mossambicus**): Another important species in aquaculture, Mozambique Tilapia is a hardy fish that can survive in a wide range of salinities, from freshwater to highly saline environments. It is an omnivore, feeding on a mix of plant material, algae, and small invertebrates. Optimal water temperatures are between 22°-30°C, with a pH range of 6.0-8.5. While less commonly farmed than Nile Tilapia due to slower growth rates, Mozambique Tilapia is important in areas with brackish or saline water conditions.

Family: Notopteridae (Featherbacks)

Indian Featherback (*Notopterus notopterus*): This species is a carnivorous fish native to rivers, lakes, and ponds in South Asia. It feeds primarily on small fish and invertebrates. Indian Featherback prefers water temperatures of 24°-30°C and slightly alkaline water with a pH range of 6.5-8.0. It is an important species in traditional fisheries and is sometimes cultured for local markets due to its high market value.

Bronze Featherback (*Chitala chitala*): The Bronze Featherback is a large, carnivorous fish prized in aquaculture for its flesh quality. It feeds on small fish, crustaceans, and insects and prefers warm water conditions of 26°-30°C, with a pH range of 6.5-8.0. It is cultivated in South Asia, primarily in pond systems, and is also used in the ornamental fish trade due to its unique shape and appearance.

Family: Mastacembelidae (Spiny Eels)

Peacock Spiny Eel (*Macrognathus aral*): The Peacock Spiny Eel is a small, nocturnal fish native to freshwater bodies in South Asia. It feeds on small crustaceans, worms, and insects. This species prefers water temperatures of 22-28°C and a neutral pH. It is often kept in aquariums due to its striking appearance and is sometimes used in local fisheries.

Tire-track Spiny Eel (***Mastacembelus armatus***): Known for the intricate tire-track pattern on its body, this species is a carnivorous spiny eel that inhabits rivers, streams, and ponds. It feeds on small invertebrates, crustaceans, and worms. It prefers water temperatures of 24°-30°C and a pH range of 6.5-7.5. Tire-track Spiny Eel is popular in ornamental aquaculture and is also used in traditional fisheries in Southeast Asia.

Family: Osphronemidae (Gouramis)

Giant Gourami (***Osphronemus goramy***): A large, omnivorous species, Giant Gourami is widely farmed in Southeast Asia for its meat. It feeds on plants, algae, and small invertebrates and thrives in water temperatures of 25°-30°C, with a pH range of 6.0-8.0. Giant Gourami is known for its ability to adapt to low-oxygen environments, making it suitable for pond-based aquaculture.

Dwarf Gourami (***Trichogaster lalius***): A smaller species, Dwarf Gourami is popular in the ornamental fish trade due to its bright coloration. It is an omnivore, feeding on algae, plant matter, and small insects. The species prefers water temperatures of 24°-28°C and a pH range of 6.0-7.5. While primarily ornamental, it plays a role in local fisheries.

Three-spot Gourami (***Trichopodus trichopterus***): This species is named for the three spots along its body and is both an ornamental and food fish. It is an omnivore, feeding on plants, algae, and small invertebrates. Three-spot Gourami thrives in water temperatures of 24°-30°C and a pH range of 6.0-7.5. It is commonly found in ponds and slow-moving rivers across Southeast Asia.

Banded Gourami (***Trichogaster fasciata***): Native to South Asia, Banded Gourami is an omnivorous species that feeds on plant matter, algae, and small insects. It prefers water temperatures of 24°-28°C and slightly acidic to neutral water (pH 6.0-7.5). It is commonly used in ornamental aquaculture and local fisheries.

Family: Belontiidae (Climbing Perches)

Climbing Perch (***Anabas testudineus***): This species, known for its ability to move over land using its pectoral fins and labyrinth organ, is found in freshwater habitats across South and Southeast Asia. It is an omnivore, feeding on small invertebrates, plant matter, and detritus. Climbing Perch prefers water temperatures of 24°-30°C and can survive in low-oxygen environments due to its accessory breathing organs. It is farmed in some regions, though it is more commonly captured from the wild, and is valued for its resilience and adaptability to harsh conditions.

Family: Nandidae (Leaf Fishes)

Nandus (***Nandus nandus***): This ambush predator is native to slow-moving rivers, lakes, and swamps in South Asia. It primarily feeds on small fish and invertebrates. *Nandus nandus* prefers a water temperature range of 24°-30°C and a slightly acidic to neutral pH (6.0-7.5). Although not widely farmed, it is valued in local fisheries for its flesh and is sometimes kept in aquaria due to its unique, leaf-like appearance and camouflage behavior.

Family: Ambassidae (Glass Perches)

Indian Glassy Fish (***Parambassis ranga***): This species is popular in the aquarium trade due to its translucent body. It is found in freshwater rivers, lakes, and swamps across South Asia, where it feeds on small invertebrates and plankton. Indian Glassy Fish prefers water temperatures between 22°-28°C and neutral pH conditions. Although primarily ornamental, it is sometimes used in local fisheries.

Giant Glassfish (***Parambassis lala***): Similar to *P. ranga*, this species has a semi-transparent body and inhabits freshwater habitats in South and Southeast Asia. It is carnivorous, feeding on small aquatic insects and zooplankton. It thrives in warm waters with temperatures between 24°-30°C and a pH of 6.0-7.5. While mostly kept as an aquarium fish, it holds some importance in local subsistence fisheries.

Family: Cobitidae (Loaches)

Zebra Loach (***Botia striata***): This small, bottom-dwelling species is popular in the aquarium trade for its distinctive zebra-like stripes. Native to fast-flowing rivers and streams in India, Zebra Loach feeds on small invertebrates, algae, and detritus. It prefers water temperatures of 24°-28°C and slightly acidic to neutral pH (6.0-7.0). It is valued primarily as an ornamental species.

Dwarf Chain Loach (***Yasuhikotakia sidthimunki***): This small, social species is known for its schooling behavior and is popular in the aquarium trade. Native to freshwater habitats in Southeast Asia, it feeds on small crustaceans and invertebrates. The Dwarf Chain Loach prefers water temperatures of 24°-28°C and neutral to slightly acidic conditions. It is primarily ornamental.

Nemacheilus Loaches (***Nemacheilus spp.***): These loaches are found in fast-flowing streams and rivers across Asia. They are bottom-dwelling omnivores, feeding on algae, small invertebrates, and organic debris. They thrive in cooler waters with temperatures of 20°-26°C and a neutral pH. They are often used in aquariums due to their active behavior and unique appearance.

Family: Clupeidae (Shads)

Indian River Shad (***Gudusia chapra***): This small, schooling fish is found in rivers, lakes, and reservoirs across South Asia. It feeds primarily on plankton and detritus and prefers warm waters with temperatures between 22°-30°C and a neutral pH. Indian River Shad is an important species in inland fisheries and is harvested in large quantities for its economic value.

Hilsa (***Tenualosa ilisha***): The Hilsa is an anadromous fish, migrating from the sea to freshwater rivers for spawning. It is one of the most commercially important fish in South Asia, particularly in the Ganges-Brahmaputra delta. Hilsa is an omnivore, feeding on plankton, algae, and detritus. It thrives in a wide range of salinities and prefers water temperatures of 24°-32°C. The species is highly valued for its flavor and is a staple in Bengali cuisine, with significant cultural and economic importance.

Family: Synbranchidae (Swamp Eels)

Asian Swamp Eel (***Monopterus albus***): This elongated, air-breathing fish is found in swampy areas, rice paddies, and slow-moving rivers across Southeast Asia. It is carnivorous, feeding on small fish, crustaceans, and insects. The Asian Swamp Eel can survive in low-oxygen environments due to its ability to breathe air and prefers water temperatures between 24°-30°C and a pH of 6.0-8.0. It is valued in local fisheries and is often caught from the wild for consumption.

Family: Sisoridae (Sisorid Catfishes)

Glyptothorax species (***Glyptothorax madraspatanum***): This genus of small, benthic catfish is found in fast-flowing rivers and streams across South Asia. Glyptothorax species are carnivorous, feeding on insects, crustaceans, and small fish. They prefer cooler, well-oxygenated waters with temperatures of 20°-25°C and a neutral to slightly alkaline pH. These fish are important in local fisheries and are often used in traditional medicine.

Goonch (***Bagarius bagarius***): Goonch is a large, predatory catfish found in rivers across South Asia. It is known for its strong, muscular body and large size, making it a prized catch for anglers. Goonch feeds on fish, crustaceans, and sometimes small mammals. It prefers cool, fast-flowing waters with temperatures between 18°-26°C and a neutral to slightly alkaline pH. It is an important species in local fisheries but is also threatened by habitat loss and overfishing.

Family: Aplocheilidae (Killifish)

Striped Panchax (***Aplocheilus lineatus***): This small, freshwater fish is native to slow-moving waters and rice paddies in South Asia. It is an insectivore, feeding on aquatic insects, larvae, and small invertebrates. The species thrives in warm waters with temperatures of 24°-30°C and a neutral pH. Striped Panchax is valued in the aquarium trade for its bright coloration and peaceful nature.

Blue Panchax (***Aplocheilus panchax***): Another popular aquarium species, Blue Panchax is found in freshwater rivers and ponds across South and Southeast Asia. It is a surface-dwelling predator, feeding on insects and small invertebrates.

It prefers water temperatures between 24°-30°C and neutral to slightly alkaline water. This species is primarily ornamental but is sometimes used in mosquito control due to its feeding habits.

Family: Poeciliidae (Livebearers)

Guppy (***Poecilia reticulata***): Known for its vibrant colors and adaptability, the guppy is one of the most popular aquarium fish species globally. Native to northern South America, guppies are omnivores, feeding on algae, plant material, small invertebrates, and detritus. They prefer water temperatures between 22°-28°C, with a pH range of 6.8-7.8. Due to their high reproductive rate and tolerance to varying environmental conditions, guppies are often used in mosquito control programs as well as in aquariums for decorative purposes.

Molly (***Poecilia sphenops***): Mollies are hardy, freshwater fish known for their ability to adapt to various water conditions, including brackish environments. They are omnivores, feeding on algae, invertebrates, and plant material. The species thrives in water temperatures between 24°-28°C and a pH of 7.0-8.5. Mollies are widely used in aquaculture for ornamental purposes due to their ease of breeding and variety of color morphs.

Swordtail (***Xiphophorus hellerii***): Native to Central America, swordtails are named for the elongated lower lobe of the males' caudal fin. They are omnivorous, consuming algae, plant material, and small invertebrates. Swordtails prefer warm water with temperatures ranging from 22°-28°C and a pH of 7.0-8.0. They are popular in the aquarium trade due to their vibrant colors and active swimming behavior.

Platy (***Xiphophorus maculatus***): The platy is a small, brightly colored livebearer native to Central America. It is an omnivore, feeding on algae, detritus, and small aquatic organisms. Platys thrive in water temperatures between 22-26°C and a pH range of 7.0-8.0. Their peaceful temperament, ease of breeding, and wide range of color varieties make them a staple in the ornamental fish trade.

Family: Eleotridae (Sleepers)

Tank Goby (***Glossogobius giuris***): Tank goby is a widely distributed species found in freshwater and brackish waters of Asia, Africa, and Australia. It is a carnivorous species, feeding on small fish, crustaceans, and invertebrates. This species thrives in tropical waters with temperatures between 24°-30°C and slightly alkaline water (pH 7.0-8.0). It is of moderate importance in local fisheries, and in some areas, it is consumed as food.

Family: Gobiidae (Gobies)

Freshwater Goby (***Rhinogobius giurinus***): Native to East Asia, this species inhabits freshwater rivers and streams and is known for its small size and benthic lifestyle. It primarily feeds on small invertebrates and organic matter. The freshwater goby prefers water temperatures of 20°-26°C and a pH range of

6.0-7.5. It has limited economic value but is sometimes kept in aquariums for its unique behavior and appearance.

Family: Cyprinodontidae

Striped Killifish (***Aplocheilus panchax***): This species is commonly found in slow-moving freshwater bodies, such as ponds, streams, and rice fields, in South and Southeast Asia. It is carnivorous, feeding primarily on insects, small crustaceans, and larvae. The striped killifish prefers temperatures between 24°-30°C and slightly acidic to neutral pH levels (6.0-7.5). It is a popular choice in the aquarium trade for its striking patterns and peaceful nature.

Family: Adrianichthyidae (Ricefish)

Indian Ricefish (***Oryzias dancena***): This small, hardy species is found in freshwater bodies, including rice paddies, across South and Southeast Asia. It is omnivorous, feeding on algae, small invertebrates, and detritus. Indian Ricefish thrives in water temperatures between 22°-28°C, with a neutral pH of 7.0-7.5. While primarily ornamental, it is also important in mosquito control and as a model organism in scientific research due to its simple breeding requirements.

Family: Xenentodontidae (Needlefish)

Freshwater Needlefish (***Xenentodon cancila***): This elongated, predatory species is native to freshwater rivers and lakes in South and Southeast Asia. It is carnivorous, feeding on smaller fish, insects, and crustaceans. Freshwater needlefish prefer warm water temperatures of 24°-28°C and a neutral to slightly alkaline pH of 7.0-8.0. While not widely used in aquaculture, it is sometimes kept in aquariums for its unique appearance.

Family: Atherinidae (Silversides)

Indian Silverside (***Atherina boyeri***): This small, schooling fish is found in freshwater, brackish, and marine environments across the Indian subcontinent. It is an omnivore, feeding on plankton, small invertebrates, and detritus. Indian Silverside prefers warm waters with temperatures between 22°-28°C and a neutral pH range of 6.5-7.5. The species holds some local commercial value and is sometimes captured for food or aquarium trade.

Family: Schilbeidae (Schilbeid Catfish)

Butter Catfish (***Ompok bimaculatus***): This slender, freshwater catfish is found in rivers and ponds across South and Southeast Asia. It is carnivorous, feeding on small fish, insects, and crustaceans. The species thrives in water temperatures of 24°-30°C, with a slightly acidic to neutral pH of 6.0-7.5. Butter catfish is valued for its high-quality flesh and is an important species in inland fisheries.

Family: Anguillidae (Eels)

Indian Mottled Eel (***Anguilla bengalensis***): This long, slender, and snake-like fish is a catadromous species, migrating from freshwater to the sea for spawning. It is distributed across rivers and streams of South Asia. Indian Mottled Eels are carnivorous, feeding on small fish, crustaceans, and aquatic insects. They prefer warm tropical waters with temperatures of 25°-30°C and slightly alkaline pH levels. These eels are highly prized in local fisheries for their flesh and are also farmed in some regions.

Family: Anabantidae (Climbing Gouramis)

Climbing Gourami (***Anabas testudineus***): Also referred to as the climbing perch, this species is noted for its ability to move across land using its pectoral fins and specialized breathing organ. Found across South and Southeast Asia, *Anabas testudineus* is an omnivore, feeding on invertebrates, plant matter, and detritus. It thrives in tropical waters with temperatures of 24°-30°C and can survive in low-oxygen environments due to its labyrinth organ. This species is important in small-scale fisheries and is also valued for its resilience in aquaculture.

Endemic Species

Chela cachius (***Chela cachius***): Also known as the Glass Barb, this small schooling species is found in freshwater rivers and lakes across South Asia. It is omnivorous, feeding on zooplankton, small insects, and detritus. This species thrives in warm waters with temperatures of 24°-30°C and a slightly acidic to neutral pH range (6.0-7.5). *Chela cachius* is sometimes kept in ornamental aquariums for its translucent body and active behavior.

Dawkinsia assimilis (***Dawkinsia assimilis***): This species is endemic to the southern rivers of India and is part of the barb family. It is a detritivore, feeding primarily on organic material and algae. It prefers warm water temperatures of 24°-28°C and a neutral pH range. *Dawkinsia assimilis* is occasionally kept in aquariums but plays a more significant role in local subsistence fisheries.

Labeo boga (***Labeo boga***): Found in rivers of the Indian subcontinent, *Labeo boga* is a benthopelagic fish that primarily feeds on algae, detritus, and other organic matter. This species prefers slow-moving waters with temperatures ranging from 24°-28°C and a slightly alkaline pH. It is important in subsistence fisheries and plays a role in maintaining ecosystem balance by controlling algae growth. Although not widely farmed, it has potential for aquaculture in certain regions.

Labeo dero (***Labeo dero***): This freshwater cyprinid is native to the rivers and streams of the Himalayan region. It primarily consumes algae, aquatic plants, and organic detritus. It prefers cold to temperate waters with temperatures ranging from 18°-24°C and a neutral pH. Although not widely farmed, it is harvested in local fisheries and has potential for cold-water aquaculture in mountain regions.

***Neolissochilus hexagonolepis* (*Chocolate Mahseer*)**: This mahseer species is found in the clear, fast-flowing rivers of the Himalayan foothills. It is an omnivorous fish, consuming a wide variety of aquatic plants, algae, insects, and small invertebrates. *Neolissochilus hexagonolepis* prefers cooler waters with temperatures ranging from 18°-24°C and a neutral to slightly alkaline pH of 7.0-8.0. It is considered a valuable sport fish and is important for aquaculture due to its high market value and slow-growing nature.

Osteobrama belangeri (*Osteobrama belangeri*): Native to the Manipur region in northeastern India, this species is an important endemic fish in local fisheries. It is a herbivorous species, mainly consuming phytoplankton, algae, and detritus. *Osteobrama belangeri* thrives in subtropical waters with temperatures of 18°-28°C and slightly acidic to neutral pH levels (6.5-7.5). It is a commercially important species in Manipur, consumed as a food fish and in regional cuisines.

Pseudosphromenus cupanus (*Pseudosphromenus cupanus*): Commonly known as the Spotted Paradise Fish, this species is native to the freshwater bodies of southern India and Sri Lanka. It is an omnivore, feeding on small invertebrates, plant matter, and detritus. *Pseudosphromenus cupanus* thrives in tropical waters with temperatures between 24°-28°C and a pH of 6.5-7.5. It is sometimes kept in aquariums for its unique appearance and hardy nature.

***Puntius denisonii* (*Sahyadria denisonii*)**, also known as the Red Line Torpedo Barb): Native to fast-flowing rivers and streams of the Western Ghats in India, this species is known for its striking red line along the body. It is omnivorous, feeding on small invertebrates, algae, and detritus. It thrives in cool, well-oxygenated waters with temperatures between 22°-26°C and a pH of 6.5-7.5. *Puntius denisonii* is a popular ornamental species, though its population in the wild has declined due to overcollection for the aquarium trade.

Freshwater shell fish species

Apple Snail (*Pomacea canaliculata*): The Apple Snail, *Pomacea canaliculata*, from the family Ampullariidae, is a freshwater gastropod native to South America but now widespread in tropical and subtropical regions due to its invasive nature. It thrives in slow-moving freshwater habitats, such as ponds, lakes, and rice paddies, with optimal water temperatures between 25°C and 30°C and a pH range of 6.5 to 8.0. Apple Snails are voracious herbivores, feeding on aquatic plants and algae, and are known to cause significant damage to crops like rice. Despite their pest status, they are also cultivated in some regions for food, especially in Southeast Asia. Managing water quality is crucial to preventing the spread of this invasive species in aquaculture systems.

Freshwater Crab (*Potamon potamios*): *Potamon potamios*, known as the Freshwater Crab, belongs to the family Potamidae and is native to freshwater habitats across the Middle East and the Mediterranean region. This species inhabits rivers, streams, and lakes with clear, oxygen-rich water. Optimal conditions for *P. potamios* include water temperatures of 20°C to 25°C, pH levels between 7.0 and 8.0, and dissolved oxygen levels above 6 mg/L. Freshwater crabs are omnivores, feeding on detritus, small invertebrates, and plant matter. Although not extensively farmed, they are sometimes collected for food and are of ecological importance as scavengers in their native habitats.

Freshwater Mussel (*Lampsilis siliquoidea*): The Fatmucket Mussel, *Lampsilis siliquoidea*, belongs to the family Unionidae and is a freshwater bivalve native to North America. It is typically found in rivers, lakes, and streams with moderate to fast flow, where it burrows in sandy or gravelly substrates. This species thrives in water temperatures ranging from 10°C to 25°C, with a pH of 6.5 to 8.0 and dissolved oxygen levels above 5 mg/L. *L. siliquoidea* is a filter feeder, consuming plankton and detritus suspended in the water column. It plays an important ecological role in water filtration and nutrient cycling. While not widely farmed, freshwater mussels like *L. siliquoidea* are vital for maintaining healthy freshwater ecosystems and are used in pearl production.

Freshwater Prawn (*Macrobrachium malcolmsonii*): *Macrobrachium malcolmsonii*, commonly known as the Indian River Prawn, belongs to the family Palaemonidae. Native to freshwater systems across South Asia, particularly in the rivers and estuaries of India, Bangladesh, and Myanmar, this species thrives in tropical freshwater environments. Optimal water temperatures for *M. malcolmsonii* range from 26°C to 32°C, with a pH level between 7.0 and 8.5. It requires dissolved oxygen levels above 5 mg/L for optimal growth, and though primarily a freshwater species, it can tolerate brackish water during early larval stages, with salinity levels below 5 ppt. *M. malcolmsonii* is an omnivorous species, feeding on detritus, algae, small invertebrates, and organic matter. It is a significant species for aquaculture in South Asia due to its fast growth and high market value, reaching a harvestable size of 100-150 grams within 6 to 8 months. It is typically cultured in ponds or river systems and is a valuable source of income for rural communities in its native range.

Freshwater Prawn (*Macrobrachium rosenbergii*): *Macrobrachium rosenbergii*, commonly known as the Giant Freshwater Prawn, belongs to the family Palaemonidae and is one of the largest species of freshwater prawns. Native to South and Southeast Asia, this species thrives in tropical and subtropical freshwater environments, with optimal water temperatures ranging from 28°C to 31°C and a pH of 7.0 to 8.5. The ideal dissolved oxygen levels are above 5 mg/L, and salinity should remain under 5 ppt, as *M. rosenbergii* is primarily a freshwater species but can tolerate brackish conditions during larval stages. It is an omnivorous species, feeding on plant matter, small invertebrates, and detritus. Commercially farmed for its rapid growth and high market value, this prawn can reach harvest size within 6 to 8 months.

Zebra Mussel (*Dreissena polymorpha*): The Zebra Mussel, *Dreissena polymorpha*, belongs to the family Dreissenidae and is native to the freshwater lakes and rivers of Eastern Europe and Western Asia. However, it has become a highly invasive species in North America and other regions. This species thrives in a wide range of freshwater environments, from lakes to slow-moving rivers, with optimal water temperatures between 18°C and 25°C and a pH of 7.0 to 8.5. Zebra Mussels are filter feeders, consuming phytoplankton and organic particles from the water. Although considered a nuisance species due to their propensity to clog water intake pipes and outcompete native mussels, they also significantly impact water clarity and nutrient cycling in their invaded ecosystems. Control measures in aquaculture focus on preventing their spread.

CHAPTER 19

BRACKISHWATER FIN FISH AND SHELL FISH SPECIES

Asian Moon Scallop (*Amusium pleuronectes*): The Asian Moon Scallop, *Amusium pleuronectes*, belongs to the family Pectinidae and is commonly found in the shallow coastal waters of Southeast Asia. This species inhabits sandy or muddy substrates in areas with salinities ranging from 25-35 ppt and water temperatures between 24°C and 28°C. The optimal pH range for culture is between 7.5 and 8.5. *A. pleuronectes* is a filter feeder, subsisting on phytoplankton and other suspended organic particles. Moon Scallops are highly prized for their tender, sweet flesh, making them a sought-after seafood item in both local and international markets. Aquaculture of this species typically involves bottom culture or suspended net methods, with scallops reaching market size (7-10 cm) within 12-18 months.

Asian Seabass (Barramundi) (*Lates calcarifer*): Asian Seabass, commonly known as Barramundi, is a bony fish (teleost) of significant commercial value in aquaculture. Belonging to the family Latidae, order Perciformes, this euryhaline species can thrive in a wide range of salinities (0-35 ppt), making it suitable for both brackish and marine water systems. Optimal water temperature for Barramundi ranges from 26°C to 32°C, with a pH of 7.5 to 8.5. The species is carnivorous, requiring a protein-rich diet, and typically reaches market size (500-1000g) within 6-12 months under ideal conditions. Asian Seabass are protandrous hermaphrodites, initially maturing as males and later transforming into females.

Banana Shrimp (*Penaeus merguiensis*): The Banana Shrimp, *Penaeus merguiensis*, also known as the White Banana Prawn, is a penaeid shrimp species found in tropical and subtropical coastal waters of the Indo-Pacific. *P. merguiensis* thrives in salinities ranging from 10-35 ppt and prefers water temperatures between 26°C and 32°C. The ideal pH range for culture is between 7.5 and 8.5. This species is omnivorous, feeding on plankton, detritus, and small invertebrates. Banana Shrimp have a fast growth rate, reaching market size (20-30g) within 4-6 months under intensive farming conditions. They are commercially important in wild capture fisheries and are also cultured in Southeast Asia for export, particularly to markets in Japan and Australia.

Banded Archerfish (*Toxotes jaculatrix*): The Banded Archerfish is known for its unique ability to shoot water jets to capture prey. Belonging to the family Toxotidae, this species thrives in brackish water habitats, often found in estuaries and mangrove swamps. Optimal water conditions for *Toxotes jaculatrix* include salinities ranging from 5-20 ppt, water temperatures between 25°C to 30°C, and a neutral to slightly alkaline pH (7.0-8.0). This species is primarily carnivorous, feeding on insects and small aquatic organisms, and is popular in the ornamental fish trade.

Barramundi Cod (*Cromileptes altivelis*): The Barramundi Cod, or Humpback Grouper, is a species of high commercial value, particularly in Southeast Asian mariculture. It belongs to the family Serranidae, subfamily Epinephelinae. This fish thrives in brackish water with salinities between 15-30 ppt and requires water temperatures of 25°C to 30°C for optimal growth. The pH should be maintained between 7.5 and 8.5. *Cromileptes altivelis* is a carnivorous species, feeding on small fish and crustaceans. It grows slowly, requiring up to 2-3 years to reach marketable size (1-2 kg), but commands a high price in the live fish trade due to its unique appearance and quality meat.

Black Tiger Shrimp (*Penaeus monodon*): The Black Tiger Shrimp is a significant species in global aquaculture, especially in brackish water systems. Belonging to the family Penaeidae, this shrimp is known for its large size and striking black stripes. Optimal culture conditions include salinity levels of 15-25 ppt, water temperatures between 28°C and 30°C, and a pH range of 7.5 to 8.5. Black Tiger Shrimp are omnivorous, feeding on a mixture of plant and animal material, and reach market size (30-50g) within 4-6 months under favorable conditions. However, they are susceptible to viral diseases like White Spot Syndrome, necessitating stringent biosecurity practices.

Blood Cockle (*Anadara granosa*): The Blood Cockle, *Anadara granosa*, is a bivalve mollusk of the family Arcidae, found in brackish and estuarine waters, particularly in Southeast Asia. This species thrives in low-salinity environments, typically between 5-20 ppt, and prefers water temperatures of 24°C to 30°C. The ideal pH for its culture ranges from 7.0 to 8.5. *A. granosa* is a deposit feeder, consuming organic matter and detritus from the sediment. Blood Cockles are named for their red-colored hemolymph, which contains hemoglobin. They are an important aquaculture species in many Asian countries, valued for their nutritious, iron-rich meat. Blood Cockles are typically harvested from natural beds, but aquaculture practices are increasingly adopted to meet rising demand. Market size (2.5-4 cm) is generally reached within 6-12 months, depending on environmental conditions.

Blue Swimming Crab (*Portunus pelagicus*): The Blue Swimming Crab, *Portunus pelagicus*, is a species of crab from the family Portunidae, commonly found in shallow coastal waters, estuaries, and brackish lagoons across the Indo-Pacific region. This species thrives in salinities ranging from 25-35 ppt and prefers water temperatures between 24°C and 30°C. The optimal pH for Blue Swimming Crab culture is between 7.5 and 8.5. *P. pelagicus* is omnivorous, feeding on small fish, mollusks, algae, and detritus. Known for its striking blue color and sweet-tasting meat, this crab is a highly sought-after seafood product, particularly in Southeast Asia. Blue Swimming Crabs are harvested from the wild and farmed in aquaculture systems, with a growing emphasis on sustainable management to prevent overexploitation. Market size (100-150 g) is typically reached within 6-8 months in culture settings.

Brackish Water Catfish (*Plotosus canius, Arius spp.*): Brackish water catfish, such as *Plotosus canius* and *Arius spp.*, are members of the families Plotosidae and Ariidae, respectively, and are well-adapted to estuarine and coastal environments. These species tolerate a broad range of salinities (5-30 ppt) and prefer water temperatures between 26°C and 32°C. The optimal pH range for their culture is slightly acidic to neutral, between 6.5 and 7.5. Brackish water catfish are bottom-dwelling carnivores, feeding on small fish, crustaceans, and detritus. They are popular in small-scale aquaculture due to their hardiness and ability to thrive in low-oxygen conditions. Growth rates vary, but they can reach market size (500g-1kg) within 12-18 months under proper conditions.

Brackish Water Snail (*Cerithidea obtusa*): The Brackish Water Snail, *Cerithidea obtusa*, is a gastropod mollusk found in mangroves, mudflats, and brackish water environments across Southeast Asia and the Indo-Pacific. This species belongs to the family Potamididae and thrives in salinities ranging from 10-30 ppt. The optimal temperature for its culture is between 24°C and 30°C, with a pH range of 7.0 to 8.5. *Cerithidea obtusa* is a detritivore, feeding on organic matter and algae found in the sediment. Although not commonly farmed, this snail plays an important role in nutrient cycling within mangrove ecosystems. In some regions, it is harvested for food or used as bait in fisheries.

Dusky Grouper (*Epinephelus marginatus*): The Dusky Grouper, belonging to the family Serranidae, is a valuable species for both commercial fishing and aquaculture, particularly in the Mediterranean and Atlantic coasts. This carnivorous fish thrives in brackish and marine environments with salinities ranging from 15-35 ppt. Optimal water temperatures for culture are between 22°C and 28°C, with a pH range of 7.5 to 8.5. Dusky Grouper is a slow-growing species, taking up to 3-5 years to reach market size (3-5 kg), but is highly prized for its firm flesh and excellent taste. The species primarily feeds on fish and crustaceans, making it a high-value candidate for intensive aquaculture operations.

Estuarine Stingray (*Dasyatis fluviorum*): The Estuarine Stingray, *Dasyatis fluviorum*, is a brackish water species belonging to the family Dasyatidae, found along the coasts and estuaries of Australia. This species is well adapted to a range of salinities, from 5-30 ppt, and thrives in water temperatures between 22°C and 28°C. The optimal pH for their culture is between 7.0 and 8.5. *Dasyatis fluviorum* is a benthic feeder, preying on crustaceans, mollusks, and small fish. It is of interest for small-scale aquaculture and research, though it is more commonly encountered in local fisheries. The Estuarine Stingray can grow up to 1.5 meters in diameter, making it a large and valuable species for certain markets, though its slow growth and specialized diet present challenges for aquaculture.

Fiddler Crab (*Uca spp.*): Fiddler Crabs, belonging to the genus *Uca*, are small crabs found in brackish and estuarine environments. These crabs are known for the extreme size difference between their two claws, with males possessing one large claw used for communication and defense. *Uca spp.* thrive in salinities ranging from 10 to 30 ppt and prefer water temperatures between 24°C and 30°C. The pH range for their habitat is typically between 7.0 and 8.5. Fiddler Crabs are detritivores, feeding on organic matter and algae in the mudflats. Although not a major aquaculture species, they play an important ecological role in sediment turnover and are sometimes cultured for the ornamental trade or as bait for fishing.

Flathead Grey Mullet (*Mugil cephalus*): The Flathead Grey Mullet, a widely distributed species in coastal and estuarine areas, is a member of the family Mugilidae. This euryhaline species can tolerate a wide range of salinities, from fresh to full-strength seawater (0-35 ppt). Optimal water temperatures for mullet culture are between 24°C and 28°C, with a pH of 7.0 to 8.5. Flathead Grey Mullet are omnivorous, feeding primarily on detritus, algae, and small invertebrates. They are commonly used in polyculture systems due to their ability to utilize natural food sources efficiently. Mullets can reach market size (500g-1kg) in 12-18 months under ideal farming conditions.

Fourfinger Threadfin (*Eleutheronema tetradactylum*): The Fourfinger Threadfin is a species found in coastal and estuarine waters across the Indo-West Pacific. Belonging to the family Polynemidae, this species thrives in brackish water environments with salinities ranging from 10-30 ppt and water temperatures between 25°C and 30°C. The optimal pH for their culture is between 7.0 and 8.5. Fourfinger Threadfin are carnivorous, feeding on small fish, crustaceans, and polychaetes. Due to their excellent meat quality and relatively fast growth, they have potential for commercial aquaculture, reaching market size (500g-1kg) within 12 months.

Ghost Shrimp (*Trypaea australiensis*): Ghost Shrimp, *Trypaea australiensis*, also known as Yabbies, are burrowing crustaceans found in sandy or muddy coastal waters and estuaries. Belonging to the family Callianassidae, these shrimp are well adapted to brackish water environments with salinities ranging from 10-30 ppt. The optimal temperature for Ghost Shrimp culture is between 20°C and 28°C, with a pH range of 7.0 to 8.5. *Trypaea australiensis* are filter feeders, consuming organic particles and plankton from the sediment. Although not widely farmed for food, Ghost Shrimp are popular as bait in recreational fishing and are sometimes used in laboratory studies due to their burrowing behavior and ecological importance in sediment dynamics.

Giant African Snail (*Achatina achatina*): The Giant African Snail, *Achatina achatina*, originally from the family Achatinidae, is native to West Africa but has become an invasive species in many tropical and subtropical regions worldwide. While typically associated with terrestrial habitats, some populations have adapted to brackish environments in coastal areas. In these brackish habitats, *A. achatina* can tolerate salinities up to 10 ppt and temperatures between 24°C and 30°C. The optimal pH for its survival is between 6.5 and 7.5. These snails are herbivorous, feeding on a wide range of plants, algae, and decaying organic matter. Although *A. achatina* is considered a pest in many regions due to its voracious appetite for crops, it is also cultivated for its meat, which is a source of protein in some African communities. However, its invasive nature requires careful management to prevent further ecological damage.

Giant Freshwater Prawn (*Macrobrachium rosenbergii*): The Giant Freshwater Prawn, *Macrobrachium rosenbergii*, is a large species of prawn belonging to the family Palaemonidae. It is primarily cultured in freshwater but can tolerate brackish water environments with salinities up to 10-15 ppt. Optimal water temperatures for culture range from 28°C to 31°C, and the ideal pH is between 7.0 and 8.5. *M. rosenbergii* is an omnivorous species, feeding on detritus, algae, and small aquatic organisms. Under optimal conditions, the Giant Freshwater Prawn can reach market size (100-150g) within 6-8 months. This species is highly valued for its large size and high market demand, particularly in Southeast Asia and South Asia, where it is farmed extensively. However, successful culture requires careful management of water quality and stocking densities.

Giant Mottled Eel (*Anguilla marmorata*): The Giant Mottled Eel is a species of eel belonging to the family Anguillidae and is found in tropical and subtropical brackish waters across the Indo-Pacific region. This catadromous species migrates from freshwater rivers to the ocean for spawning. *Anguilla marmorata* can tolerate a wide range of salinities (0-35 ppt) and prefers water temperatures between 24°C and 30°C. The pH range for optimal growth is 7.0 to 8.5. Giant Mottled Eel is carnivorous, feeding on fish, crustaceans, and mollusks. It is a slow-growing species, taking 2-3 years to reach marketable size (1-2 kg). Eel farming is popular in East Asia, particularly in Japan, where eel is considered a delicacy.

Green Mussel (Perna viridis**):** The Green Mussel, *Perna viridis*, is a large bivalve mollusk from the family Mytilidae, native to the Indo-Pacific region but now widely distributed in tropical and subtropical waters. This species thrives in brackish and marine environments with salinities ranging from 18-33 ppt and prefers water temperatures between 26°C and 30°C. The optimal pH for Green Mussel culture is between 7.5 and 8.5. *Perna viridis* is a filter feeder, subsisting primarily on phytoplankton and suspended organic matter. Green Mussels are extensively farmed in Southeast Asia due to their fast growth, reaching market size (8-12 cm) within 6-9 months. They are valued for their high protein content and are consumed as a popular seafood delicacy. Additionally, Green Mussels play a crucial role in biofiltration and can help improve water quality in aquaculture systems.

Green Tiger Shrimp (*Penaeus semisulcatus*): The Green Tiger Shrimp, *Penaeus semisulcatus*, is another commercially important species in the Indo-Pacific region, belonging to the family Penaeidae. This species inhabits brackish and marine waters with salinities ranging from 15-35 ppt and prefers water temperatures between 26°C and 32°C. The ideal pH for Green Tiger Shrimp culture is between 7.5 and 8.5. *P. semisulcatus* is an omnivorous species, feeding on detritus, plankton, and small aquatic organisms. It is known for its large size and is often farmed alongside other penaeid shrimp species. Green Tiger Shrimp can reach market size (25-30g) within 4-6 months. The species is highly valued in both domestic and export markets, particularly in Southeast Asia.

Halfbeak (*Hemiramphus brasiliensis*): The Halfbeak, *Hemiramphus brasiliensis*, is a species found in tropical and subtropical coastal waters, including estuaries and brackish lagoons. Belonging to the family Hemiramphidae, this species can tolerate a range of salinities from 10 to 30 ppt. Optimal water temperatures for Halfbeak culture range from 24°C to 28°C, and the ideal pH range is between 7.0 and 8.0. Halfbeaks are omnivorous, feeding on algae, plankton, and small invertebrates. They are relatively fast-growing, reaching a market size of 100-200g within 6-8 months. While not a major aquaculture species, Halfbeaks are important in local fisheries and are also kept as ornamental fish.

Hilsa Shad (*Tenualosa ilisha*); Hilsa Shad, a member of the family Clupeidae, is a highly prized fish in South Asia, particularly in India and Bangladesh. This anadromous species migrates from the sea to brackish and freshwater rivers for spawning. Hilsa prefer salinities between 10-25 ppt during their juvenile stages in estuaries and water temperatures between 24°C and 30°C. They are primarily plankton feeders, consuming phytoplankton and zooplankton. Due to their complex life cycle, Hilsa aquaculture is still in the experimental stages, although significant research is ongoing. Hilsa is highly sought after for its rich flavor and nutritional value, making it a key species in regional fisheries.

Horseshoe Crab (*Tachypleus gigas, Carcinoscorpius rotundicauda*): Horseshoe Crabs, including *Tachypleus gigas* and *Carcinoscorpius rotundicauda*, are ancient marine arthropods belonging to the class Merostomata. These species are found in coastal and estuarine waters of the Indo-Pacific, thriving in brackish environments with salinities ranging from 15-30 ppt and water temperatures between 24°C and 30°C. The optimal pH for their habitat is between 7.5 and 8.5. Horseshoe Crabs are bottom-dwellers that feed on worms, mollusks, and other small invertebrates. While not a major aquaculture species, they are of significant ecological importance due to their role in coastal ecosystems and their use in biomedical research. Horseshoe Crab blood contains a unique substance called Limulus Amebocyte Lysate (LAL), which is used in testing for bacterial endotoxins in medical applications.

Indian Swamp Eel (*Monopterus cuchia*): The Indian Swamp Eel, *Monopterus cuchia*, belongs to the family Synbranchidae and is native to South Asia. This air-breathing species is well adapted to stagnant, low-oxygen environments, including swamps, rice paddies, and brackish waters. It thrives in salinities up to 10-15 ppt and prefers water temperatures between 24°C and 30°C, with a pH range of 6.5 to 8.0. *Monopterus cuchia* is a carnivorous species, feeding on small fish, insects, and other invertebrates. It is highly resilient, capable of surviving in dry conditions by burrowing into mud. Though its growth rate is slow, it can reach market size (200-500g) within 12-24 months. The species is valued in local markets for its medicinal properties and high nutritional content, particularly in traditional Asian medicine.

Indian Threadfin (*Polynemus indicus*): The Indian Threadfin is a commercially important species found in coastal and brackish water environments across the Indo-Pacific region. Belonging to the family Polynemidae, this fish thrives in water with salinities ranging from 10-25 ppt and temperatures between 26°C to 32°C. The pH should be maintained at 7.0 to 8.5. Indian Threadfin are typically omnivorous, feeding on small fish, crustaceans, and plankton. With fast growth rates, they can reach market size (around 500g) within 6-9 months under optimal conditions. This species is particularly valued for its firm and tasty flesh.

Kuruma Shrimp (*Marsupenaeus japonicus*): The Kuruma Shrimp, *Marsupenaeus japonicus*, is a valuable shrimp species in aquaculture, especially in Japan and other parts of East Asia. Belonging to the family Penaeidae, this species thrives in brackish and marine waters with salinities ranging from 15-35 ppt and water temperatures between 26°C and 30°C. The optimal pH range for culture is slightly alkaline, between 7.5 and 8.5. *M. japonicus* is omnivorous, feeding on plankton, detritus, and small organisms. The species is known for its high market value due to its firm texture and delicate flavor. Kuruma Shrimp can reach market size (30-40g) within 4-6 months under optimal conditions. However, their culture is challenging due to their susceptibility to diseases such as White Spot Syndrome Virus (WSSV).

Ladyfish (*Elops saurus*): Ladyfish, also known as Tenpounders, belong to the family Elopidae and are typically found in coastal and brackish water environments, particularly in the Western Atlantic and Gulf of Mexico. *Elops saurus* thrives in salinities ranging from 5-30 ppt and prefers water temperatures between 24°C and 30°C. The species can tolerate a pH range of 7.0 to 8.5. Ladyfish are carnivorous, feeding on smaller fish and crustaceans, and are known for their fast swimming and active behavior. Although not a major target for aquaculture, they are of interest in sport fishing and are occasionally used in stocking programs. Growth to market size (1-2 kg) can take 12-18 months.

Mangrove Crab (*Neosarmatium meinerti*): The Mangrove Crab, *Neosarmatium meinerti*, is a species of sesarmid crab commonly found in mangrove forests along the coasts of the Indo-Pacific region. This species thrives in salinities ranging from 15-30 ppt and prefers water temperatures between 24°C and 30°C. The pH range for optimal growth is between 7.5 and 8.5. *Neosarmatium meinerti* is primarily herbivorous, feeding on mangrove leaves and detritus, making it an important species in the recycling of nutrients within mangrove ecosystems. While not a major aquaculture species, Mangrove Crabs are of ecological significance and are sometimes harvested locally for food.

Mangrove Oyster (*Crassostrea spp.*): Mangrove Oysters, belonging to the genus *Crassostrea*, are a group of bivalve mollusks typically found in brackish water environments, particularly in the mangroves and estuaries of the Indo-Pacific region. These oysters thrive in salinities ranging from 10-30 ppt and prefer water temperatures between 24°C and 30°C. The optimal pH range for Mangrove Oyster culture is between 7.5 and 8.5. Like other oysters, *Crassostrea spp.* are filter feeders, consuming plankton and organic particles from the water. Mangrove Oysters are highly valued for their economic importance, providing both food and livelihoods in coastal communities. They are cultivated using various methods, including rack-and-bag systems, and can reach market size (5-10 cm) within 6-12 months. Additionally, they contribute to ecosystem health by filtering large volumes of water and enhancing water clarity.

Mangrove Whipray (*Urogymnus granulatus*): The Mangrove Whipray, a member of the family Dasyatidae, is typically found in shallow coastal waters and mangroves across the Indo-Pacific region. *Urogymnus granulatus* is a euryhaline species, tolerating salinities from 10-35 ppt, and prefers water temperatures between 24°C and 30°C. The species can tolerate a pH range of 7.0 to 8.5. Mangrove Whipray primarily feeds on benthic invertebrates, including crustaceans and mollusks. Though not widely cultured, this species is occasionally targeted in artisanal fisheries for its meat and cartilage. Due to its large size, reaching up to 1.8 meters in diameter, the Mangrove Whipray is also of interest to ecotourism and conservation efforts.

Milkfish (*Chanos chanos*): Milkfish, the only living species of the family Chanidae, is a key species in brackish water aquaculture across Southeast Asia. This species is highly adaptable to different salinity levels, thriving in both coastal and inland brackish water ponds. Milkfish grow best at water temperatures between 26°C and 30°C and require dissolved oxygen levels above 5 mg/L for optimal growth. Their diet consists primarily of phytoplankton and detritus, making them an efficient species for integrated farming systems. Milkfish can reach market size (500g-1kg) in about 6-12 months, depending on water quality and feeding practices.

Mud Clam (*Polymesoda erosa*): The Mud Clam, *Polymesoda erosa*, is a bivalve mollusk found in brackish waters and estuarine environments across the Indo-Pacific region. This species belongs to the family Cyrenidae and thrives in salinities ranging from 5-20 ppt. The optimal temperature for Mud Clam culture is between 24°C and 30°C, with a pH range of 7.0 to 8.5. *P. erosa* is a filter feeder, consuming plankton and detritus from the water. Mud Clams are of commercial importance in local fisheries and are often harvested for food, particularly in Southeast Asia. They are known for their sweet, tender meat and are also used in traditional medicine.

Mud Crab (*Scylla serrata*): The Mud Crab, *Scylla serrata*, also known as the Mangrove Crab, is a large crustacean from the family Portunidae, commonly found in mangrove forests, estuaries, and brackish water environments in the Indo-Pacific region. This species thrives in salinities ranging from 10-35 ppt and prefers water temperatures between 25°C and 30°C. The optimal pH for its culture is between 7.0 and 8.5. *S. serrata* is an omnivorous scavenger, feeding on small fish, mollusks, plant material, and detritus. Mud Crabs are highly prized for their sweet, succulent meat, making them a valuable commodity in seafood markets. They are typically farmed in extensive or semi-intensive systems, including mangrove pens or brackish water ponds, with market size (500-1,000 g) typically reached in 6-12 months. However, the species is vulnerable to overfishing and habitat loss, making sustainable aquaculture practices essential for its long-term viability.

Mud Lobster (*Thalassina anomala*): The Mud Lobster, *Thalassina anomala*, is a burrowing crustacean belonging to the family Thalassinidae. This species is found in mangroves and mudflats along the coasts of Southeast Asia and the Indo-Pacific region. Mud Lobsters thrive in brackish water environments with salinities ranging from 10-25 ppt and prefer water temperatures between 24°C and 30°C. The optimal pH for their habitat is between 7.0 and 8.5. *T. anomala* is an omnivorous detritivore, feeding on organic matter in the mud. Although not commonly farmed, Mud Lobsters are of ecological significance due to their burrowing activity, which helps aerate the soil and improve nutrient cycling in mangrove ecosystems. In some regions, they are harvested for food, although their meat is considered a delicacy in limited markets.

Mullet (*Mugil cephalus*, Liza spp.): Mullets, including *Mugil cephalus* and species of the genus *Liza*, are highly versatile fish found in brackish water systems such as estuaries, lagoons, and coastal ponds. They belong to the family Mugilidae and are euryhaline, tolerating a wide range of salinities (0-35 ppt).

Mullets are typically detritivorous, feeding on algae, organic detritus, and small invertebrates. Optimal water temperatures for mullet culture range from 24°C to 28°C, with pH levels between 7.0 and 8.5. They are used in polyculture systems due to their ability to thrive on natural food sources and can reach market size (around 500g) within 12-18 months.

Pearl Spot (*Etroplus suratensis*): The Pearl Spot, also known as the Green Chromide, is a brackish water fish native to India and Sri Lanka. Belonging to the family Cichlidae, *Etroplus suratensis* is a hardy species that thrives in salinities ranging from 5-30 ppt, making it highly adaptable to estuarine and coastal environments. Optimal water temperatures for culture are between 26°C and 30°C, and the ideal pH range is slightly alkaline, from 7.0 to 8.5. Pearl Spot is an omnivorous species, feeding on algae, detritus, and small invertebrates, and is often used in integrated farming systems. Growth to market size (200-500g) typically takes 8-12 months under favorable conditions. The species is valued for its delicate taste and is a significant component of local aquaculture in India.

Pomfret (*Pampus argenteus*): Pomfret, particularly the Silver Pomfret (*Pampus argenteus*), is a valued species in brackish water aquaculture due to its delicate flavor and high market demand. This species belongs to the family Stromateidae and is typically found in coastal waters and estuaries. Silver Pomfret thrive in salinities ranging from 20-30 ppt, with optimal water temperatures between 25°C to 29°C. The species prefers slightly alkaline water with a pH range of 7.5 to 8.5. *Pampus argenteus* is omnivorous, feeding on plankton, small invertebrates, and algae, and requires around 9-12 months to reach a marketable size of 300-500g.

Scat (*Scatophagus argus*): Scat, particularly the Spotted Scat (*Scatophagus argus*), is a popular brackish water species in both aquaculture and the aquarium trade. Belonging to the family Scatophagidae, this species thrives in environments with salinities ranging from 5-25 ppt and water temperatures between 24°C and 30°C. Scats are omnivorous, with a diet consisting of algae, detritus, and small invertebrates, making them useful in polyculture systems for controlling algae and maintaining water quality. The species prefers slightly alkaline water with a pH range of 7.5 to 8.5. Scats are relatively slow-growing, reaching market size (200-300g) within 12-24 months.

Sheepshead (*Archosargus probatocephalus*): The Sheepshead, a member of the family Sparidae, is found in coastal and brackish waters of the western Atlantic and Gulf of Mexico. This species is highly valued for its mild, sweet-tasting flesh and is recognized for its distinctive black and white vertical stripes. Sheepshead thrive in salinities between 10-30 ppt and prefer water temperatures ranging from 20°C to 28°C. The optimal pH for their culture is between 7.5 and 8.5. Sheepshead are omnivorous, feeding on crustaceans, mollusks, and small fish. They are known for their strong, protruding teeth, which are used to crush shellfish. Growth to market size (500-800g) typically takes around 12-18 months under aquaculture conditions.

Silver Perch (*Bidyanus bidyanus*): Silver Perch, a native Australian species, belongs to the family Terapontidae and is increasingly being cultured in both freshwater and brackish water systems. This species tolerates a wide range of salinities, up to 10-15 ppt, and prefers water temperatures between 20°C and 28°C for optimal growth. The ideal pH for Silver Perch culture is slightly alkaline, ranging from 7.0 to 8.0. Silver Perch are omnivorous, feeding on aquatic plants, detritus, and small invertebrates. Under optimal conditions, they can reach marketable size (500g-1kg) within 12-18 months. The species is also valued for its adaptability to intensive farming systems and resilience to environmental stress.

Snakehead (*Channa striata*): Snakehead, specifically *Channa striata*, is a popular species in brackish and freshwater aquaculture across Southeast Asia. Belonging to the family Channidae, this air-breathing fish is highly resilient, capable of surviving in low-oxygen conditions and tolerating a wide range of salinities (up to 15 ppt). Optimal water temperatures for Snakehead culture are between 26°C and 32°C, with a pH range of 6.5 to 8.0. *Channa striata* is a carnivorous species, feeding on fish, frogs, and crustaceans. It grows rapidly, reaching marketable size (500g-1kg) within 6-12 months. Snakehead is highly valued for its nutritional benefits, particularly for its high protein content and medicinal properties.

Snapper (*Lutjanus argentimaculatus*): The Mangrove Red Snapper, *Lutjanus argentimaculatus*, is a popular species in brackish and marine aquaculture, belonging to the family Lutjanidae. This carnivorous fish thrives in salinities between 15-30 ppt and water temperatures of 26°C to 30°C, with a pH range of 7.5 to 8.5. The species is known for its high growth rate, reaching market size (600g-1kg) within 12-18 months under favorable conditions. Mangrove Red Snapper feeds on small fish, crustaceans, and mollusks, making it a high-value species for both food and sport fishing. The fish is also known for its resilience to environmental changes, making it suitable for intensive farming.

Spotted Scat (*Scatophagus argus*): Spotted Scat, commonly found in estuaries and mangroves, belongs to the family Scatophagidae. This euryhaline species can tolerate a wide range of salinities from 5 to 25 ppt and prefers water temperatures between 24°C and 30°C. The pH range for optimal growth is between 7.5 and 8.5. *Scatophagus argus* is omnivorous, feeding on algae, detritus, and small invertebrates, which makes it suitable for polyculture systems where it can help control algae and improve water quality. The fish is slow-growing, reaching market size (200-300g) within 12-24 months. Spotted Scat is popular in both the ornamental fish trade and as a food fish, especially in Southeast Asia.

Tilapia (*Oreochromis mossambicus, Oreochromis niloticus*): Tilapia, specifically *Oreochromis mossambicus* and *Oreochromis niloticus*, are among the most widely cultured species in brackish and freshwater aquaculture. These species belong to the family Cichlidae and are highly adaptable to various environmental conditions, tolerating salinities up to 25 ppt, although they grow best in salinities between 10-15 ppt. The optimal water temperature for tilapia culture ranges from 26°C to 30°C, with a pH range of 6.5 to 8.5. These herbivorous and omnivorous fish thrive on a diet of algae, plankton, and formulated feeds. Nile Tilapia (*O. niloticus*) is preferred in commercial farming due to its faster growth rates and resistance to diseases, reaching market size within 6-8 months.

Whiteleg Shrimp (*Penaeus vannamei*): The Whiteleg Shrimp, *Penaeus vannamei*, also known as Pacific White Shrimp, is a prominent species in global aquaculture due to its fast growth rate, adaptability to various environmental conditions, and disease resistance. Native to the Pacific coast of Central and South America, this shrimp species is now farmed extensively in tropical and subtropical regions worldwide. *L. vannamei* thrives in brackish waters with salinities ranging from 10-40 ppt and water temperatures between 26°C and 30°C. The optimal pH for Whiteleg Shrimp culture is between 7.5 and 8.5. They are omnivorous, feeding on algae, small invertebrates, and detritus. Whiteleg Shrimp grow rapidly, reaching market size (15-25 g) in 3-4 months. Due to its high productivity, low feed conversion ratio, and relatively low susceptibility to diseases compared to other shrimp species, *P. vannamei* dominates global shrimp aquaculture, particularly in Asia and Latin America.

CHAPTER 20

MARINE AQUACULTURE SPECIES

Finfish

Atlantic Salmon (Salmo salar): Atlantic salmon is one of the most significant species in marine aquaculture due to its high market demand. Native to the North Atlantic Ocean, this anadromous species can thrive in both freshwater and marine environments during different life stages. Optimum rearing conditions in aquaculture include water temperatures of 10-15°C, dissolved oxygen levels above 7 mg/L, and salinities between 30-35 ppt for marine phases. Atlantic salmon requires a protein-rich diet and efficient feed conversion, making it an economically valuable species. The species is known for its streamlined growth performance, reaching market size within 18 to 36 months.

Barramundi (Lates calcarifer): Also known as Asian seabass, Barramundi is a euryhaline species found in both marine and freshwater habitats across the Indo-Pacific region. It thrives in temperatures between 26°-30°C and salinities ranging from freshwater to full seawater (~35 ppt). The species has fast growth rates, typically reaching 400-600 g in 6-9 months. Barramundi are carnivorous, requiring diets high in protein and lipids. Their adaptability to various environmental conditions and their suitability for intensive farming systems make them a prominent species in marine aquaculture.

Cobia (Rachycentron canadum): Cobia, a fast-growing, carnivorous marine fish, is native to tropical and subtropical waters. It thrives in water temperatures of 25°-30°C and salinities of 25-35 ppt. Known for its robust growth, cobia can reach 4-6 kg within 12 months under optimal aquaculture conditions. Cobia is prized for its white, firm flesh, and it is typically farmed in sea cages or recirculating aquaculture systems (RAS). It requires a protein-rich diet and efficient feed management for economic viability. Its resistance to diseases makes it a promising candidate for intensive marine aquaculture.

Dusky Kob (Argyrosomus japonicus): Dusky kob, also known as mulloway, is a carnivorous, euryhaline species native to coastal waters of southern Africa and Australia. It thrives in water temperatures of 18-25°C and tolerates salinity ranges from 20 to 35 ppt. The species is known for its fast growth, reaching market size (1-2 kg) within 12-18 months under good farming conditions. Dusky kob are farmed in sea cages or land-based recirculating systems, and they require high-protein diets for optimal performance.

European Sea Bass (Dicentrarchus labrax): This carnivorous species is native to the Mediterranean Sea and the eastern Atlantic Ocean, favoring temperate waters between 13-22°C. European sea bass is commonly farmed in sea cages or land-based tanks and prefers a salinity range of 30-40 ppt. It requires high-protein feeds and optimal water quality, with dissolved oxygen levels above 5 mg/L for healthy growth. The species is known for its relatively slow growth, taking around 18-24 months to reach a market size of 300-400 g.

Gilthead Sea Bream (Sparus aurata): Native to the Mediterranean and the eastern Atlantic, the gilthead sea bream is a euryhaline, protandrous hermaphrodite species, meaning individuals typically change from male to female during their life cycle. This species thrives in water temperatures of 18-26°C and salinities between 30-38 ppt. Gilthead sea bream is one of the most farmed fish in the Mediterranean, owing to its adaptability, moderate growth rate, and resilience in aquaculture systems. It requires a diet rich in protein, often reaching a marketable size of 400 g within 12-18 months.

Groupers (Epinephelus spp.): Groupers are carnivorous, demersal fish commonly found in warm waters of the Indo-Pacific region. Groupers thrive in temperatures ranging from 22°-30°C and require salinities of 30-35 ppt. They are typically slow-growing species, taking 18-24 months to reach market size (800 g to 1.5 kg). In aquaculture, groupers require high-protein diets and are often farmed in sea cages or recirculating aquaculture systems (RAS).

Halibut (Hippoglossus hippoglossus): Atlantic halibut is a cold-water, bottom-dwelling species found in the North Atlantic. In aquaculture, it thrives at water temperatures of 5-12°C and salinities of 28-35 ppt. Halibut has a long growth cycle, typically taking 3-4 years to reach market size (4-6 kg). It requires well-oxygenated water and a carnivorous diet rich in fishmeal and fish oil. Halibut farming is challenging due to its slow growth, but it is highly valued in seafood markets due to its large size and high-quality white flesh.

Japanese Flounder (Paralichthys olivaceus): Native to the northwest Pacific, Japanese flounder is a bottom-dwelling species highly prized for its delicate flesh. It thrives in cold, temperate waters with temperatures between 14-18°C and salinity levels of 30-35 ppt. The species requires well-oxygenated environments (above 6 mg/L of dissolved oxygen) and takes about 18-24 months to reach market size (500 g to 1.5 kg). Japanese flounder is mainly cultured in land-based tanks or recirculating aquaculture systems due to its benthic nature.

Milkfish (Chanos chanos): Milkfish is a highly tolerant euryhaline species, farmed extensively in Southeast Asia. It can thrive in a wide range of salinities from freshwater to hyper-saline conditions (~50 ppt), and optimal growth occurs in temperatures between 26-32°C. Milkfish are herbivorous and feed primarily on algae and detritus, making them suitable for polyculture systems with lower feed costs. It takes about 5-8 months for milkfish to reach 500 g under good farming conditions.

Pacific Cod (Gadus macrocephalus): Native to the northern Pacific, this cold-water species is a commercially important finfish in both capture fisheries and aquaculture. It thrives in water temperatures between 5-12°C, requiring well-oxygenated water with salinities of 30-35 ppt. Pacific cod is a demersal fish that grows to market size (1-2 kg) in about 18-24 months. Its farming is usually done in sea cages or bottom enclosures, and it requires a protein-rich diet for optimal growth.

Red Snapper (Lutjanus campechanus): Native to the Gulf of Mexico and the western Atlantic, the red snapper is a popular species in aquaculture. It thrives in warm waters between 22-28°C and requires salinities of 30-35 ppt. Red snapper is a carnivorous species that feeds on smaller fish and crustaceans, and it requires a diet high in protein. Growth rates are moderate, with individuals reaching marketable sizes (1-2 kg) in about 18-24 months. It is typically farmed in sea cages or large enclosures in coastal areas.

Striped Bass (Morone saxatilis): Striped bass, native to North American coastal waters, is a highly adaptable species used in aquaculture. It thrives in temperate waters ranging from 18-24°C and salinities from freshwater to full seawater (~35 ppt). Striped bass are carnivorous and require high-protein diets to grow efficiently. In commercial aquaculture, they are typically raised in ponds or tanks and can reach market size (500 g to 1 kg) in about 12-18 months.

Turbot (Scophthalmus maximus): A cold-water, carnivorous flatfish native to European waters, the turbot is renowned for its firm, white flesh, making it highly valued in aquaculture. Optimal farming conditions include water temperatures between 14-18°C and salinity levels of 30-35 ppt. Turbot typically reach market size (1-2 kg) within 18-24 months. Due to its benthic nature, turbot requires specialized tank designs that mimic natural habitats for growth and welfare.

Yellowtail Kingfish (Seriola lalandi): Also known as the greater amberjack, this species is highly valued in aquaculture for its rapid growth and high-quality flesh. Native to temperate and subtropical waters, yellowtail kingfish thrives at water temperatures of 20-26°C and salinity levels between 30-36 ppt. It can reach a marketable size (3-5 kg) within 18-24 months. Kingfish are carnivorous and require high-protein diets, making their farming capital-intensive but commercially rewarding. It is typically farmed in sea cages, requiring excellent water quality management to maintain healthy stock.

Shellfish

Abalone (Haliotis spp.**)**: Abalone are marine gastropods prized for their tender meat and decorative shells. They thrive in cool, temperate waters with temperatures between 12-20°C and salinities of 30-35 ppt. Abalone farming is typically conducted in tanks or sea cages, where they are fed a diet rich in macroalgae like kelp. Abalone take 3-4 years to reach market size (70-120 mm shell length). Their high market value makes them a lucrative species for aquaculture, but they require careful water quality management and disease prevention measures.

Blue Mussel (Mytilus edulis**)**: Native to the North Atlantic, blue mussels are filter-feeding bivalves that thrive in cold, nutrient-rich waters with temperatures between 10-18°C and salinities of 25-35 ppt. Blue mussels are often farmed on ropes suspended in the water column, where they efficiently filter phytoplankton from the water. They reach market size (60-80 g) in 12-18 months and are prized for their high meat content and sustainability in aquaculture.

Clams (Venerupis philippinarum, Mercenaria mercenaria**)**: The Manila clam (*Venerupis philippinarum*) and the hard clam (*Mercenaria mercenaria*) are widely farmed bivalves in aquaculture. Clams thrive in estuarine environments with salinities between 20-35 ppt and water temperatures ranging from 15-25°C. These filter feeders are typically farmed in sandy or muddy substrates in coastal regions, where they consume phytoplankton and organic matter. Clams grow to market size (50-70 g) within 12-24 months. They are highly valued for their tender flesh and are commonly harvested using dredging or manual collection methods.

Cockles (Cerastoderma edule**)**: Native to European coastal waters, cockles are filter-feeding bivalves found in intertidal sandy or muddy substrates. They thrive in temperate regions with water temperatures between 10-18°C and salinity ranges from 15-35 ppt. Cockles are known for their high filtration capacity and are typically farmed in managed coastal beds or harvested from natural populations. Growth to market size (30-50 g) can take 12-18 months, and they are a popular seafood item in Europe, prized for their briny flavor.

European Flat Oyster (Ostrea edulis**)**: Native to European coastal waters, the European flat oyster is a slow-growing bivalve that prefers water temperatures between 12-18°C and salinities of 28-35 ppt. It is filter-feeding and relies on phytoplankton for sustenance. This species typically takes 3-4 years to reach a market size of 80-120 g. European flat oysters are prized for their distinct flavor and texture, making them valuable in the high-end seafood market.

Geoduck (Panopea generosa**)**: The geoduck is a large, burrowing clam native to the Pacific Northwest. It thrives in cold waters with temperatures of 10-14°C and salinities between 30-35 ppt. Geoducks are farmed in subtidal zones where they burrow deep into sandy or muddy substrates. Due to their slow growth, it takes 4-6 years to reach market size (1-2 kg). Geoducks are highly valued in Asian markets for their sweet, firm meat and are considered a delicacy, often sold at premium prices.

Green Mussel (Perna viridis**)**: Found in the Indo-Pacific region, green mussels thrive in warm waters with temperatures of 26°-30°C and salinities of 20-35 ppt. They are filter feeders and are typically farmed using longline or raft culture methods. Green mussels are fast-growing, reaching market size (70-100 g) in 6-12 months, making them a popular species in tropical aquaculture.

Pacific Oyster (Crassostrea gigas**)**: Native to the Pacific coast of Asia, the Pacific oyster is one of the most widely farmed oyster species worldwide. It thrives in estuarine environments with salinities ranging from 20-35 ppt and water temperatures between 10-20°C. Pacific oysters are filter feeders, requiring nutrient-rich waters with high plankton levels. They are usually farmed in suspended systems and can reach market size (70-100 g) in 18-24 months under optimal conditions.

Razor Clams (Ensis directus**)**: Razor clams, found in temperate coastal waters, are burrowing bivalves that thrive in sandy or muddy substrates. They prefer water temperatures between 10-18°C and salinities of 25-35 ppt. Razor clams are typically harvested from natural populations, though farming methods using subtidal bed management are being explored. They reach market size (80-100 g) within 18-24 months and are known for their sweet, delicate flavor, making them a popular seafood item.

Scallops (Pecten maximus, Argopecten irradians**)**: Scallops are filter-feeding bivalves that thrive in cold, nutrient-rich waters with temperatures between 10-15°C and salinities of 30-35 ppt. Pecten maximus, commonly known as the king scallop, and Argopecten irradians, the bay scallop, are highly prized for their sweet, delicate flesh. Scallops are typically farmed using suspended cages or on-bottom culture methods, where they filter plankton from the water. Growth to market size takes 18-24 months, and they are harvested when their shells reach around 10-12 cm in diameter. Scallop farming requires high water quality and well-oxygenated environments for optimal growth and survival.

Crustaceans

American Lobster (Homarus americanus**)**: The American lobster, found along the Atlantic coast of North America, is a cold-water species that thrives in temperatures between 5-15°C and salinities of 30-35 ppt. Similar to its European counterpart, the American lobster is slow-growing and can take up to 5-7 years to reach market size (500 g to 1 kg). Lobster aquaculture is challenging due to the species' complex life cycle and aggressive behavior, but experimental farming methods are being explored to supplement wild stocks.

Banded Coral Shrimp (Stenopus hispidus): A popular species in marine aquariums, the banded coral shrimp is a small, colorful crustacean native to tropical reef environments. It thrives in warm waters with temperatures of 24°-28°C and salinities of 30-35 ppt. Banded coral shrimp are omnivorous, feeding on detritus, algae, and small invertebrates. Though not farmed commercially for food, they are bred in captivity for the aquarium trade, where they serve as "cleaners" by removing parasites from fish.

Black Tiger Shrimp (Penaeus monodon): Black tiger shrimp is a tropical species native to the Indo-Pacific region, farmed extensively in Asia. It prefers water temperatures of 26°-30°C and salinities between 15-25 ppt. The species grows rapidly, with a production cycle of 3-6 months to reach a marketable size of 30-40 g. Black tiger shrimp are carnivorous, requiring high-protein diets, and are often cultured in semi-intensive and intensive farming systems.

Blue Crab (Callinectes sapidus): Native to the western Atlantic and the Gulf of Mexico, the blue crab is a commercially important species. It thrives in estuarine environments with salinities of 10-30 ppt and temperatures between 18-26°C. Blue crabs are omnivorous, feeding on small fish, invertebrates, and plant matter. They are typically farmed in coastal ponds or enclosures and can reach market size (100-150 g) in 12-18 months.

European Lobster (Homarus gammarus): The European lobster is a large, clawed crustacean found in the rocky coastal waters of the North Atlantic. It thrives in cool, temperate waters with temperatures between 10-18°C and salinities of 30-35 ppt. European lobsters are slow-growing, taking 4-5 years to reach market size (500 g to 1 kg). They are carnivorous, feeding on mollusks, fish, and other crustaceans, and require complex habitats with ample hiding spaces. Lobster farming, although limited, is conducted in sea cages or tanks, with careful management of water quality and diet.

Mud Crab (Scylla serrata): Mud crabs are found in tropical and subtropical regions of the Indo-Pacific and are valued for their large size and meat quality. They thrive in brackish waters with salinities of 15-25 ppt and temperatures of 25°-30°C. Mud crabs are carnivorous and require a protein-rich diet. They are often cultured in ponds or mangrove areas and can reach market size (500-1,000 g) in 6-12 months under optimal conditions.

Red King Crab (Paralithodes camtschaticus): Red king crab, native to the Bering Sea and North Pacific, is one of the most valuable crab species in the seafood industry. It thrives in cold waters with temperatures of 2-7°C and salinities of 30-35 ppt. Red king crabs are slow-growing and can take 4-5 years to reach market size (2-5 kg). They are typically farmed in cold-water sea cages or collected from the wild. Their high market value and large size make them a sought-after delicacy, particularly in Asia and the U.S.

Rock Lobster (Panulirus spp.): Rock lobsters, also known as spiny lobsters, are clawless crustaceans found in warm, tropical waters. They thrive in temperatures of 22-28°C and salinities of 30-35 ppt. Rock lobsters are typically farmed in sea cages or coastal ponds, where they feed on mollusks, fish, and plant matter. They take 2-3 years to reach market size (1-2 kg) and are highly valued for their sweet, firm meat. Farming rock lobsters is labor-intensive due to their specific dietary needs and slow growth.

Spiny Lobster (Panulirus argus): Spiny lobster is a tropical species native to the Caribbean and Western Atlantic regions. It thrives in warm waters with temperatures between 24°-28°C and salinities of 30-35 ppt. Spiny lobsters are omnivorous, feeding on a wide range of plant and animal matter. They are farmed in sea cages or raceway systems and can take 18-24 months to reach market size (500 g to 1 kg). Spiny lobsters are highly prized in seafood markets, particularly in Asia.

Seaweeds and Marine Plants

Caulerpa (Caulerpa lentillifera): Commonly known as sea grapes, Caulerpa lentillifera is a green algae species cultivated in tropical and subtropical waters, primarily in Southeast Asia. It thrives in warm, shallow coastal areas with temperatures between 25°-30°C. This seaweed is farmed in ponds or lagoons, where it grows rapidly under nutrient-rich conditions. Caulerpa is highly valued as a food source due to its high nutritional content, particularly its richness in vitamins, minerals, and omega-3 fatty acids. It is harvested within 4-6 weeks of cultivation and is consumed fresh in salads or as a garnish.

Eucheuma (Eucheuma spp.): Eucheuma is a red algae commonly farmed in tropical and subtropical waters for its carrageenan content, a gelatinous substance used in food processing. It grows best in warm waters with temperatures between 25°-30°C and is cultivated using longline or off-bottom farming methods. Eucheuma is harvested every 45-60 days, and its rapid growth makes it an important resource in the global seaweed industry, particularly in Southeast Asia.

Gracilaria (Gracilaria spp.): Gracilaria is a red algae commonly farmed in tropical and subtropical regions, primarily for its agar content. It thrives in warm waters between 25°-30°C and can grow in both marine and brackish environments. Gracilaria is often farmed using raft or longline methods, and it is harvested several times a year. It is used for agar production and as a raw material in food and cosmetic industries.

Irish Moss (Chondrus crispus): Also known as carrageen, Irish moss is a red algae species found in cold waters of the North Atlantic, thriving in temperatures between 10-18°C. It grows in intertidal zones on rocky substrates and is harvested for its carrageenan content, a polysaccharide used as a gelling agent in food, cosmetics, and pharmaceuticals. Irish moss is typically wild-harvested, but cultivation techniques, such as rope farming in coastal waters, are being developed. Harvesting occurs 60-90 days after planting, and its demand in the food industry is significant due to its natural thickening properties.

Kelp (Saccharina latissima, Laminaria japonica): Kelp is a fast-growing brown algae farmed in cold, nutrient-rich waters, typically below 15°C, with a salinity of 30-35 ppt. Kelp is grown in suspended systems, requiring little to no input after initial seeding. It is used in food, cosmetics, and pharmaceuticals due to its rich content of alginates, iodine, and other nutrients.

Nori (Porphyra spp.): Nori is a red algae extensively cultivated in the intertidal zones of East Asia, especially Japan and Korea. It grows best in nutrient-rich waters with temperatures ranging from 10-20°C. Nori farming involves nets suspended in the water, and the seaweed can be harvested multiple times a year. It is a vital component of sushi, known for its high nutritional value, particularly in vitamins, minerals, and essential fatty acids.

Sargassum (Sargassum spp.): Sargassum is a brown macroalga that grows in tropical and subtropical waters, often forming large floating mats. It thrives in nutrient-rich waters with temperatures between 20°-30°C and salinities of 30-35 ppt. Sargassum is farmed in coastal areas using off-bottom or floating line methods. It is used for its high content of alginates, which are extracted for use in food processing, pharmaceuticals, and industrial applications. Additionally, Sargassum is increasingly being studied for its potential in carbon sequestration and its role in combating coastal eutrophication. It is harvested approximately every 45-60 days.

Sea Lettuce (Ulva lactuca): Sea lettuce is a green algae that thrives in both marine and brackish waters, often found in nutrient-rich environments. It grows best in water temperatures between 15-25°C and salinities ranging from 20-35 ppt. Ulva is widely farmed for its nutritional value, containing high levels of protein, vitamins, and minerals. It is used as food for humans, in animal feed, and in bioremediation projects due to its ability to absorb excess nutrients from the water. Sea lettuce grows rapidly, with harvesting occurring every 30-60 days, depending on environmental conditions. Its potential as a sustainable food source is gaining attention in aquaculture.

Ulva (Ulva spp.): Commonly known as sea lettuce, Ulva is a green algae found in both marine and brackish waters. It thrives in nutrient-rich environments with temperatures between 15-25°C. Ulva is cultivated in coastal ponds or tanks and is valued for its high protein and vitamin content. It is used in human food, animal feed, and as a biofilter in aquaculture systems due to its ability to absorb excess nutrients.

Wakame (Undaria pinnatifida): Wakame is a brown seaweed native to the temperate coastal areas of East Asia. It grows best in cool, nutrient-rich waters with temperatures between 5-18°C. Wakame is commonly farmed in suspended systems where the fronds can grow rapidly, and multiple harvests are possible within a year. It is rich in vitamins, minerals, and antioxidants, and is widely used in soups and salads in Asian cuisine.

Ornamental Species

Angelfish (Pomacanthus spp., Centropyge spp.): Marine angelfish are found in tropical reefs of the Atlantic, Indian, and Pacific Oceans. They prefer water temperatures between 24°-28°C and salinities of 30-35 ppt. Angelfish are herbivorous or omnivorous, feeding on sponges, algae, and small invertebrates. They are popular in the aquarium trade for their bright colors and striking patterns. Larger species, such as those from the Pomacanthus genus, can reach up to 45 cm, while dwarf angelfish (Centropyge spp.) typically grow to about 10 cm. Angelfish are known to be territorial, so careful tank setup and management are needed in aquaculture and ornamental fishkeeping.

Butterflyfish (Chaetodon spp.**)**: Butterflyfish are small, vibrantly colored marine fish known for their striking patterns and are commonly found in coral reefs of the Indo-Pacific and Atlantic Oceans. They thrive in water temperatures between 24°-28°C and salinities of 30-35 ppt. Butterflyfish are omnivorous, feeding on coral polyps, small invertebrates, and algae. They are popular in the aquarium trade due to their beauty, but they can be challenging to keep in captivity because of their specialized diet and sensitivity to water quality. Butterflyfish grow to around 12-22 cm and require well-established reef tanks with plenty of hiding spots.

Cardinalfish (Pterapogon kauderni**)**: The Banggai cardinalfish is a small, striking species native to the Banggai Archipelago in Indonesia. It thrives in water temperatures of 24°-28°C and salinity levels of 30-35 ppt. Known for its distinctive black and white banding, it is a popular aquarium species. Cardinalfish are mouthbrooders, with males carrying the eggs and young in their mouths until they are ready to be released. They are carnivorous, feeding on small crustaceans and zooplankton. Cardinalfish reach sizes of 5-8 cm and are known for their relatively easy breeding in captivity, though they are vulnerable to overharvesting in the wild.

Cleaner Shrimp (Lysmata amboinensis**)**: Cleaner shrimp are known for their symbiotic relationships with fish, where they clean parasites and dead tissue from their hosts. Native to coral reefs in the Indo-Pacific, they thrive in water temperatures between 24°-28°C and salinity levels of 30-35 ppt. Cleaner shrimp are omnivorous, feeding on detritus, algae, and the parasites they remove from fish. In captivity, they are popular in reef aquariums for their cleaning behavior and are relatively easy to breed. They grow to around 5-6 cm and require a stable, well-maintained environment to thrive.

Clownfish (Amphiprioninae**)**: Clownfish, popular in the aquarium trade, are small marine fish found in tropical coral reefs, often in symbiosis with sea anemones. They thrive in water temperatures of 25-28°C and salinity levels of

30-35 ppt. Clownfish are bred in captivity, making them one of the first marine species to be successfully cultured for aquariums. They are known for their bright coloration and hardy nature, making them easy to care for. Clownfish feed on zooplankton and algae and are usually raised in tanks with artificial reefs or anemone analogs. They grow to a size of 5-10 cm and are popular for their adaptability in home aquariums.

Damselfish (Pomacentridae spp.): Damselfish are small, colorful marine fish found in tropical and subtropical coral reefs, with a wide distribution across the Indo-Pacific and Atlantic Oceans. They thrive in water temperatures between 24°-28°C and salinities of 30-35 ppt. Damselfish are hardy, adaptable, and relatively easy to breed in captivity, making them popular in the aquarium trade. They are omnivorous, feeding on zooplankton, algae, and small invertebrates. Damselfish are typically 6-15 cm in size and are known for their territorial behavior. Their ability to withstand a range of water conditions makes them ideal for both beginner and advanced aquarists.

Mandarinfish (Synchiropus splendidus): The mandarinfish is a small, vibrantly colored marine fish native to the Pacific Ocean, often found in coral reefs. It thrives in tropical waters with temperatures between 24°-28°C and salinities of 30-35 ppt. Mandarinfish are popular in the aquarium trade due to their striking appearance, though they require specific care, including a diet of live copepods and amphipods. They are difficult to breed in captivity, and maintaining water quality and habitat complexity is essential for their well-being.

Seahorses (Hippocampus spp.): Seahorses are unique marine fish known for their upright posture and horse-like head. They are found in shallow, coastal waters and thrive in water temperatures of 22-28°C and salinity levels of 28-35 ppt. Seahorses are slow-moving and feed on small crustaceans such as copepods. They are difficult to breed in captivity due to their complex life cycle, though successful aquaculture has been developed to reduce pressure on wild populations. Seahorses are farmed for the aquarium trade and traditional medicine, and their populations are vulnerable to overharvesting and habitat degradation.

Mollusks (Other than Shellfish)

Cuttlefish (Sepia officinalis): Cuttlefish are cephalopods found in shallow coastal waters of the Mediterranean, North Sea, and Atlantic Ocean. They thrive in temperatures between 15-25°C and salinity levels of 30-35 ppt. Cuttlefish are carnivorous, feeding on small fish, crustaceans, and other mollusks. Known for their ability to change color rapidly, cuttlefish are also a key species in cephalopod aquaculture. They have a relatively short lifespan of about 1-2 years and reach market size quickly. Cuttlefish are harvested for both food and their internal shells (cuttlebone), which are used in bird cages for calcium supplementation.

Octopus (Octopus vulgaris): The common octopus is a highly intelligent cephalopod found in temperate and tropical waters globally. It thrives in water temperatures between 15-25°C and salinity levels of 30-35 ppt. Octopuses are carnivorous, feeding on crabs, mollusks, and fish. They are known for their problem-solving abilities, camouflaging capabilities, and complex behaviors. In aquaculture, octopuses are cultured for food in parts of Europe and Asia, with growing interest in sustainable farming practices. Octopus farming requires precise control of water quality and temperature, as well as specialized tanks to accommodate their behavior and prevent escape.

Squid (Loligo spp.): Squid are fast-swimming cephalopods found in both temperate and tropical waters worldwide. They thrive in water temperatures between 10-25°C and salinity levels of 30-35 ppt. Squid are carnivorous, feeding on fish, crustaceans, and other squid. In aquaculture, squid farming is still in its early stages, with challenges in larval rearing and lifecycle management. Squid are harvested primarily for food, with species like *Loligo vulgaris* and *Loligo pealei* being important in commercial fisheries. Squid grow rapidly and are known for their use in seafood dishes, particularly in the Mediterranean and Asian cuisines.

Echinoderms

Sea Cucumber (Holothuria scabra): Sea cucumbers are echinoderms found in shallow coastal waters and estuaries, with optimal water temperatures between 24°-28°C and salinity levels of 28-35 ppt. They play a key ecological role by recycling nutrients through sediment bioturbation. In aquaculture, sea cucumbers are farmed primarily in Asia for their high demand as a delicacy and their perceived medicinal properties. They are typically grown in polyculture systems with other marine species like shrimp or fish. Sea cucumbers reach market size (300-500 g) within 12-18 months and are valued for their ability to improve water quality in integrated aquaculture systems.

Sea Urchins (Strongylocentrotus **spp.**): Sea urchins are spiny echinoderms found in temperate and tropical coastal waters. They thrive in water temperatures between 10-18°C and salinity levels of 30-35 ppt. Sea urchins are herbivorous, feeding primarily on macroalgae, especially kelp. They are farmed for their roe, which is considered a delicacy in many cuisines, particularly in Japan. Sea urchin aquaculture is challenging due to their slow growth rates and specific dietary requirements. They are also valued for their role in controlling algal growth in marine ecosystems, making them an important species in both wild fisheries and aquaculture.